Principles and Practice

of

Managing Soilborne Plant Pathogens

Edited by Robert Hall
University of Guelph
Guelph, Ontario

APS PRESS
The American Phytopathological Society
St. Paul, Minnesota

Chapters 1–13 are revised versions of papers presented at the
Sixth International Congress of Plant Pathology, held at Montreal,
July 28–August 6, 1993.

This book has been reproduced directly from computer-generated
copy submitted in final form to APS Press by the editor of the volume.
No editing or proofreading has been done by the Press.

Reference in this publication to a trademark, proprietary product,
or company name by personnel of the U.S. Department of Agriculture
or anyone else is intended for explicit description only and does not
imply approval or recommendation to the exclusion of others that
may be suitable.

Library of Congress Catalog Card Number: 96-79341
International Standard Book Number: 0-89054-223-6

Printed in the United States of America on acid-free paper

The American Phytopathological Society
3340 Pilot Knob Road
St. Paul, Minnesota 55121-2097, USA

CONTRIBUTORS

Alabouvette, C., I.N.R.A. - Centre de Microbiologie du Sol et de l'Environnement, Laboratoire de recherches sur la Flore pathogène du Sol, B.V. 1540, 17 rue Sully, 21034 Dijon Cedex, France

Baker, R., Department of Plant Pathology and Weed Science, Colorado State University, Fort Collins, CO 80523, USA (Deceased)

Boehm, M.J., Department of Plant Pathology, Ohio Agricultural Research and Development Center, The Ohio State University, Columbus, OH 43210, USA

Campbell, C.L., Department of Plant Pathology, North Carolina State University, Raleigh, NC 27695–7616, USA

Canfield, M., Department of Botany and Plant Pathology, Oregon State University, Corvallis, OR 97331–2902, USA

Evans, K., Entomology and Nematology Department, Institute of Arable Crops Research, Rothamsted Experimental Station, Harpenden, Hertfordshire, AL5 2JQ, England

Hall, R., Department of Environmental Biology, University of Guelph, Guelph, Ontario, Canada N1G 2W1

Hoitink, H.A.J., Department of Plant Pathology, Ohio Agricultural Research and Development Center, The Ohio State University, Wooster, OH 44691, USA

Hornby, D., Plant Pathology Department, Institute of Arable Crops Research, Rothamsted Experimental Station, Harpenden, Hertfordshire, AL5 2JQ, England

Katan, J., Department of Plant Pathology and Microbiology, The Hebrew University of Jerusalem, Rehovot 76100, Israel

Kerry, B.R., Entomology and Nematology Department, Institute of Arable Crops Research, Rothamsted Experimental Station, Harpenden, Hertfordshire, AL5 2JQ, England

Lemanceau, P., I.N.R.A. - Centre de Microbiologie du Sol et de l'Environnement, Laboratoire de recherches sur la Flore pathogène du Sol, B.V. 1540, 17 rue Sully, 21034 Dijon Cedex, France

Madden, L.V., Department of Plant Pathology, Ohio Agricultural Research and Development Center, The Ohio State University, Wooster, OH 44691, USA

Moore, L.W., Department of Botany and Plant Pathology, Oregon State University, Corvallis, OR 97331–2902, USA

Neher, D.A., Department of Plant Pathology, North Carolina State University, Raleigh, NC 27695–7616, USA

Paulitz, T.C., Department of Plant Science, Macdonald Campus of McGill University, Ste Anne de Bellevue, Quebec, Canada H9X 3V9

Postma, J., DLO Research Institute for Plant Protection (IPO-DLO), Wageningen, The Netherlands

Steinberg, C., I.N.R.A. - Centre de Microbiologie du Sol et de l'Environnement, Laboratoire de recherches sur la Flore pathogène du Sol, B.V. 1540, 17 rue Sully, 21034 Dijon Cedex, France

Thomashow, L.S., Root Disease and Biological Control Research Unit, USDA-ARS, 362 Johnson Hall, Washington State University, Pullman, WA 99164–6430, USA

Tronsmo, A., Department of Biotechnological Sciences, Agricultural University of Norway, N-1432 Ås, Norway

Trudgill, D.L., Scottish Crop Research Institute, Invergowrie, Dundee DD2 5DA, Scotland

Van Vuurde, J.W.L., DLO Research Institute for Plant Protection (IPO-DLO), Wageningen, The Netherlands

Weller, D.M., Root Disease and Biological Control Research Unit, USDA-ARS, 362 Johnson Hall, Washington State University, Pullman, WA 99164–6430, USA

CONTENTS

Section III: Cases

Section IV: Epilogue

INTRODUCTION

This is the seventh in a series of books on soilborne plant pathogens that began with the publication in 1970 of the proceedings of an international symposium that dealt with factors determining the behavior of plant pathogens in soil and was held at the University of California, Berkeley, 7 to 13 April 1963. The series continued with five books based on papers given at the first five International Congresses of Plant Pathology, held at five-year intervals from 1968 to 1988.

The first 13 chapters in this book are based on eight symposium papers and five keynote addresses presented at the 6th International Congress of Plant Pathology, held in Montreal, Canada, 28 July to 6 August 1993. The papers were given at three symposia that examined detection and management of soilborne plant pathogens (S1), management of soilborne plant pathogens by cultural practices (S15), and plant pathogens and microbial interactions in soil (S17). The keynote addresses introduced five discussion sessions on bacterial diseases, ecophysiology of nematodes in plants, cultural management of soilborne fungi, *Trichoderma* and biological control, and biological control of soilborne plant pathogens. Papers and posters related to soilborne plant pathogens were widely distributed among the 25 symposia and 56 discussion sessions that made up the core of the scientific program in Montreal. This book provides a taste of the subjects discussed and views presented.

Two of the symposia represented in this book (S1 and S17) were organized by the Soilborne Plant Pathogens Committee of the International Society for Plant Pathology, which was invited by the National Organizing Committee of the congress to serve as the Soilborne Pathogens subject matter committee. The third symposium (S15) was arranged by the Environment and Disease Management subject matter committees of the congress. Discussion sessions were developed around abstracts and posters submitted by congress delegates and were introduced by invited keynote speakers.

The chapters are arranged into four sections. The first section, entitled "Prologue", contains one chapter that sets the historical, social, and scientific contexts of the book. The second section, "Fundamentals", includes five chapters that offer a conceptual framework for understanding the biology and control of soilborne plant pathogens. The third section, "Cases", consists of seven chapters that focus on recent developments in practical approaches to the control of these pathogens. Biological control has been a recurring theme in the series to which this book belongs and is retained here as an important component of the theory and practice of integrated management of soilborne plant pathogens. The final section, "Epilogue", comprises one chapter that considers how principles and practice interrelate and lead to innovations in management of soilborne plant pathogens.

Hornby (Chapter 1) sets the stage by noting (p. 2) that "it is still the invention of new techniques, or the speeding up of old ones, that drives the process of discovery". The current climate for research includes environmentalism, cutting of positions, and reduction in curiosity-oriented research. Methods and concepts considered by Hornby as they relate to management of soilborne plant pathogens include detection of pathogen or disease, cultural control, biological control, diagnosis, fungal vectors of viruses, interactions with mycorrhizae, soil suppressiveness, root disease, soilborne inoculum, inoculum potential, computers, modeling, decision making, expert systems, precision farming, molecular biology, immunology, and root function. A closing section on managing disease in a changing world comments on (i) the need to link progressive areas to achieve sustainable disease management, (ii) registration of biological control agents (BCAs), (iii) integrated pest management, (iv) the importance of field research, and (v) public pressures influencing management technologies.

Campbell and Neher (Chapter 2) suggest that epidemiology can set the strategy for disease management (p. 21). They note the need for methods to quantify inoculum, assess disease, and design effective studies, and comment on opportunities for modeling root disease, linking soil ecology to ecology of root pathogens, comparing and classifying epidemics, and setting strategies for disease management. The suggested strategies are to reduce inoculum, to reduce the rate of primary infection or secondary infection, and to alter host density. They emphasize the importance of establishing a sound theoretical framework, including mathematical models and simulation models, and the need to present strategies for root disease management based on host resistance, natural processes to regulate pathogen populations, and cultural practices. It is also important to develop practical systems for analyzing root epidemics. The systems would include practical protocols for sampling soil and roots for pathogens, assessing intensity of root disease, ensuring procedures are reliable, analyzing spatial and temporal patterns of pathogens and disease, and comparing epidemics of root disease.

Baker and Paulitz (Chapter 3) focus on the theoretical basis of biological control of soilborne plant pathogens and its importance for determining future directions of research. They call for the elaboration of "mind-stretching theory" (p. 50) and identify three management strategies: (i) protect the infection court (fixed and motile), (ii) reduce inoculum potential in sites not necessarily associated with the infection court, and (iii) induce host resistance. Methods for implementing these strategies include treating seeds with BCAs, adding carbon amendments such as cellulose to immobilize nitrogen by increasing the carbon:nitrogen ratio, managing irrigation to sustain bacterial antagonism, adding antagonists to soil, increasing competition for iron, rotating crops to reduce inoculum density, using mycoparasites, adding crop residues or organic matter, and inducing resistance by exposing plants to nonpathogens or other inducers.

Thomashow and Weller (Chapter 4) discuss the molecular basis of pathogen suppression by antibiosis in the rhizosphere. They examine how the methods of molecular biology can be used to (i) study the genetics and regulation of antibiosis as a mechanism of antagonism of root pathogens in the rhizosphere, and (ii) improve biological control. New methods overcome limitations of old methods in elucidating mechanisms. Improved knowledge of mechanisms of antibiosis helps to optimize performance of existing agents, rationalizes development of new strains, and assists selection of new strains. Suggested principles for selecting superior BCAs include (i) transfer genes among strains, (ii) combine agents known to differ in antagonistic mechanisms, and (iii) use biosynthetic genes as probes to identify earlier, enhanced, or novel production of antibiotics.

Van Vuurde and Postma (Chapter 5) organize methods for detecting viable bacteria and fungi in soil into three groups: isolation, serology, and gene probes. Each group is referred to as a principle of detection, and the characteristics and limitations of each principle are described. Criteria are provided for selection of the most suitable method.

Trudgill (Chapter 6) also provides clear examples of the development of principles and their application to root pathogen management. He describes a thermal time basis for understanding the ecology of pathogens and the epidemiology of diseases they cause. He starts with the principle that temperature is a rate-determining factor for growth and multiplication of organisms. Numerous other principles and hypotheses are then developed that relate temperature to development. Knowing how temperature affects the activities of organisms enables us to predict their behavior under particular conditions. This, in turn, enables us to plan pest management strategies and tactics, such as predicting the threat of introduced pathogens, estimating hatching dates, and harvesting early to minimize pest damage. New applications of the theory are suggested, including selecting environments for testing plant response to pathogens, assessing risks posed by introduced pathogens, and classifying pathogens according to their survival strategies.

Kerry and Evans (Chapter 7) describe new strategies and tactics for managing plant parasitic nematodes. After noting the need for alternatives to fumigant pesticides, they consider integrated management of potato cyst nematodes as a case study. Control strategies were based on a detailed knowledge of the biology of the causal nematodes, *Globodera pallida* and *Globodera rostochiensis*. The two nematodes are morphologically similar and it was assumed that tactics developed to control *G. rostochiensis* would control *G. pallida*. However, it was discovered that biological differences between the nematodes dictated different management tactics. This is turn emphasized the need for correct identification of the species. The authors thus describe advances in serological and molecular techniques for identification of nematodes, and new components of integrated management of nematodes, such as transgenic plants and BCAs. Strategies developed for managing the two nematodes are the same—rotation, fumigation, and cultivar resistance—but the tactics differ. Control of *G. pallida* requires more methods, longer rotations, different fumigants, and different rotation of cultivars.

Moore and Canfield (Chapter 8) connect management of crown gall to the biology of the causal agent (biovars of *Agrobacterium*). They emphasize the importance of understanding diversity in the pathogen, and thus the importance of information on taxonomy, classification, detection, characterization, identification, mechanisms of variation, pathogenicity, and host range. Identification techniques based on physiology, pathogenicity, serology, and molecular biology are described. They also stress the importance of understanding the ecology of the pathogen. Management strategies discussed include biological control, heat therapy, chemical treatment of roots, fumigation of soil, pathogen-free plants, and resistant rootstocks.

Alabouvette et al. (Chapter 9) consider the biology of biological control of Fusarium wilts and the application of this knowledge to the development of a commercial biological control product. They observe that biological control of soilborne plant pathogens has been studied for a long time, often because there are few other effective management options. The characteristics of soils suppressive to Fusarium wilts are described. Organisms responsible for suppression of the diseases include fluorescent pseudomonads and nonpathogenic forms of *Fusarium oxysporum*. Mechanisms of disease suppression include induced resistance in the plant, and competition for infection sites, carbon, and iron. Several principles are enunciated. One is that strains of BCAs differ in their control efficacy. Another is that "Competition for nutrients occurs only among microorganisms having the same requirements and sharing the same ecological niches" (p. 202). Thus one criterion for selecting a BCA is that it should have rhizosphere competence. A third principle identified is that the efficacy of a BCA is determined by the ratio of BCA to pathogen. Two strategies arise from this principle: (i) use a BCA in situations where the amount of rooting medium is limited, e.g. plants grown in small containers such as pots, and (ii) use a BCA and production methods that enable the generation of large amounts of BCA inoculum. The authors show how biological information, principles, and strategies led to the development of a commercial biological control product.

Tronsmo (Chapter 10) examines biological control with *Trichoderma harzianum* in the soil, the phyllosphere, and stored products, and notes how principles and strategies can be transferred among systems. Principles discussed include selection of BCAs, production of BCAs, specificity of BCAs, and integrated control with BCAs. The observation that antagonism can occur at low temperatures indicates the potential for control of diseases that occur at low temperatures, such as storage rots and snow molds. Studies on biological control in the phyllosphere show the role of prior colonization in protection, a principle that can be tested in other habitats. The recognition that some biological control is effected through metabolites of BCAs suggests that disease control could be achieved by incorporating genes conditioning production of these substances into plants or other microorganisms.

Hoitink et al. (Chapter 11) explore biological control with organic amendments. The introduction of new analytical methods, and an increased interest by society in composting as a method of waste management, permitted rapid advances in this area. The authors discuss the fate of BCAs during composting, and mechanisms of biological control with composts, especially organic matter decomposition level and biological energy availability in soil and amendments. They present an important new principle: decomposition level of composts affects suppression of plant disease. The compost must be neither immature nor mature for maximum effectiveness. They suggest possibilities for developing a test to determine the suppressiveness of compost. One test might be based on estimating microbial activity by measuring hydrolysis of fluorescein diacetate.

Katan (Chapter 12) reviews the use of soil solarization to control soilborne plant pathogens. He starts with basic principles, e.g. that disinfesting soil improves plant health, and that heat can kill pathogens and therefore reduce disease. He also notes that new materials enable the development of new techniques; in this case, the availability of plastic sheets permitted the development of modern solarization techniques. He provides an extensive summary of principles of soil solarization (pp. 253–255), and examples of tactics for implementing the technology. The mechanisms of solarization, including physical, chemical and biological changes in solarized soil, thermal inactivation of pathogens, microbe interactions, and improved plant growth, are discussed. Strategies for improving soil solarization are presented, and the value of simulative and predictive models of soil heating is described.

In chapter 13, I relate inoculum dynamics of *Fusarium solani* f. sp. *phaseoli* to management of Fusarium root rot of bean. Attention is drawn to the critical importance of methods for identification, isolation, and enumeration of the fungus; the central role of a suitable selective medium; the impact of the disease on bean yield; and the use of the pathosystem as a model. Qualitative and quantitative features of chlamydospores and their relation to the disease are discussed. One principle enunciated is that disease severity increases with inoculum density, but the relationship is strongly influenced by the environment. The strategy derived is that disease can be reduced by lowering inoculum density and modifying the environment. Tactics of disease management can be based on observations that (i) inoculum density declines in the absence of the preferred host (bean), and (ii) effective modifications to the environment include relieving stresses (e.g. compacted soil, drought), and adding organic materials with high carbon:nitrogen ratios.

In the final chapter, I present a model facilitating innovation, and discuss how theoretical, strategic, and tactical modes can be applied to the planning, implementing, observing, and evaluating phases of research. Potential uses of the innovation model are illustrated using examples from the previous chapters.

I would like to thank the authors for contributing to this book, the members of the committees mentioned above for organizing the symposia, Stellos M. Tavantzis, Pam Johnson, and Gay Phillips for editorial assistance, and APS Press for publishing this material and so continuing a tradition that was established 26 years ago.

R. Hall
Chair 1988–93
Soilborne Plant Pathogens Committee
International Society for Plant Pathology

Chapter 1

KEEPING UP WITH SOCIETY AND CATCHING UP WITH TECHNOLOGY

D. Hornby

The brief for the symposium on Detection and Management of Soilborne Plant Pathogens at the 6th International Congress of Plant Pathology (6th ICPP) in Montreal in 1993 was to focus on the more applied aspects of "soil pathology", such as diagnosis, prediction and control (biological, chemical and integrated). This expanded version of the remarks with which I opened that symposium is intended to set the scene for the chapters to follow and, hopefully, it will help readers to put recent developments in the study of soilborne plant pathogens into perspective. I have observed the study of soilborne plant pathogens for over a quarter of a century as a researcher into root diseases of arable crops, a participant in all the ICPPs to date, and a founder member and first chair (1978–88) of the Soilborne Plant Pathogens Committee (SBPPC) of the International Society for Plant Pathology. In writing this paper I drew on these experiences, but since there is much emphasis on impact and publicity in science and agriculture today I have also quoted from a wider range of sources than I am accustomed to doing in order to capture some of the perceptions that affect our activities.

TRENDS AND THEMES

The title for this chapter arose because more than ever before external trends keep much, if not all, of what we do reactive rather than proactive. Perhaps the biggest trend is environmentalism—"the greatest trend of the late 20th century" (37)—which makes the public, certainly in developed countries, increasingly less disposed to leave plant pathologists (or indeed others involved in agriculture) to their own devices. Another trend is well exemplified in the UK, where government funding of agricultural research continues to be cut and, since the shortfall is not being made up by industry, posts continue to be lost.   Much of the plant pathological research that remains is directed at priorities laid down by funding agencies and is tightly planned, so that opportunities to indulge in "interesting" leads are much fewer than hitherto.   Perhaps the most exciting feature of modern biological research is the way in which studies in one area can unexpectedly have consequences for another (35). However, in the current massive technological trend, advances continue to flood in from many different fields, adding to the number with as yet unproven application or with limited or uncertain usefulness in plant pathology.   Some plant pathologists feel that there is enough technology, we now need to apply it.   Whereas most of the traditional compartments of study have broken down, it is still the invention of new techniques, or the speeding up of old ones, that drives the process of discovery (35). These are themes I wish to explore in this chapter.

As a program theme, management of soilborne plant pathogens last had prominence in Section 5 (soilborne plant pathogens) of ICPPs at the 4th ICPP in Melbourne in 1983.   Although the early detection of a pathogen or disease is likely to be an important part of most disease management strategies, detection came to prominence as a program theme only recently at the 6th ICPP.   My introductory chapter (22) to the volume of proceedings of the 4th ICPP contained observations on the contemporary situation and predicted some of what the future might bring (Tables 1 to 3 list the points).   I tried to indicate the kinds of problems and challenges we faced by: (i) identifying some goals, some contentions and some examples of areas relying on few, well-used and less-than-perfect methods (Table 1); (ii) predicting the coming to prominence of new themes (Table 2) and the further development of several that were long-running (Table 3); (iii) pointing out some areas in which we were our own worst enemies (Table 3); and (iv) suggesting that some of our cherished concepts (e.g. inoculum potential and rhizoplane) were flawed.

Some of these issues remain relevant, some have been ignored and new ones have come to the fore, heralded or otherwise.

Although fertilizers as a cultural control of disease and biological control continued to be strong topics during the subsequent 5 years, rapid methods of diagnosis of soilborne plant pathogens, fungal vectors of viruses, and interactions with mycorrhizae were topics of increasing importance (23).   By the time of the 6th ICPP further areas of change, progress and quiescence were identifiable.   The entries in Table 1 may no longer all be major aspirations within our discipline, but it was recognized at the Plenary Session of the 6th ICPP that a sustainable system of crop production requires a holistic approach (32).

Table 1.   Study of soilborne plant pathogens: some aspirations, contentions and examples of areas relying on few well-used and less-than-perfect methods in 1985.

A. Aspirations

1. Better definition and reduction of losses as a means to increase crop yields
2. Multidisciplinary cooperation
3. Holistic systems approach

B. Contentions

1. There had been no major changes in direction
2. Agriculture had changed more than our preoccupations
3. Outside influences dictated new topics
4. Increasing skills, improved techniques, and growing sophistication enabled response with few conceptual upheavals
5. Preservation of traditional barriers should be questioned
6. There were few new techniques for studying diseases in soil
7. Too little was known about soil microbial ecology

C. Areas relying on few well-used and less-than-perfect methods

1. Measuring soil suppressiveness
2. Assessing root disease
3. Quantifying soilborne inoculum

The traditional program format was abandoned at the 6th ICPP in Montreal and the organizers of Section 5 involved the SBPPC in drawing up the soilborne plant pathogen's program and sought contributions from nematologists (Table 1, item B5).   This was a move towards breaking down traditional boundaries, but since it is easy to find other under-represented areas, there is much scope for further initiatives. Interactions between protozoa and fungi ought perhaps to be given more attention in the future, particularly since protozoa may be beneficial biological control agents (BCAs) or detrimental competitors of inocula of useful organisms in soil (38). Also, little is known about how pathogens affect natural plant populations, even though biotic forces may be of primary importance in determining species abundance and diversity within major biomes. The large body of knowledge gained from agricultural systems, however, may not be that helpful in understanding natural systems. It is claimed (3) that lack of involvement by agricultural funding agencies in basic research and the concentration of plant pathologists in agricultural departments have added to the separation of pure and applied aspects of the subject. The study of soilborne inoculum has not received the attention I expected (Table 1). Twenty years ago inoculum potential was a term at once derided, pervasive, widely used, not lightly dismissed, and suggesting concepts that writers were trying to express (46). Little has changed and it is still appearing in titles of publications in the 1990s and is more in vogue than hitherto in papers dealing with VA mycorrhizae.

COMPUTERS AND PRECISION FARMING

By seeing the computer's contribution mainly in the context of modeling (Table 2), I was much too restrictive. Certainly models continue, as in many disciplines, to serve the useful purpose of focusing clear attention on deficiencies in our knowledge (27). However, since 1985, the computer has been used more for identification and decision making through expert systems and in assisting in visualising data sets in meaningful ways. What is variously called precision farming, computer-aided farming, or prescription farming is already with us (8,9). The environmental impact of farming is on farmed and nonfarmed ecosystems, human health and socio-economic effects. Wasted inputs, waste products, inappropriate objectives and risk, or the perception of risk, make the rejection of farming "by averages" in favor of precision farming both attractive and persuasive. The goals are increased efficiency in the management of soils and crops, more economic use of resources and less environmental impact.

The technology for precision farming includes Global Positioning Systems (GPSs), Geographic Information Systems (GISs), Management Information Systems (MISs) and Decision Support Systems (DSSs). As in the case of molecular biology, plant pathologists are still discovering how all this new technology can help them. Although there has been progress in applying this technology to cereal production systems, there has been little progress with crops like grass, sugar beet and potato. Combine harvesters fitted with various kinds of location-finding equipment (usually a satellite-based GPS, using microwave, or a differential

Table 2. Study of soilborne plant pathogens: themes predicted to come into prominence after 1985.

---

1. Minor pathogens/plant growth-promoting rhizobacteria (PGPR)
2. Similarities rather than compartmentalisation
   (mycostasis/suppressive soils)
3. Modeling (its value, stochastic models, disease progress in time and
   space)
4. Phloem translocated fungicides (awaiting progress)
5. Genetic manipulation (awaiting progress)
6. Integrated pest management (IPM)
7. Experimental designs and procedures for testing BCAs and PGPR
8. Better experimentation for endemic diseases that insidiously decrease
   yield

GPS which also uses a fixed reference point on the ground) and various kinds of yield measuring devices (e.g. gamma absorption sensors, capacitative grain mass flow sensors), which may, for example, sample yield every 3 seconds over the width of the combine, are already in use. It is claimed that such a combine's position in the field can be located to within 50 cm using a GPS and base stations. The information is transmitted to a suitable computer and yield contour maps may be produced using various kinds of statistical procedures. Interpretation of such maps is not necessarily easy, but with the help of a GIS using overlays and trends etc., it is possible to arrive at treatment maps which may incorporate information on such things as location, quantity and timing of treatments. Some farmers in southern England are now making use of commercial services that sample fields intensively to create pH and yield maps, which usually reveal huge variations within fields and demonstrate the inaccuracy of the conventional "W" sampling pattern for soils. This makes it possible to adjust crop husbandry using spatially varied input applications of seed, fertilizers, growth regulators, fungicides, nitrification inhibitors etc. on the basis of the treatment map for the field. Farmers seem favorably disposed to this approach, arguing that knowledge of the variability of a field is of immediate benefit, giving a better insight into which areas are more productive than others. Most see ways of making savings on inputs on the basis of pH and nutrient maps of their fields, or they can use them to identify problems, or land for set-aside. However, several years of data will be required to identify the intrinsic variability of a field and to develop a reliable guide to the soil's potential. As yet it is not known how yield contours will vary among crops, species and cultivars, or what are the seasonal or rotational variations. There is little data at present to show if ultimately there will be net financial benefits from full precision farming.

These advances bring exciting possibilities for crop protection research and more efficient disease control. Apart from the question of a suitable resolution (suggested resolutions are 5 to 50 m for fertilizers, 5 m for yield, 1 m for herbicides and where overlap avoidance is important < 0.1 m) a major limitation at present is getting sufficient, appropriate disease data at the right time. For instance, how are the current year's weed and disease problems to be identified in order to produce a treatment map and at what cost? Some years ago (44) infestation by *Verticillium dahliae* was mapped in potato fields and the information used in conjunction with information on nematode numbers and different seed lots as the basis for decisions about expensive fumigation and which fields to utilize for seed tuber production. However, collecting the data required laborious soil sampling and host-infection assays, or monitoring disease in a preceding susceptible crop (usually cotton). Weeds maps are currently generated by field walking; the advent of hand-held dataloggers with a GPS will facilitate this, but in general there is currently little technology for this sort of problem. The way forward may ultimately come through linking precision farming technology with the new technology for rapid detection of soilborne pathogens.

Diseases continue to be a major and costly factor in field crop production. Many, particularly those caused by soilborne plant pathogens, often manifest themselves as patches of poorly growing and/or discolored plants in more or less complex patterns within the crop. The information in these patches is rarely exploited in farm management systems or used efficiently in national disease surveys and yield predictions. Perhaps it is time to explore possibilities for bringing together expertise in (i) remote sensing of plant diseases using infrared aerial photography, (ii) yield mapping, and (iii) plant disease assessment and epidemiology to improve the control and management of crops and their diseases and the prediction of yield shortfalls. For a disease like take-all of cereals, the initial aim might be to establish the extent to which damaging disease extends beyond visible patches (24) and to establish the strength of correlations between disease maps produced by remote sensing and yield maps.

MOLECULAR BIOLOGY

We are experiencing the "DNA revolution" and molecular biology has come to the fore and is evolving rapidly. However, only one in 10 biotechnology companies is applying biotechnology to food and agriculture, partly because of uncertainties about market prospects and consumer backlash (13). Plant pathologists are required to function in a scientific milieu where molecular biology currently is omnipresent (16), which has drawbacks: "molecular biology dominates grant applications and awards; molecular biologists are rarely exponents of chemistry or biology and there is a tendency to view all problems through a genetic, molecular biology filter" (11). The benefits of molecular biology in our own discipline are to be seen in all areas from detection and identification (18,40) through to biological control. There are undoubtedly benefits in allied disciplines: the numbers of oral presentations on molecular approaches to fungal systematics at meetings of the Mycological Society of America rose from six in 1989 to 34 in 1992 (43) and dialogue between molecular biologists and ecologists is expected to advance soil ecology, especially where molecular tools are used in conjunction with traditional methods (14). In plant studies the most important form of genetic manipulation (Table 2, item 5) is gene "add-ons", which are just reaching the stage of becoming commercially valuable (35). Many crop plants, including the main cereals, have now yielded to this technique and there are examples of resistance to insect pests and viruses engineered into plants. Field testing and safety concerns are now hurdles that have to be overcome. Some of my own, vicarious, experience of molecular biology has been encouraging, despite the time it is taking to develop methods for the field, or indeed methods to advance our epidemiological and ecological work. The control of plant pathogens depends much on the quality and speed of our diagnostic techniques. At Rothamsted we have begun to make progress with developing rapid tests based on DNA probes, restriction fragment length polymorphisms (RFLPs) and the polymerase chain reaction (PCR), the most important technical advance in molecular biology in the last few years, for discriminating amongst fungi of the *Gaeumannomyces-Phialophora* complex (6,48,49,50,51). Others (10) have provided genetic proof that enzymic detoxification of a preformed antifungal defence substance (a saponin, avenacin) is essential for the pathogenicity of *Gaeumannomyces graminis* var. *avenae* to oats.

During the last 10 years, a new, patch-forming disease has occurred in a few, scattered crops of winter cereals in the UK (Figures 1A, 1B). It has been attributed to the Basidiomycete fungus *Omphalina pyxidata*

(Figure 1D), the basidiomata of which frequently occur amongst the poorly growing plants in spring. An *Omphalina* sp., possibly damaging to wheat, was isolated from pasture soil in South Australia in the 1950s (47). *Omphalina pyxidata* has proved problematical in conventional handling and we have been able to culture extremely slow-growing colonies only from spore prints on agar. Inocula from these cultures were not infectious to roots of cereal seedlings in tests. Consequently it was not possible to fulfil Koch's postulates and demonstrate that the fruiting bodies, sclerotia on roots (Figure 1C) and sclerotia found in the soil were all the same fungus and that this fungus was responsible for the poor growth in cereal crops. Further, collections of basidiomata have differed in appearance and in the mean length of basidiospores. DNA studies by E. Ward at Rothamsted have enabled us to progress beyond this conventional impasse. In the absence of fresh basidiomata, herbarium material of *O. pyxidata* and *Omphalina hepatica* and material from three of our collections (cultures, sclerotia, and preserved roots with sclerotia) were used. A relationship between sclerotia (on roots and in soil) from a collection made in Cumbria in 1993 and a culture from spores of a collection of basidiomata from Suffolk in 1992 (Figure 1D) was established. There were differences between cultures from the Suffolk collection and a collection from Herefordshire made in 1989. None of the recent collections were identical to the herbarium specimens and the closest relationship was between the Herefordshire collection and *O. hepatica*. This work has demonstrated that at least one of the collections of basidiomata was of the same fungus as that forming sclerotial structures on stunted plants, although what we currently call *O. pyxidata* is quite variable genetically and morphologically.

Immunological detection techniques for agricultural applications have progressed in a way I failed to foresee in 1985 and the advent of diagnostic kits for use by farmers and growers to detect fungal pathogens of plants (31) is a major achievement. This advance arose primarily through the development of monoclonal antibody techniques and by improved purification methods to produce polyclonal antisera with enhanced specificity and sensitivity. Quantitative immunoassays have been used to monitor changes in populations of fungal pathogens in crops, providing accurate diagnosis and detecting early increases (31). This has important management implications for optimizing fungicide inputs, implementing other control measures and minimizing environmental impact. It is predicted that "agricultural production will benefit from the availability of immunoassays for the detection of plant pathogens affecting a wide range of crops" (31). Because many effective pesticides have been

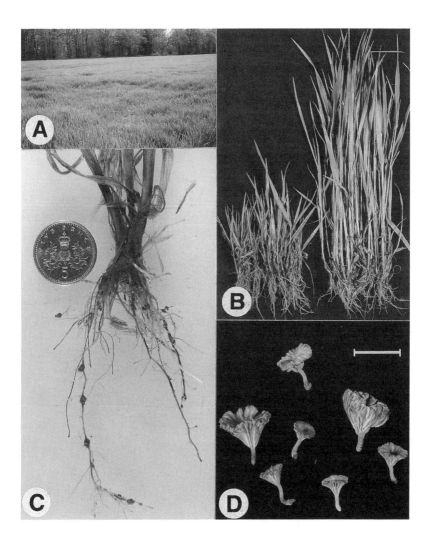

Figure 1. (A) Winter barley in a field near Ipswich, Suffolk showing a patch of poor growth 29 April 1992. (B) Plants from inside (left) and outside (right) the patch. (C) Roots of a plant from within the patch showing sclerotia. Coin is 1.8 cm in diameter. (D) Collection of basidiomata of a fungus tentatively identified as *Omphalina pyxidata* (Bull. ex Fr.) Quél. from within the patch. Scale line represents 2 cm.

lost, and because legislative and environmental pressures are increasing, rapid disease diagnosis is becoming the key to modern crop protection. American-made commercial kits, now available in the UK for pathogens such as *Pythium, Phytophthora* and *Rhizoctonia*, were heralded (1) as the start of a new era of disease management. There are, as yet, no simple monoclonal antibody ELISA kits for economically important *Armillaria* spp. and nucleic acid hybridization techniques have not been adapted for use in the field (19).

ROOTS AND THE SOIL

Because we still have insufficient knowledge about the properties of roots and root-soil relationships, and do not know what fraction of the root system is active in taking up materials (27), alternative approaches to assessing root disease such as measuring or estimating functional tissue instead of diseased tissue (Table 1, item C2) seem no nearer. My hopes that in situ observation would aid inoculum studies (Table 1, item C3) have not materialized. However, the goal of being able to study plants with root disease and changes in the soil without destructive sampling is coming closer through new, minimally invasive or noninvasive techniques and equipment. Time Domain Reflectometry (TDR), introduced to soil science in 1980, can give instantaneous volumetric water content of the adjacent soil. A further advantage is rapid measurement of water content in the top 10 to 15 cm of soil, a zone where previous methodology has severe limitations (12). The dielectric constant of soil is a good measure of its volumetric water content. TDR measures the velocity of propagation of a high frequency electric signal or voltage pulse; this velocity is decreased in proportion to the inverse of the square root of the dielectric constant. Computer-assisted tomography is being used experimentally for studies of moisture characteristics around roots and provides "an exciting new method of nondestructive imaging within a solid matrix, with considerable potential for studying soil behavior and soil-plant-water relations in space and time" (4). Improvements in scanning geometry, counting electronics and software as well as reduction in scanning times are needed to realize the potential of the technique.

Several other techniques offer promise. Magnetic resonance imaging may also by used for looking at roots in situ. Meters measuring the spectral transmittance of leaves can give readings rapidly that are proportional to the amount of chlorophyll in a leaf. Although this is perhaps more useful in deciding on the need for essential nutrients, we have used it to measure leaf yellowing to study the distribution of take-all patches. Work with portable radiothermometers and a thermal infrared camera has indicated the possibility of detecting attacks of *Heterodera avenae* in large fields of wheat by remote sensing (36). Videomicrography has been used to explore mycoparasitism in the laboratory (28) to great effect, despite the time required for recordings and the limitation that imposes on replication for experimental purposes. Perhaps it will not be too long before we will be able to watch microbial interactions on roots in vivo in the soil.

Many traditional or conventional approaches have much still to offer. An example is the most probable number (MPN) method applied to quantifying microorganisms in soil. One innovation has been its application to both the beet necrotic yellow vein virus *and* its vector *Polymyxa betae* (45). Another has been its link to molecular methods in studies of actinorhizal symbionts (*Frankia* spp.) (33). Here plant bioassays were used to estimate "nodulation units" (units physiologically capable of nodulation) and "polymerase chain reaction MPN" used to detect genomic units (one unit being the amount of *Frankia* containing a single genome), which should be proportional to the total population of *Frankia*. At the same time, statistical advances have allowed correction of bias in estimates of density from dilution counts (21,42). There has also been an increase or a resurgence of interest in old themes (Table 3, item B) such as the rhizosphere (30), typified by the formation of a Rhizosphere Ecology Group at Rothamsted (2). This group declared an interest in sampling and characterizing soil microbiota through bulk extraction of microorganisms from soil by aqueous two-phase partitioning (41), in addition to identifying and quantifying soilborne viruses and their vectors, identifying soilborne plant pathogens, quantifying rhizobia, and exploiting interactions between organisms in soil.

MANAGING DISEASE IN A CHANGING WORLD

Many developing countries have made remarkable increases in agricultural production in the last decade, sometimes by adopting new and unstable systems (34,39). In Europe, however, overproduction of several commodities is a problem. Here, the debate has turned towards production systems that maximize profits within strategies that have a neutral or high concern for the environment and an awareness of the concerns of ecologists and the public's perception of risk. The official view in the UK (20) may be taken as an example. High levels of agricultural support under the Common Agricultural Policy (CAP) were seen to have stimulated further intensification of production, with consequent adverse effects on the environment, including habitat loss and problems of water pollution in some areas, particularly from nitrates and farm wastes. Following the 1992 CAP reform, UK polices were to: promote environmentally friendly farming, conserve wildlife and habitats, protect and restore landscapes, reduce pollution from agricultural inputs and wastes, minimize use of pesticides and conserve nonrenewable resources. Less intensive arable systems incorporating decreased usage of fertilizers and pesticides, with proponents in Europe, contrast with the "maximum yield research-best management practices-maximum economic yield" school (29), where the focus is on intensification, increasing efficiency and more inputs. The ongoing arguments for and against these approaches and their intermediates compound the task of developing disease management strategies, because different systems tend to differ in their soilborne disease problems.

Progress frequently requires the linking of progressive areas. In considering what technical and political options there were to decrease the use of chemicals in cereals in Western Europe and to rescue farmers from the "endless spiral of intensification", it was stated (7) that substantial reductions in pesticide usage could be achieved by encouraging farmers to use DSSs such as EPIPRE. Eyespot disease of cereals has been included in more than one DSS. However, for most cereal growers, DSSs have not been very interesting economically, because they require laborious field observations and involve difficult diagnoses of, for example, stem base diseases. This latter point may be eased by new developments in diagnostic kits, which will need to be easy to use, qualitative and, in many cases, quantitative. Such kits will be valuable, not only for DSSs, but for the selection of resistant varieties of host plant.

Registration of environmentally safe products such as BCAs should be quicker and environmentally safe methods should be supported by subsidies (7). For BCAs to succeed on arable crops careful selection of suitable agents, a thorough understanding of the ecology of BCAs and pathogen, integration of biological and other methods of management and close collaboration between research workers and industry are needed (26).

Having progressed from initial scepticism, through outright hostility and grudging recognition, integrated pest management (IPM, Table 2, item 6) now finds growing acceptance, and its philosophy is increasingly applied to pest, disease and weed management (52). The example of integrated fruit management in Europe emphasized the importance of gaining consumer acceptance and the need for political and social initiatives to make IPM an established part of both food production and

Table 3. Study of soilborne plant pathogens in 1985.

A. Ways in which progress was being hindered

1. Premature publicity (e.g. biological control)
2. Uncritical acceptance
3. Generalization

B. Several long-running themes predicted to develop further

1. Soil microorganisms
2. Soil environment
3. Root exudates, root environment, rhizosphere
4. Pathogenesis and resistance
5. Mechanism of antagonism, biological control
6. Soil inoculum, population dynamics, quantification, epidemiology
7. Interactions (soil, microbe, plant)

---

marketing (52).

In some areas, management has not progressed as much as one would have hoped. New opportunities to manage take-all have been few and far between, despite much effort on natural biological control, introduced BCAs and, more recently, seed-treatment fungicides. At Rothamsted, we have spent many years developing field methods for testing and evaluating putative controls and the development of these revealed how the efficiency of soil-applied fungicides increased with increased disease pressure, whereas that of BCAs did not (5,25).

Experience has taught me that facing up to the great intractable problems in plant pathology, several of which are in the realms of soilborne plant pathogens, means accepting more complexity than is currently acknowledged in many research projects. I subscribe to the view that one is not really a plant pathologist until one has done a field experiment. Consequently, I often find myself in vigorous debate about pot experiments in take-all research. The proponents argue that in a complex system it is necessary to study component processes in simple or isolated conditions to make progress. My stance has a parallel in ecology; in considering whether we can make ecological predictions at a scale appropriate to the problems that need attention, it was pointed out that in exploring in progressively more detail the minutiae of ecological processes, possible complicating factors operating over a larger scale tend to be ignored (17). In the 5 years since the 5th ICPP in Kyoto, where biological control was in the spotlight, new technology has not yet come up with the answers. Furthermore, excessive shedding of traditional skills in favor of the molecular biological approach and/or relying on computer simulation will seriously disturb the balance of our science. This concern for subjects out of balance is not unique; fungi are increasingly employed as tools for investigating biochemical processes, frequently for industrial exploitation, and at the expense of attention to the whole fungus (15).

In many countries there is uncertainty about what politicians and the public want with regard to agriculture. Political and economic considerations leading to schemes such as set-aside shift emphasis with regard to underlying plant pathological challenges. Changes in circumstances may lead to problem avoidance (if cereals were not grown frequently in Europe, take-all would cease to be a problem), but what would we store up for the future by not solving the pathological problem?

LITERATURE CITED

1. Abel, C. 1994. SAC's Alert kit spots disease in 10 minutes. Farmers Weekly, 6 May, p. 53.
2. Anonymous. 1993. The roots of research at Rothamsted. Interrelationships between soil, soil organisms and plants. AFRC News, April, pp. 22–25.
3. Antonovics, J. 1994. The interplay of numerical and gene-frequency dynamics in host-pathogen systems. Pages 129–145 in: Ecological Genetics. L.A. Real, ed. Princeton University Press, Princeton, NJ.
4. Aylmore, L.A.G. 1990. Use of computer-assisted tomography in studying water movement around plant roots. Adv. Agron. 49:1–54.
5. Bateman, G.L., Hornby, D., Payne, R.W., and Nicholls, P.H. 1994. Evaluation of fungicides applied to soil to control naturally-occurring take-all using a balanced-incomplete-block design and very small plots. Ann. Appl. Biol. 124:241–251.
6. Bateman, G.L., Ward, E., and Antoniw, J.F. 1992. Identification of *Gaeumannomyces graminis* var. *tritici* and *G. graminis* var. *avenae* using a DNA probe and non-molecular methods. Mycol. Res. 96:737–742.

7.   Bigler, F., Forrer, H.R., and Fried, P.M. 1992. Integrated crop protection and biological control in cereals in Western Europe. Pages 95–116 in: Biological Control and Integrated Crop Protection: Towards Environmentally Safer Agriculture. J.C. Van Lenteren, A.K. Minks, and O.M.B. de Ponti, eds. Pudoc Scientific Publishers, Wageningen, The Netherlands.

8.   Blackmore, S., ed. 1994. Precision Farming: Issues and Opportunities. Proceedings, 5th Silsoe Link Conferences, Silsoe Link, Silsoe, UK. 38 pp.

9.   Blake, A. 1994. Yields all mapped out. Farmers Weekly, 22 April, p. 55.

10. Bowyer, P., Bryan, G., Lunness, P., Clarke, B., Daniels, M., and Osbourn, A. 1994. Saponin-detoxifying enzymes in plant pathogenic fungi. Abstract 1761 in: Abstracts, 4th International Congress of Plant Molecular Biology, 19–24 June, Amsterdam, The Netherlands.

11. Burgess, J. 1993. Soggy tomatoes and bumps on the head. New Scientist, 24 April, pp. 47–48.

12. Cassel, D.K., Kachanoski, R.G., and Topp, G.C. 1994. Practical considerations for using a TDR cable tester. Soil Technol. 7:113–126.

13. Coghlan, A. 1993. Engineering the therapies of tomorrow. New Scientist, 24 April, pp. 26–31.

14. Coleman, D.C., Dighton, J., Ritz, K., and Giller, K.E. 1994. Perspectives on the compositional and functional analysis of soil communities. Pages 261–271 in: Beyond the Biomass. Compositional and Functional Analysis of Soil Microbial Communities. K. Ritz, J. Dighton, and K.E. Giller, eds. John Wiley & Sons, New York, NY.

15. Cooke, R.C., and Whipps, J.M. 1993. Ecophysiology of Fungi. Blackwell Scientific Publications, Oxford, UK. 337 pp.

16. Cussenot, M., contact. 1993. Techniques moléculaires de détection et d'identification des micro-organismes (PCR, RFLP). INRA Mensuel, 69:18.

17. Edwards, P.J., May, R.M., and Webb, N.R., eds. 1994. Large-Scale Ecology and Conservation Biology. Blackwell Scientific Publications, Oxford, UK. 375 pp.

18. Fox, R.T.V. 1990. Rapid methods for diagnosis of soil-borne plant pathogens. Soil Use and Management 6:179–184.

19. Fox, R.T.V., Manley, H.M., Culham, A., Hahne, K., and Tiffin, A.I. 1994. Methods of detecting *Armillaria mellea*. Pages 119–133 in: Ecology of Plant Pathogens. J.P. Blakeman and B. Williamson, eds. CAB International, Wallingford, UK.

20. G.B. Parliament. 1994. Sustainable Development. The UK Strategy. Command Paper 2426, HMSO, London, UK. 267 pp.

21. Haas, C.N. 1989. Estimation of microbial densities from dilution count experiments. Appl. Environ. Microbiol. 55:1934–1942.

22. Hornby, D. 1985. The study of soilborne plant pathogens: changing outlook or more of the same? Pages 3–6 in: Ecology and Management of Soilborne Plant Pathogens. C.A. Parker, A.D. Rovira, K.J. Moore, P.T.W. Wong, and J.F. Kollmorgen, eds. The American Phytopathological Society, St. Paul, MN.

23. Hornby, D., compiler. 1990. Soil Use and Management 6:167–217.

24. Hornby, D. 1994. Aspects of the autecology of the take-all fungus. Pages 209–226 in: Ecology of Plant Pathogens. J.P. Blakeman and B. Williamson, eds. CAB International, Wallingford, UK.

25. Hornby, D., Bateman, G.L., Payne, R.W., Brown, M.E., Henden, D.R., and Campbell, R. 1993. Field tests of bacteria and soil-applied fungicides as control agents for take-all in winter wheat. Ann. Appl. Biol. 122:253–270.

26. Kerry, B.R. 1992. The biological control of soil-borne pests and diseases. Pages 117–123 in: Biological Control and Integrated Crop Protection: Towards Environmentally Safer Agriculture. J.C. Van Lenteren, A.K. Minks, and O.M.B. de Ponti, eds. Pudoc Scientific Publishers, Wageningen, The Netherlands.

27. Klepper, B., and Rickman, R.W. 1990. Modeling crop root growth and function. Adv. Agron. 44:113–132.

28. Laing, S.A.K., and Deacon, J.W. 1991. Video microscopical comparison of mycoparasitism by *Pythium oligandrum, P. nunn* and an unnamed *Pythium* species. Mycol. Res. 95:469–479.

29. Ludwick, A.E. 1994. Maximum yield research: friend of the environment. Pages 2–12 in: Research for Maximum Yield in Harmony with Nature. Transactions of the 15th World Congress of Soil Science. Acapulco, Mexico, July 1994, Satellite Symposium Sponsored by PPI. J.D. Etchevers, ed. The International Society of Soil Science and The Mexican Society of Soil Science, Mexico.

30. Lynch, J.M., ed. 1990. The Rhizosphere. John Wiley & Sons, New York, NY. 458 pp.

31. Miller, S.A., Rittenburg, J.H., Petersen, F.P., and Grothaus, G.D. 1992. From the research bench to the market place: development of commercial diagnostic kits. Pages 208–221 in: Techniques for the Rapid Detection of Plant Pathogens. J.M. Duncan and L. Torrance, eds. Blackwell Scientific Publications, Oxford, UK.

32. Morrissey, J.B. 1993. Perspective on sustainability from the Canadian agricultural research community. Page 1 in: Abstracts, 6th International Congress of Plant Pathology, Montreal, Canada. National Research Council of Canada, Ottawa, Canada.

33. Myrold, D.D., Hilger, A.B., Huss-Danell, K., and Martin, K.J. 1994. Use of molecular methods to enumerate *Frankia* in soil. Pages 127–136 in: Beyond the Biomass. Compositional and Functional Analysis of Soil Microbial Communities. K. Ritz, J. Dighton, and K.E. Giller, eds. John Wiley & Sons, New York, NY.

34. Nene, Y.L. 1993. Sustainable agriculture: future hope for developing countries. Page 1 in: Abstracts, 6th International Congress of Plant Pathology, Montreal, Canada. National Research Council of Canada, Ottawa, Canada.

35. Newmark, P. 1993. Biology. Pages 223–236 in: World Science Report 1993. Anon., ed. UNESCO Publishing, Paris, France.

36. Nicolas, H., Rivoal, R., Duchesne, J., and Lili, Z. 1991. Detection of *Heterodera avenae* infestations on winter wheat by radiothermometry. Rev. Nématol. 14:309–315.

37. Pleydell-Bouverie, J. 1994. Cotton without chemicals. New Scientist, 24 September, pp. 25–29.

38. Pussard, M., Alabouvette, C., and Levrat, P. 1994. Protozoan interactions with soil microflora and possibilities for biocontrol of plant pathogens. Pages 123–146 in: Soil Protozoa. J.F. Darbyshire, ed. CAB International, Wallingford, UK.

39. Saunders, D.A., and Hettel, G.P., eds. 1994. Wheat in Heat-Stressed Environments: Irrigated, Dry Areas and Rice-Wheat Farming Systems. CIMMYT, Mexico, D.F. 402 pp.

40. Schots, A., Dewey, F.M., and Oliver, R., eds. 1994. Modern Assays for Plant Pathogenic Fungi: Identification, Detection and Quantification. CAB International, Wallingford, UK. 267 pp.

41. Smith, N.C., and Stribley, D.P. 1994. A new approach to direct extraction of microorganisms from soil. Pages 127–136 in: Beyond the Biomass. Compositional and Functional Analysis of Soil Microbial Communities. K. Ritz, J. Dighton, and K.E. Giller, eds. John Wiley & Sons, New York, NY.

42. Strijbosch, L.W.G., Does, R.J.M.M., and Albers, W. 1990. Multiple-dose design and bias-reducing methods for limiting dilution assays. Statistica Neerlandica 44:241–261.

43. Taylor, J.W. 1993. A contemporary view of the holomorph: nucleic acid sequence and computer databases are changing fungal classification. Pages 3–13 in: The Fungal Holomorph: Mitotic, Meiotic and Pleomorphic Speciation in Fungal Systematics. D.R Reynolds and J.W. Taylor, eds. CAB International, Wallingford, UK.

44. Tsror, L., Nachmias, A., and Krikun, J. 1985. Improvement of potato profitability through field mapping of *Verticillium dahliae* inoculum potential in the soil. Phytopathology 75:1365. (Abstr.).

45. Tuitert, G. 1994. Quantifying *Polymyxa betae* and BNYVV in soil. Pages 161–167 in: Ecology of Plant Pathogens. J.P. Blakeman and B. Williamson, eds. CAB International, Wallingford, UK.

46. Van der Plank, J.E. 1975. Principles of Plant Infection. Academic Press, New York, NY. 216 pp.

47. Warcup, J.H. and Talbot, P.H.B. 1962. Ecology and identity of mycelia isolated from soil. Trans. Br. Mycol. Soc. 45:495–518.

48. Ward, E. 1994. Use of the polymerase chain reaction for identifying plant pathogens. Pages 143–160 in: Ecology of Plant Pathogens. J.P. Blakeman and B. Williamson, eds. CAB International, Wallingford, UK.

49. Ward, E., and Akrofi, A.Y. 1994. Identification of fungi in the *Gaeumannomyces-Phialophora* complex by RFLPs of PCR-amplified ribosomal DNAs. Mycol. Res. 98:219–224.

50. Ward, E., and Bateman, G. L. 1994. Identification of fungi in the *Gaeumannomyces-Phialophora* complex associated with take-all of cereals and grasses using DNA probes. Pages 127–134 in: Modern Assays for Plant Pathogenic Fungi: Identification, Detection and Quantification. A. Schots, F.M. Dewey, and R.P. Oliver, eds. CAB International, Wallingford, UK.

51. Ward, E., and Gray, R.M. 1992. Generation of a ribosomal DNA probe by PCR and its use in identification of fungi within the *Gaeumannomyces-Phialophora* complex. Plant Pathol. 41:730–736.

52. Wearing, C.H. 1993. IPM and the consumer: the challenge and opportunity of the 1990s. Pages 21–27 in: Pest Control and Sustainable Agriculture. S.A. Corey, D.J. Dall, and W.M. Milne, eds. CSIRO, Canberra, Australia.

Chapter 2

CHALLENGES, OPPORTUNITIES, AND OBLIGATIONS IN ROOT DISEASE EPIDEMIOLOGY AND MANAGEMENT

C.L. Campbell and D.A. Neher

Root diseases cause extensive damage to crops and forest trees and result in significant losses in the production of food, forage and fiber worldwide. This damage is often not recognized because of the unseen nature of many root diseases and the fact that the visible foliar symptoms often belie the true nature of the damage that occurs below ground. Also, perhaps due to the circumstance that root diseases often remain the "hidden enemy" and because of the complexity and challenges of working with such diseases, losses in potential production due to root diseases remain largely unquantified.

The habitat in which roots grow is quite complex and provides significant challenges to researchers who delve into the mysteries of soilborne pathogens and root diseases. Within the physically and chemically variable milieu of soil, potential root pathogens survive and grow in competition with the myriad other microbes of the soil food web. Competition for nutrients, especially nitrogen, is intense and as roots grow through soil, they provide a primary source of nutrients through exudation, damage to fragile epidermal cells, and the sloughing of dead cells. Root turnover also contributes to the pool of scarcely available nutrients. Parasitism, predation, and omnivory are common in the soil food web. Some microbes survive well as saprophytes, only becoming root pathogens when specific opportunities arise. The soil environment, although seemingly well buffered, shifts continuously to favor one group of organisms or another within microsites on and between soil particles.

Given the complexity of soil ecosystems and root disease epidemics, substantial progress has been achieved in understanding the epidemiology of root diseases and in providing effective strategies and practices for their management (22). There remains, however, much to be accomplished, and, as the true significance of roots diseases in reducing food, forage and fiber production is recognized, there will be an even greater demand for practical, economical and environmentally safe management options for root diseases. If epidemiology is to set the strategy for disease management (106), root disease epidemiologists are faced with a number of challenges, opportunities and obligations.

Our goal in this chapter is to identify some of these challenges, opportunities and obligations that face root disease epidemiologists. Through this process, we seek to compel our colleagues (and ourselves!) to continue to think about root diseases with a critical but innovative view. We also hope to entice new researchers to join the discipline of root disease epidemiology as they perceive the possibilities for advancing the science and solving meaningful problems.

CHALLENGES

Researchers continue to be challenged by many aspects of the ecology of soilborne pathogens and the epidemiology of root diseases because each pathosystem presents its unique challenges. In this section, we concentrate on three challenges applicable to many root disease systems: quantifying inoculum, assessing disease, and designing effective studies. Others will certainly wish to add to the list; however, if researchers will accept and resolve the challenges presented, we will have made tremendous progress!

Quantifying Inoculum

Propagules of soilborne pathogens are associated intimately with soil particles, organic residues, and other organisms in soil. Successful quantification of inoculum requires the initial separation, isolation or selection of propagules that will be effective in infecting host roots from the associated soil, organic matter and soil biota. The propagules must then be captured physically and identified. Propagules can be identified and quantified visually, or through a species-specific assay, or by growth on a culture medium by morphological, biochemical, or microscopic criteria. Finally, to quantify effective inoculum density in the soil ecosystem in which the propagules reside, there must be some mechanism to determine what proportion of the propagules obtained are viable and capable of infecting host roots. Ideally, there should also be means of determining what proportion of the potentially effective propagules have been recovered from the soil and how representative the soil sample is for the area (e.g. plot, field, county or state) of interest. The process is not a simple task and as Benson (12) indicates, a large proportion of a research budget must often be expended for materials and time to assay inoculum in soil.

Many methods for quantifying inoculum of soilborne plant pathogens are available (12,98). The primary methods include direct counts, often after soil sieving (9,10,65), bioassays (10,89), and soil assays with baits or selective or semiselective media (25,51,98), enzyme-linked immunosorbent assay (ELISA) in commercial kits (3,62,67,96,101), and ELISA with monoclonal antibodies (44,100). Presence of a fungal pathogen in soil can also be determined through substrate colonization (68,85,102) or through the use of amplification procedures and species-specific DNA probes (19,20,43).

Which method is "best" for quantifying inoculum must be judged in relation to the original purpose for which it was developed and the purpose for which the method is to be used. Regardless of the method selected, there should be a critical assessment of the efficacy of that method prior to its use. Some comparative studies on the reliability (i.e. relative ability to identify the target pathogen correctly), precision (i.e. relative ability to provide the same result when the assay is performed repeatedly for a given soil sample), accuracy (i.e. the closeness of the value obtained to the true value) and efficiency (i.e. the cost per unit of information obtained) of some assays for quantifying inoculum have been performed (28,58,80,86,90,99). Data related to inoculum quantification is important in understanding the epidemiology of root diseases; more studies on inoculum quantification are needed.

The statistical issues of sampling, coupled with the biological reality of spatially aggregated propagules, often pose an apparent dilemma in quantifying the inoculum of soilborne pathogens. Often an investigator cannot obtain and assay a sufficient number of samples to obtain data with the desired degree of precision, especially for inoculum that is highly aggregated or clustered. Aggregation generally increases sample variance compared to a situation where propagules have a random or uniform spatial pattern. Yet, funds are often not available to conduct all of the assays required to obtain the desired degree of precision. Reasons for this include: (i) the relative expense (in terms of material and personnel costs) of most assays; (ii) the overwhelming desire to use a small number of samples to represent a relatively large area; and (iii) the physical, chemical and biological variation that occurs naturally in soil. One resolution to the dilemma, which is invoked all too often, is for the investigator simply to decide how much time and money can be spent on assays, calculate the time and cost per assay and use the quotient of the two quantities to provide the sample number without due consideration of the statistical consequences. A better solution is to perform the needed preliminary studies with the target organism and intended assay to quantify the components of variance associated with the various sampling and assay procedures and then to calculate an optimum sample allocation plan and sample size (28,79). Even with this procedure, it may not be possible to obtain the optimum number of samples; however, at least an informed decision on resource allocation can be made prior to the actual study to quantify propagules.

The degree of success we achieve in relating data on inoculum density to other components of the epidemic depends primarily on the quality of the data obtained (21). As a result, a quality assurance plan should be developed for each study that will involve inoculum quantification. A quality assurance plan will probably include, whenever possible, the inclusion of the assay of a certain number of "known" samples (or calibration standards) during the actual performance of the assay to insure that all procedures are being followed and performed correctly. If a sample from a reservoir of soil with a known inoculum density is included in each "batch" of soil samples, the same propagule number (within whatever limits of measurement error that are established) should be obtained each time.

To meet the challenges of quantifying inoculum from soil, answers to a series of questions should be obtained prior to the collection of data with a particular method. Such questions include:

- How reliable, precise, accurate and efficient is the proposed assay?
- What are the critical steps and likely sources of error in performing the proposed assay?
- What allocation of resources during sampling and assay performance will provide the best quality data?
- What quality assurance procedures are in place to insure that the data are of the best quality possible?

Assessing Disease

Disease assessment is one of the most important and often most challenging tasks in the study of plant diseases. With root diseases, assessment is even more of a challenge than with most foliar diseases, because the host parts on which we desire to assess disease, i.e. roots and other subterranean plant parts, are "hidden" in the soil. This means that symptoms, and even the extent of the host tissue to be assessed, can not be evaluated readily on roots. Additional or alternative steps must be taken to complete disease assessment.

Whether roots or shoots are the appropriate host part to be sampled and evaluated will often depend on the purpose of the assessment. If the purpose is to describe temporal progress of the disease on roots, then roots are the logical plant part for evaluation. If the purpose is to relate disease severity or incidence to yield, then the portion of the plant harvested for yield will be a determining factor in selecting the plant part to be evaluated. For example, with soil rot or pox (*Streptomyces ipomoea*) of sweet potato, severity of symptoms on the fleshy storage roots will be related directly to yield quality and quantity, whereas severity of symptoms on fibrous roots may be related only indirectly to yield (92). In contrast, when fruit are harvested from the above-ground plant parts, the severity of symptoms on shoots may influence yield more than the severity of root symptoms as in the case of Phytophthora root rot of processing tomato (78).

Various procedures or alternatives are available for the assessment of root or shoot symptoms associated with root diseases. We have presented specific methods for estimating severity and incidence of root diseases previously (27) and will not further discuss those methods here. Rather, the challenge to be considered here is how to choose and observe the most representative and meaningful sample of plant material for disease assessment while causing the least possible disturbance to the epidemic and the pathosystem. Three specific options have been utilized by root disease epidemiologists as a surrogate for assessment of root disease: in situ observation of roots; removal of plants (and roots) from soil; and assessment of foliar or shoot symptoms.

Rhizotrons or root observations boxes or tubes can be constructed and placed in the field (17,52,60,61,103,104). This option can be quite effective for a small sample of plants but is impractical and cost-prohibitive for large areas of fields or with many fields. Additionally, there is some disturbance of the soil system with placement of the observation ports and, if glass or plastic surfaces are used for observation ports, a modified environment is created for those roots and organisms being observed.

Plants can be excavated and excess soil removed so that the roots are exposed for assessment (26,29,52,77,93). This option requires that the sample unit be destroyed during the assessment, thereby eliminating the possibility of repeated observation on the same sample unit or plant. Also, it is labor intensive (particularly for larger annual plants and certainly for many large perennials!), causes significant disturbances in the soil ecosystem and limits the number of times assessments can be made, particularly if yield data are required from the same study. Another challenge with root excavation for disease assessment is that those roots which have the greatest amount of disease may actually be sloughed prior to sampling or lost during the removal process due to their weakened physical structure. Because of this likely loss of severely diseased roots and the possibility that an important symptom is the stunting of the root system, disease assessments should be made in comparison to a healthy root system. Assessment of root area or volume in relation to healthy plants may be as significant a portion of a disease assessment as the determination or estimation of the area or volume of root tissue occupied by lesions.

A third option is to assess shoot symptoms that develop as a result of the root disease. This option is the least labor intensive and results in the least disturbance to the soil ecosystem. Although this approach has been used successfully in some pathosystems (24,31,50,53,94,95), it also results in assessment data that are not necessarily representative of the true progress of the root disease. Rather, root disease probably develops to a certain stage, which may be dependent on weather conditions, before any visible shoot symptoms are apparent (78). Inference of the extent of root disease by observing only foliar symptoms such as yellowing, epinasty, chlorosis or wilting should be done with caution (47). In some diseases for which foliar or shoot wilting occurs rapidly after root infection, e.g. Phymatotrichum root rot of cotton (50), evaluation of foliar symptoms may be quite appropriate. With other host plants, however, foliar symptoms may not be sufficiently sensitive to reflect actual damage occurring on roots or the appearance and severity of symptoms may be confounded with factors such as weather and host genotype.

Another challenging aspect of assessing disease severity on excavated roots is the relative location of the disease within the root system. In order to determine the physiological effects of disease on plants, the position and depth of lesions on roots should be evaluated (47). Morphometric root analysis systems are available (33–36) for the specification of root order (first, second or third) and type (lateral, tap). Lesions on tap roots would likely reduce water flow to stems more than lesions on lateral roots such that shoot symptoms would be present when the tap root has lesions but not when lesions are restricted to lateral roots (52,73).

Finally, the challenge of selecting the most appropriate number of samples to quantify how much disease is present enters into disease assessment in much the same way it did for quantifying inoculum of soilborne pathogens. The resolution of this issue can be achieved with the procedures identified previously. An added challenge is that if plants are dug so that roots can be examined, adjacent plants must be excluded from future assessments because of the extensive disturbance to the soil ecosystem and possible alteration of plant competition. Thus, studies must be planned so that the number of specified, randomly selected samples can be chosen within the constraint of missing plants and a shrinking population from which to select future samples.

Designing Effective Field Studies

Root diseases pose a special challenge in designing effective field studies, because the initial inoculum is usually present in soil prior to the initiation of host growth or is introduced with the host. This is certainly true for most annual crops and for some perennials. Although the influx of inoculum during the course of a growing season from sources outside a field or adjacent area should not be discounted, compared to the situation with most foliar diseases, the continuous or regular influx of inoculum is a relatively rare event for root diseases. These factors imply that the spatial dimension of inoculum pattern is of primary importance in selecting the initial experimental design for field studies of root disease epidemiology.

Spatial scale is a factor that should receive more critical consideration in the design of field studies for root diseases. The spatial scale of concern often occurs in the horizontal plane of a field. However, the vertical pattern of propagule occurrence, which involves propagule distribution and root growth within the soil profile, has been considered for several pathogens (1,16,18,63,66,72) and should not be ignored.

Ecologists partition spatial scale into extent, the overall area encompassed by a study, and grain, the size of individual units of observation such as quadrats (2,108). Extent and grain define the upper and lower limits of resolution of a study, because inferences about patterns or processes of events cannot be made legitimately beyond the extent or below the grain of a study. Also, because our ability to discern and interpret biological and environmental effects on spatial processes is dependent on extent and grain of the study (83,108), investigators should consider these items in designing field studies carefully.

With agronomic and horticultural crops, field size or farm size may influence root disease epidemics and thus, determine the extent of the epidemic. However, a wide range of extent values have been used by researchers working with soilborne pathogens and root diseases (23). Other biological and ecological criteria for determining extent will have to be identified for forests, riparian areas, or other natural ecosystems where no specific management practices, such as cultivation, define the boundaries of the ecosystem. The determination of an appropriate grain size for studying a root disease may be more challenging than the determination of an appropriate extent. Investigators have used grain sizes from < 1 to > 100 m^2 (23). Actual grain selection should be based upon: (i) biological factors such as cluster size and potential dispersal distance for a soilborne pathogen; (ii) soil factors such as soil map unit, soil type or obvious physical attributes of an area; and (iii) cultural factors such as row spacing and pattern of cultivation. The determination of grain, therefore, requires a fair amount of prior knowledge about the pathogen, host and the experimental site, including climatic influences.

Another factor of importance in designing field studies for root diseases is that the soil has a successional status and a degree of ecosystem stability that extends among seasons and from year to year. A complex and integrated food web of organisms is present in virtually every soil and soilborne pathogens are only one component in that food web (69). The interactions among pathogens and the other living residents of agricultural, forest, wetland, and prairie soils, particularly the species composition and function of microbes, play a vital role in determining the suppressiveness or conduciveness of soils for the development of root diseases and should be considered in designing field studies (105).

A third factor of importance in designing field studies concerns the physical and chemical properties of the soil itself. These compositional factors can significantly influence the environment to which roots are exposed. Differences in soils within fields and among fields, even in the absence of cultural and biological factors, thus represent a potential variation in environment that can affect the development of root disease epidemics.

A fourth factor of importance is the temporal scale at which the soil environment changes. Because of the chemical buffering capacity and the biological complexity and ecosystem stability of most soils, the soil environment changes rather slowly. However, cultural practices and weather patterns can affect the soil ecosystem among seasons or years. The addition of specific soil amendments and use of crop rotations can influence the soil environment and, thus, conditions for root disease development from year to year (105). In some cases, multiple year studies may be required to characterize the effects of specific treatments or factors on root disease epidemics.

* * * * *

The challenges we have discussed are fundamental to the science of root disease epidemiology. These challenges in quantifying inoculum, assessing disease and designing effective field studies can be resolved and become successes. In doing so, they will serve to advance both the practical application and the theoretical framework of root disease epidemiology.

These challenges should not be viewed as impediments to research with soilborne pathogens and root diseases. Rather, they should be embraced as aspects of our experimental science that we are attempting to improve continuously. They are aspects of the discipline of root disease epidemiology that should be considered when any study is being planned and implemented. No single study will resolve all the challenges; however, even the consideration of the challenges accompanied with a serious consideration of the question "How can I quantify inoculum, assess disease and design experiments better?" must be viewed as a success.

OPPORTUNITIES

There is a growing cadre of knowledge about the population dynamics and ecology of soilborne pathogens and the epidemiology of root diseases. The epidemiological approach has been used to investigate the temporal and spatial aspects of a range of pathosystems, primarily for annual crops in agronomic and horticultural settings. A cohesive and innovative approach to the theoretical aspects of the temporal and spatial dynamics of soilborne pathogens and root disease epidemics is also developing (37,39,40,42,49).

With the increasing interest among root disease epidemiologists, the knowledge base concerning these diseases and recognition of losses in potential yield caused by root diseases has expanded. Thus, new opportunities for significant contributions to the fundamental and practical understanding of root disease epidemiology become available to researchers. Specific opportunities for innovative approaches arise in subjects such as the modeling of the components of root disease epidemics, forging the linkage between soil ecology and the ecology of root pathogens, comparing and classifying epidemics and setting strategies for disease management. Researchers can take full advantage of such opportunities, in part, by maintaining a keen awareness of research progress in similar areas with foliar pathosystems and in microbial ecology. However, the greatest progress will be made through the development of innovative approaches and the realization that root disease epidemics may not have the same fundamental components and developmental pathways as foliar epidemics. Because the soil ecosystem is more complex than, and quite different from, the ambient ecosystem of the phyllosphere, and because of the distinct differences in structure and function between roots and leaves or shoots, there may be little reason to expect epidemics of diseases of roots and shoots to develop similarly. If researchers will avail themselves of the opportunities identified for root disease epidemics, we will continue to make progress toward understanding and managing these epidemics.

Modeling Root Disease Components

As Jeger (49) noted, "the modeling of root diseases has received rather less attention than that of foliar diseases". There has, however, been significant progress since that time in defining and modeling the components of root diseases and, in all probability, the seminal work of Gilligan (37–40,42) and Jeger (49) has provided a foundation in modeling and understanding of root disease epidemics that is even more complete and cohesive than that available for foliar diseases. As a result, the opportunities now exist to further explore, via modeling and empirical studies, the roles of primary and secondary infection, of root growth and of inoculum dynamics, including the survival of inoculum and the interactions of pathogens and other microorganisms in soil ecosystems. The empirical data can be evaluated with respect to current models and will serve as a basis for further modeling efforts. The understanding provided by these further modeling and empirical explorations will also present the opportunity to develop more rational and comprehensive strategies for managing root diseases.

Idealized disease progress and inoculum dynamics curves have provided the starting point for much of the mathematical analysis of root disease epidemics. The modeling efforts have been based on the pathogen or the host. Probability models have been proposed for the pathogen-based approach which relate infection to inoculum density in soil (38). Models based strictly on symptom expression on aboveground plant parts have also been proposed (21) (although the possible hazards of employing such models has been alluded to in the section on Challenges – Assessing Disease). For either the pathogen- or host-based approaches, monotonic curves, in which disease increases progressively toward some upper, asymptotic level, have dominated work in the epidemiology of root diseases (42).

The shapes of cumulative curves of disease intensity for root diseases have been a source of interest to many researchers and have provided insight into the interrelations among the biological components of these diseases. Although the increase of plant disease intensity during epidemics caused by root pathogens often appears sigmoid over time, the expectation is for a monomolecular curve (25,64) to describe an epidemic which is monocyclic or "simple interest" sensu Vanderplank (106). The conclusion of mechanisms of disease increase based solely on the shape of the disease progress curve is inappropriate (25,71,88); however, the hypothesis of mechanisms that result in sigmoidal curves for disease development over time has been the basis of much of the recent mathematical or analytical modeling for root disease systems. Some unusual disease progress forms, such as double sigmoidal (4,45), have been described for foliar diseases but have not yet been reported for root diseases.

Models of cumulative disease curves and essential disease components have been proposed. Jeger (49) explored models based upon symptom expression by above-ground plant parts and then proposed detailed models to combine root growth and increase of lesions on roots with and without lesion expansion on roots. In an elegant and masterful sequence of publications, Gilligan (38,40–42) examined the interactive relationship between host and pathogen for components such as host infection by soilborne fungi, rate of contact of inoculum and roots, the dynamics of inoculum production and survival, the growth of roots, the occurrence of primary infections, the transmission of infection by root-to-root spread of pathogens, the role of root density, death of roots, latent and infectious periods for root pathogens, and antagonistic interactions between pathogens and other microbes in soil.

Presently, the shortage of empirical data imposes practical constraints on the analysis of root disease epidemics. For example, more experimental work is needed on the dynamics of inoculum survival in soil. How long does inoculum survive in soil, how long is it infective, and what do the curves of inoculum density and of infective inoculum density (i.e. those propagules that are actually infective) actually look like—e.g. are they monotonic or cyclic? More studies are needed on the interactive effects of pathogens and roots during epidemics. The density of roots certainly affects the dynamics of infection, but little is known about the effects of disease on root growth (42). Also, the dynamics of infection of roots is only one component of the overall disease cycle; more information is needed on the length of time between root infection and when the roots become infectious (i.e. the latent period) and how long the roots remain infectious (i.e. the infectious period). The models proposed by Gilligan and Jeger are, thus, an excellent starting point for determining the mechanisms of root disease epidemics; however, there are many excellent opportunities for additional modeling studies and for empirical studies to evaluate the validity of currently proposed models and to serve as the stimulus for new, more comprehensive models.

Forging The Linkage Between Soil Ecology
And The Ecology Of Root Pathogens

The beneficial role of nonpathogenic soil invertebrates is largely unexplored by plant pathologists, who usually just consider plant-pathogenic bacteria, fungi, and nematodes in soils. However, the nonpathogenic organisms in soil far outnumber the pathogenic ones! There are, for example, an average of 10^6 to 10^7 free-living nematodes that feed on bacteria, fungi, algae, and other nematodes, 10^4 to 10^5 enchytraeids (pot worms), 10^3 to 10^4 mollusks (slugs, snails), 10^2 to 10^3 myriapods (millipedes, centipedes), 10^2 isopods (wood lice), 10^2 Araneidae (spiders), 10^4 Collembola (springtails), and 10^5 Acranaria (mites) per square meter of soil (87). Microfauna and mesofauna in soils play important roles in decomposition of organic matter and nutrient cycling. Microinvertebrates, such as nematodes (7) and protozoa (109), contribute directly to nitrogen cycling by excreting nitrogenous wastes, which are released mostly as ammonium ions (48). Microinvertebrates also enhance soil fertility directly by depositing feces and existing as a reservoir of nutrients, which are released when they die. Microarthropods contribute to decomposition of organic matter indirectly by fragmenting detritus and increasing surface area for further microbial attack (13). Subsequently, soil invertebrates graze upon microbes, and thereby alter nutrient availability, affect microbial growth and metabolic activities by selective grazing, and alter the composition of microbial communities. In addition, soil fauna also transport bacteria, fungi, and protozoa (in gut or on the cuticle) across regions of soil impenetrable by microbiota, and thus enhance microbial colonization of organic matter (70).

Plant pathologists often look at the negative effects of soil invertebrate mesofauna (14) and do not fully consider the beneficial aspects of these organisms (8). The rhizosphere-inhabiting collembolans, *Proisotoma minuta* and *Onychiurus encarpatus,* graze preferentially upon the root pathogen, *Rhizoctonia solani*, on cotton seedlings in the presence of three well known biological control fungi, *Laetisaria arvalis, Trichoderma harzianum* and *Gliocladium virens* (32). Giant amoebae of the Vampyrellidae (*Arachnula, Thecamoeba, Saccamoeba, Vampyrella*) perforate conidia of *Cochliobolus sativus*, a fungus causing root rot of barley (82). Orabatid mites prefer feeding on pigmented fungi over nonpigmented fungi, which implies their potential for destroying pigmented pathogens such as *R. solani* and *C. sativus*. Larvae of *Bradysia coprophila* (dark-winged fungus gnat) prefer sclerotia of *Sclerotinia sclerotiorum,* the cause of lettuce drop in muck soils of Quebec, as a food source (5). *Bradysia coprophila* secretes chitinase in its saliva while feeding on sclerotia of *S. sclerotiorum*, thereby disrupting germination of the pathogen (6). The larvae failed to survive when provided the mycoparasitic fungus, *Trichoderma viride*, as a food source (6). Microarthropods create problems as pests usually because a preferred food source is absent. For example, root-grazing injury by species of the collembolan *Onychiurus* on sugar beet is caused by the rubbing of their bristled bodies against the root tissue. However, if certain types of weed species and certain kinds and amounts of organic matter are present providing the preferred microbial food supply, root injury decreases (32). These examples provide evidence that common species of small-animal communities can consume sufficient pathogen inoculum to lower disease incidence and suggest the need for a new look at the mechanisms underlying biological control and soils suppressive to root diseases. A better understanding of the interactions among soil flora and fauna would complement our understanding of root disease epidemics and should lead to a better understanding of the mechanisms of disease management strategies, particularly biological control.

Species composition and function may be more important than species diversity in determining disease suppression (105). Importance of functional groups in relation to disease suppression is exemplified by a positive correlation between suppression of corky root of tomato and the Shannon-Weaver diversity index (97) for functional groups of Actinomycetes isolated from rhizospheres of tomato seedlings grown in organically and conventionally managed soils (110). Disease suppression may depend on communities of microorganisms associated with a specific substrate of a certain quality under certain environmental and management conditions. For example, disease suppression is often enhanced by incorporation of organic amendments in soil. The effectiveness depends on the specific material used, the time elapsed since incorporation, and the pathogen under study. Fresh debris sometimes increases plant disease by providing a food base for facultative saprophytic pathogens. For example, *R. solani* can utilize cellulose as a sole source of carbon (11), and thus thrives in fresh or immature compost material relatively high in cellulose content. A biological control agent, *Trichoderma* sp., degrades cellulose rapidly (46). However, if cellulose levels are high, free glucose concentrations accumulate and may repress synthesis of chitinase involved in hyperparasitism of *R. solani* by *Trichoderma* sp. (30).

Forging the linkage between soil ecology and the ecology of root pathogens will require that root disease researchers expand their view of the interactions between pathogens and roots. It will necessitate the examination of soil food webs to ascertain the expected or "normal" composition of the soil microflora and micro- and mesofauna for certain types of soil ecosystems in specific areas. It may also require that reference sites of reference systems be defined for comparison. For example, soil ecosystems with perennial hosts such as pasture species may be appropriate reference sites in a region for comparison of the diversity and abundance of soil organisms with those sites with annual hosts such as many agricultural crops (76). Such examinations will be time-consuming and expensive; however, the result of this expanded view will be a much improved understanding of the ecology of soilborne pathogens. This, in turn, will allow a more complete examination of the epidemiology of root diseases.

Comparing And Classifying Epidemics

Epidemiologists compare the progress curves of root diseases over time and the patterns of root disease in space in order to gain fundamental knowledge about the factors that influence the course of disease progress in time and space and, ultimately, to establish strategies for management of plant diseases. Vanderplank (106) identified the need for comparison of disease progress curves and provided an initial, simple framework for making such comparisons over time. He proposed the comparison of disease progress curves based upon two parameters associated with several disease progress models—initial disease and the apparent infection rate. Kranz (54–57) recognized the greater wealth of information contained in disease progress curves and proposed a multivariate approach in establishing the subdiscipline of comparative epidemiology. The recognition of a larger number of parameters available for comparison and the use of multivariate, statistical techniques for epidemic analysis and classification provides a more realistic view of the factors involved in disease progress over time than the simpler, two-parameter approach proposed by Vanderplank. The goals of comparative epidemiology are thus twofold: (i) to examine the interrelationships among descriptors of epidemics and ascertain a minimum set of descriptors needed to characterize an epidemic; and (ii) to classify epidemics into a number of meaningful and interpretable classes or categories.

Although a few studies are available that apply the methodology of comparative epidemiology to the temporal (24,26,29) and spatial (81) aspects of epidemics caused by soilborne pathogens, there is much more to be learned in this area. Comparative epidemiological studies are needed that address both the temporal and spatial aspects of root diseases. The major challenge in making progress with such studies is the lack of data sets with information on a relatively large number (6 to 10) of observations and characteristics for epidemics of root diseases. The requisite data sets will be relatively expensive and difficult to assemble; however, the opportunities available for real advancement of our knowledge of root disease epidemiology through comparative studies will more than compensate for the costs associated with data acquisition.

An initial benefit of such comparative studies will be a better understanding of the parameters that are needed to characterize root disease epidemics. For example, there should be some degree of similarity in the things which need to be measured to characterize epidemics of wilt diseases caused by soilborne species of *Phytophthora* or *Fusarium* on annual hosts. Once a relatively small number of essential parameters is established for the characterization of certain types of epidemics, the cost and difficulty of data collection should be reduced. The set of essential parameters will, of necessity, contain elements that describe, or are surrogates for, host or root growth, inoculum dynamics, and disease development. Other elements may be needed to describe key ecological and environmental factors in the soil.

Once a fundamental understanding is obtained concerning the parameters that are needed to describe root disease epidemics, specific fundamental and immediately practical questions concerning the development and management of root diseases can be addressed. Questions to be answered might include: are there a limited number of types or categories of root disease epidemics that occur in agricultural ecosystems, and is the epidemic category determined primarily by the type of pathogen, host or soil characteristics? Do epidemics caused by species of *Phytophthora* and *Pythium* on agronomic or horticultural crops have more in common than epidemics caused by species of *Phytophthora* and *Rhizoctonia*? Do epidemics caused by fungi (e.g. *Aphanomyces*) differ from epidemics caused by prokaryotes (e.g. *S. ipomoea*), nematodes (e.g.

Meloidogyne incognita), or viruses (e.g. soilborne wheat mosaic virus)? Are root disease epidemics of annually harvested, herbaceous crops more similar to each other than to epidemics of perennial crops? Is the type of host plant (e.g. legumes or grasses) more a determinant of epidemic behavior than pathogen type? Do epidemics of annual and perennial crops differ only in the relative time or spatial scale or in other fundamental ways?

Setting Strategies For Disease Management

With regard to disease management, Gilligan (40,42) has proposed a series of three equations, one for infected roots, one for total roots, and one for inoculum, that account for primary and secondary infection of soilborne pathogens with allowance for root and inoculum dynamics. The parameters of the models suggest alternative epidemiological strategies for disease management that can be evaluated and compared among pathosystems. These include (42, p. 158): (i) reduction of inoculum by "removal" of initial inoculum and/or increase in the rate of decay of inoculum; (ii) reduction in the rate of primary infection; (iii) reduction of the rate of secondary infection; (iv) alteration of host density by change in initial host density and/or change in asymptotic root density and/or change in the rate of production of roots.

From his theoretical analysis of the influence of root growth and inoculum density on the dynamics of root disease epidemics, Jeger (49) was able to provide several specific suggestions for root disease management strategies. These suggestions were to reduce pathogen density, to maintain a low rate of root extension relative to root infection, and to restrict lesion expansion. These recommendations are compatible with the epidemiological methods proposed by Gilligan (40,42) and provide a starting point for the empirical evaluation of strategies for root disease management.

Thus, the opportunity exists to evaluate which of these strategies or combination of strategies may be best for managing specific types of root diseases. Comparative epidemiology may allow the extension of such an evaluation to determine if specific strategies or combinations work best for categories of root disease epidemics. The result would be the ability to provide at least an initial prescription of the type of management strategy that may be expected to work the best for a "new" disease without extensive and costly empirical evaluation of each management option. Such a prescription should be derived from the knowledge of the basic set of parameters needed to characterize disease progress temporally and/or spatially in specific categories of epidemics.

OBLIGATIONS

Root disease epidemiologists have the principal obligation of making the results of their research defensible theoretically and biologically, understandable to others with a reasonable knowledge of ecology and epidemiology, and useful to those who seek to manage root diseases in the real world. The obligations of root disease epidemiologists can be met, in part, through accepting the challenges and taking advantage of the opportunities discussed earlier. These obligations are not met easily and require not only scientific knowledge, insight and integrity, but also perseverance and perhaps some luck.

We propose the following specific obligations or goals for root disease epidemiologists.

1. Establish a sound theoretical framework for understanding root disease epidemics.
2. Present possible strategies for root disease management inherent in an ecologically based framework.
3. Provide a practical, useful framework for describing and analyzing epidemics of root diseases.

Establishing A Sound Theoretical Framework

The mathematical models proposed by Gilligan (38–40,42) and Jeger (49), combined with simulation models such as those of Bloomberg (15) and Reynolds et al. (91), have provided the basis for meeting the first goal. Further empirical work is needed, as noted by Gilligan (42), to discern how infection by soilborne pathogens, and, more importantly, disease, affects root growth so that such effects can be incorporated into mathematical models. Additional empirical and modeling efforts also need to be devoted to the dynamics of survival of soilborne pathogens in soil ecosystems. Studies that examine the survival dynamics of specific propagules of individual pathogens alone will be useful. Studies that examine the dynamics of inoculum in a range of soil ecosystems would, however, be more useful in the long term, because methods to incorporate effects of antagonistic or competitive microorganisms, which also reside in the soil ecosystem with pathogens and roots, must be developed.

There is a need to incorporate spatial aspects of pathogen dynamics and root disease development into the overall theoretical framework for root disease epidemics before it can be considered complete. Spatial attributes, at several scales from the rhizosphere of a root or whole plant to a focus of disease or a field and even a region, will be important (23). Although the modeling of the temporal attributes of root disease development is a significant challenge in itself, the incorporation of the spatial dimension into root disease models, along with the temporal dimension, will be even more challenging and may even require new approaches.

Presenting Strategies For Root Disease Management

As researchers and growers come to recognize the true significance of root diseases in reducing the yield of plants producing food, forage and fiber, there will be an even greater demand for strategies and practices to manage root diseases effectively and efficiently. Host resistance will answer part of the demand for better management of root diseases; however, there will be a significant need for additional strategies that are effective, economically feasible, and environmentally friendly to promote ecological sustainability in agricultural systems (74). This challenges root disease epidemiologists to broaden their view of soil ecosystems to encompass the view held by soil ecologists. As the ecological "world view" of root disease epidemiologists expands to encompass the many organisms that compose the soil food web, we will become aware of many more natural mechanisms for root disease control accounted for by various nonpathogenic microbes and invertebrates who graze preferentially on pathogenic fungi, bacteria, and nematodes (32,59).

In nature, there is a balance within soil communities between relative abundance of pathogens and nonpathogenic organisms. If the balance is disrupted, for example by use of general biocidal soil fumigants, it is difficult, if not impossible, for soil to regain its original diversity (111). We may benefit from comparative studies of soil communities between natural and agricultural ecosystems for identification of natural enemies, biopesticides, and other means of classical biological control (107). A more comprehensive understanding of the spatio-temporal distribution of microarthropods and microorganisms in soil and of the mechanisms of their interactions (59) is necessary so we can capitalize on these biological mechanisms in disease management. Biologically- and culturally-based strategies need to be explored further for control of root diseases. For example, cultural practices such as cultivation, fertilization, and irrigation influence the abundance and diversity of soil communities (75). The quantity and quality of fertilizer both affect soil communities. For example, composted manure and other organic materials increase the biodiversity of soils and reduce nutrient leaching more than mineral fertilizers (30,84). We need to understand the effect of various management practices on the balance between beneficial and pathogenic organisms in soils.

Providing A Practical Framework
For Monitoring Root Disease Epidemics

The development of a cohesive theory of root disease epidemics and the prescription of management strategies will be significant accomplishments and will require the efforts of many dedicated and innovative scientists over the course of many years. The results can potentially have long-lasting and significant impacts in reducing crop losses to root diseases and in increasing crop yields as part of an approach to developing sustainable agriculture in both the temperate and tropical regions of the world. However, if these results are to have the greatest impact possible, they must be translated into a series of practical, user-friendly steps that will provide an overall procedure to monitor populations of soilborne pathogens and epidemics of root diseases, to analyze the data obtained, and to apply the results to practical management measures.

There are at least five components that epidemiologists are obliged to include in a practical system for analyzing root disease epidemics. Each component adds to the overall completeness of the practical system and serves to provide a sound theoretical basis for the system.

First, protocols must be developed and evaluated for the sampling of soil and roots and the assay of such samples for determining the inoculum density of each of the major soilborne pathogens. Such protocols must include procedures for estimating variability within and among samples and means of assuring the quality of the data obtained. Such procedures should include, as far as possible, mechanisms of determining efficacy of propagules in the particular soil environment and information on specific strains or genotypes of the pathogens.

Second, optimum sampling plans need to be developed for assessing intensity of root diseases during the course of epidemics. Potential benefits and risks of protocols and plans proposed should be presented with regard to factors such as destructive versus nondestructive sampling, assessment of foliar compared to root symptoms, estimation of root biomass or healthy area (volume) duration of roots, and so on. Factors such as costs and variability associated with various sampling components should be included in such plans.

Third, data quality objectives and quality assurance plans should be specified for all data collected while studying root disease epidemics and the inoculum dynamics of associated pathogens. Data quality objectives provide preset limits that researchers seek to achieve during assays for pathogen propagule densities and estimation of disease intensity. Quality assurance plans provide opportunities to insert specific steps in assays and estimation protocols to assess whether such procedures are performed reliably.

Fourth, practical methods need to be developed and publicized for analyzing temporal and spatial pattern data for inoculum dynamics and disease progress. The methods required for such analyses should be made as simple statistically and analytically as possible. The conditions and assumptions under which these analyses can be used should be specified clearly, and any possible hazards associated with the analyses stated clearly and concisely. The availability of such methods would encourage many more researchers to monitor and investigate epidemics of root diseases and would add considerably to the databases available for modeling and for making management decisions.

Finally, practical methods, such as a series of templates or a classification scheme, need to be developed and publicized for comparing epidemics of root disease. Classes of epidemics based upon host or pathogen type would aid researchers and others in prescribing possible management options based upon what was known about the specific class of epidemics.

CONCLUSIONS

Considerable progress has been made in the last 10 years in the understanding of root disease epidemics, yet many relatively basic questions remain unanswered. We have attempted to identify challenges, opportunities and obligations for epidemiologists that will allow our understanding of root diseases to continue to increase in the decades ahead and will allow for the development and implementation of environmentally sound management procedures for root diseases.

The advent of molecular techniques such as PCR and DNA probes provides exciting new opportunities for identifying pathogens in soil; however, the application of such techniques is to date mostly qualitative and must become quantitative and less expensive to be utilized fully in monitoring pathogen dynamics in soils. The use of such techniques which may be much more specific in identifying pathogens than assays on semiselective media, for example, will not, however, eliminate the challenges of obtaining representative samples in the field and accounting for the complexities occurring in soil ecosystems.

The availability of faster microcomputers and more powerful software that can handle larger and more complex data sets provides enticing and encouraging challenges and opportunities for root disease epidemiologists. The enticement is to assemble and analyze larger and larger data sets of increasing complexity for the sake of recognition. The encouragement comes in knowing that data sets containing essential data can be analyzed. The hazard, of course, is that in the complexity of the analyses and with the ability to analyze larger and larger data sets, epidemiologists may lose themselves in the analysis, itself, and forget that the original purpose was to provide a simple and understandable interpretation of the epidemiological data.

With the ever growing need to limit losses due to root diseases in order to provide food, forage and fiber for human needs, the need to provide environmentally and ecologically sound management practices for these diseases becomes paramount. Agricultural practices for managing root diseases must be designed to optimize crop productivity and minimize the impacts on beneficial soil organisms and the environment. The success that epidemiologists have in setting the strategies for successful root disease management will depend on their understanding of ecology of soilborne pathogens and the epidemiology of the diseases they cause.

LITERATURE CITED

1. Adams, P.B. 1981. Forecasting onion white rot disease. Phytopathology 71:1178–1181.
2. Addicot, J.F., Aho, J.M., Antolin, M.F., Padilla, D.K., Richardson, J.S., and Soluk, D.A. 1987. Ecological neighborhoods: scaling environmental patterns. Oikos 49:340–346.
3. Ali-Shtayeh, M.S., MacDonald, J.D., and Kabashima, J. 1991. A method for using commercial ELISA tests to detect zoospores of *Phytophthora* and *Pythium* species in irrigation water. Plant Dis. 75:305–311.
4. Amorim, L., Filho, A.B., and Hau, B. 1993. Analysis of progress curves of sugarcane smut on different cultivars using functions of double sigmoid pattern. Phytopathology 83:933–936.

5. Anas, O., Alli, I., and Reeleder, R.D. 1989. Inhibition of germination of sclerotia of *Sclerotinia sclerotiorum* by salivary gland secretions of *Bradysia coprophila*. Soil Biol. Biochem. 21:47–52.

6. Anas, O., and Reeleder, R.D. 1988. Feeding habits of larvae of *Bradysia coprophila* on fungi and plant tissue. Phytoprotection 69:73–78.

7. Anderson, R.V., Gould, W.D., Woods, L.E., Cambardella, C., Ingham, R.E., and Coleman, D.C. 1983. Organic and inorganic nitrogenous losses by microbivorous nematodes in soil. Oikos 40:75–80.

8. Andrén, O., and Lagerlöf, J. 1983. Soil fauna (microarthropods, enchytraeids, nematodes) in Swedish agricultural cropping systems. Acta Agriculturae Scandinavica 33:33–52.

9. Barker, K.R. 1985. Nematode extraction and bioassays. Pages 19–35 in: An Advanced Treatise on Meloidogyne, Vol. 2: Methodology. K.R. Barker, C.C. Carter, and J.N. Sasser, eds. North Carolina State University Graphics, Raleigh, NC.

10. Barker, K.R., Townshend, J.L., Bird, G.W., Thomason, I.J., and Dickson, D.W. 1986. Determining nematode population responses to control agents. Pages 283–296 in: Methods for Evaluating Pesticides for Control of Plant Pathogens. K.D. Hickey, ed. APS Press, St. Paul, MN.

11. Bateman, D.F. 1964. Cellulase and the *Rhizoctonia* disease of bean. Phytopathology 54:1372–1377.

12. Benson, D.M. 1994. Inoculum. Pages 1–33 in: Epidemiology and Management of Root Diseases. C.L. Campbell and D.M. Benson, eds. Springer-Verlag, Berlin, Germany.

13. Berg, N.W., and Pawluk, S. 1984. Soil mesofaunal studies under different vegetative regimes in north central Alberta. Can. J. Soil Sci. 64:209–223.

14. Beute, M.K., and Benson, D.M. 1979. Relation of small soil fauna to plant disease. Annu. Rev. Phytopathol. 17:485–502.

15. Bloomberg, W.J. 1979. A model of damping-off and root rot of Douglas-fir seedlings caused by *Fusarium oxysporum*. Phytopathology 69:74–81.

16. Boag, B., Brown, D.J.F., and Topham, P.B. 1987. Vertical and horizontal distribution of virus-vector nematodes and implications for sampling procedures. Nematologica 33:83–96.

17. Böhm, W. 1979. Methods of Studying Root Systems. Springer-Verlag, Berlin, Germany. 188 pp.

18. Bruton, B.D., and Reuveni, R. 1985. Vertical distribution of microsclerotia of *Macrophomina phaseolina* under various soil types and host crops. Agric. Ecosyst. Environ. 12:165–169.

19. Cahill, D.M., and Hardham, A.R. 1994. Exploitation of zoospore taxis in the development of a novel dipstick immunoassay for the specific detection of *Phytophthora cinnamomi*. Phytopathology 84:193–200.

20. Cahill, D.M., and Hardham, A.R. 1994. A dipstick immunoassay for the specific detection of *Phytophthora cinnamomi* in soils. Phytopathology 84:1284–1292.

21. Campbell, C.L. 1986. Interpretation and uses of disease progress curves for root diseases. Pages 38–54 in: Plant Disease Epidemiology, Vol. 1: Population Dynamics and Management. K.J. Leonard and W.E. Fry, eds. Macmillan, New York, NY.

22. Campbell, C.L., and Benson, D.M., eds. 1994. Epidemiology and Management of Root Diseases. Springer-Verlag, Berlin, Germany. 344 pp.

23. Campbell, C.L., and Benson, D.M. 1994. Spatial aspects of the development of root disease epidemics. Pages 195–243 in: Epidemiology and Management of Root Diseases. C.L. Campbell and D.M. Benson, eds. Springer-Verlag, Berlin, Germany.

24. Campbell, C.L., Jacobi, W.R., Powell, N.T., and Main, C.E. 1984. Analysis of disease progression and the randomness of occurrence of infected plants during tobacco black shank epidemics. Phytopathology 74:230–235.

25. Campbell, C.L., and Madden, L.V. 1990. Introduction to Plant Disease Epidemiology. John Wiley & Sons, New York, NY. 532 pp.

26. Campbell, C.L., Madden, L.V., and Pennypacker, S.P. 1980. Structural characterization of bean root rot epidemics. Phytopathology 70:152–155.

27. Campbell, C.L., and Neher, D.A. 1994. Estimating disease severity and incidence. Pages 117–147 in: Epidemiology and Management of Root Diseases. C.L. Campbell and D.M. Benson, eds. Springer-Verlag, Berlin, Germany.

28. Campbell, C.L., and Nelson, L.A. 1986. Evaluation of an assay for quantifying populations of sclerotia of *Macrophomina phaseolina* from soil. Plant Dis. 70:645–647.

29. Campbell, C.L., Pennypacker, S.P., and Madden, L.V. 1980. Progression dynamics of hypocotyl rot of snapbean. Phytopathology 70:487–494.

30. Chung, Y.R., Hoitink, H.A.J., and Lipps, P.E. 1988. Interactions between organic-matter decomposition level and soilborne disease severity. Agric. Ecosyst. Environ. 24:183–193.
31. Culbreath, A.K., Beute, M.K., and Campbell, C.L. 1991. Spatial and temporal aspects of epidemics of Cylindrocladium black rot in resistant and susceptible peanut genotypes. Phytopathology 81:144–150.
32. Curl, E.A., Lartey, R., and Peterson, C.M. 1988. Interactions between root pathogens and soil microarthropods. Agric. Ecosyst. Environ. 24:249–261.
33. English, J.T., and Mitchell, D.J. 1989. Use of morphometric analysis for characterization of tobacco root growth in relation to infection by *Phytophthora parasitica* var. *nicotianae*. Plant Soil 113:243–249.
34. English, J.T., and Mitchell, D.J. 1994. Host roots. Pages 34–64 in: Epidemiology and Management of Root Diseases. C.L. Campbell and D.M. Benson, eds. Springer-Verlag, Berlin, Germany.
35. Fitter, A.H. 1982. Morphometric analysis of root systems: application of the technique and influence of soil fertility on root system development in two herbaceous species. Plant Cell Environ. 5:313–322.
36. Fitter, A.H. 1987. An architectural approach to the comparative ecology of plant root systems. New Phytol. 106 (Suppl.):61–77.
37. Gilligan, C.A. 1983. Modeling of soilborne pathogens. Annu. Rev. Phytopathol. 21:45–64.
38. Gilligan, C.A. 1985. Probability models for host infection by soilborne fungi. Phytopathology 75:61–67.
39. Gilligan, C.A. 1985. Construction of temporal models: III. Disease progress of soil-borne pathogens. Pages 67–105 in: Advances in Plant Pathology, Vol. 3. C.A. Gilligan, ed. Academic Press, London, UK.
40. Gilligan, C.A. 1990. Mathematical modeling and analysis of soilborne pathogens. Pages 96–142 in: Epidemics of Plant Diseases: Mathematical Analysis and Modeling, 2nd ed. J. Kranz, ed. Springer-Verlag, Berlin, Germany.
41. Gilligan, C.A. 1990. Antagonistic interactions involving plant pathogens: fitting and analysis of models to non-monotonic curves for population and disease dynamics. New Phytol. 115:649–665.

42. Gilligan, C.A. 1994. Temporal aspects of the development of root disease epidemics. Pages 148–194 in: Epidemiology and Management of Root Diseases. C.L. Campbell and D.M. Benson, eds. Springer-Verlag, Berlin, Germany.

43. Goodwin, P.H., English, J.T., Neher, D.A., Duniway, J.M., and Kirkpatrick, B.C. 1990. Detection of *Phytophthora parasitica* from soil and host tissue with a species-specific DNA probe. Phytopathology 80:277–281.

44. Harrison, J.G., Lowe, R., Wallace, A., and Williams, N.A. 1994. Detection of *Spongospora subterranea* by ELISA using monoclonal antibodies. Pages 23–27 in: Modern Assays for Plant Pathogenic Fungi: Identification, Detection and Quantification. A. Schots, F.M. Dewey, and R.P. Oliver, eds. CAB International, Wallingford, UK.

45. Hau, B., Amorium, L., and Filho, A.B. 1993. Mathematical functions to describe disease progress curves of double sigmoid pattern. Phytopathology 83:928–932.

46. Henrissat, B., Driguez, H., Viet, C., and Schülein, M. 1985. Synergism of cellulases from *Trichoderma reesei* in the degradation of cellulose. Bio/Technology 3:722–726.

47. Hornby, D., and Fitt, B.D.L. 1981. Effects of root-infecting fungi on structure and function of cereal roots. Pages 101–130 in: Effects of Diseases on the Physiology of the Growing Plant. P.G. Ayres, ed. Cambridge University Press, Cambridge, UK.

48. Hunt, H.W., Coleman, D.C., Ingham, E.R., Ingham, R.E., Elliot, E.T., Moore, J.C., Rose, S.L., Reid, C.P.P., and Morley, C.R. 1987. The detrital food web in a shortgrass prairie. Biol. Fertil. Soils 3:57–68.

49. Jeger, M.J. 1987. The influence of root growth and inoculum density on the dynamics of root disease epidemics: theoretical analysis. New Phytol. 107:459–478.

50. Jeger, M.J., and Lyda, S.D. 1986. Epidemics of Phymatotrichum root rot (*Phymatotrichum omnivorum*) in cotton: environmental correlates of final incidence and forecasting criteria. Ann. Appl Biol. 109:523–534.

51. Johnson, L.F. and Curl, E.A. 1972. Methods for Research on the Ecology of Soil-borne Plant Pathogens. Burgess, Minneapolis, MN. 247 pp.

52. Jones, K.J. 1990. Components of resistance in *Nicotiana tabacum* to *Phytophthora parasitica* var. *nicotianae*. Ph.D. thesis, North Carolina State University, Raleigh, NC.

53.   Kenerley, C.M., Papke, K., and Bruck, R.I. 1984. Effect of flooding on development of Phytophthora root rot in Fraser fir seedlings. Phytopathology 74:401–404.

54.   Kranz, J. 1968. Eine Analyse von annuellen Epidemien pilzlicher Parasiten. III. Über Korrelationen zwischen quantitativen Merkmalen von Befallskurven und Ähnlichkeiten von Epidemien. Phytopathol. Z. 61:205–217.

55.   Kranz, J. 1974. Comparison of epidemics. Annu. Rev. Phytopathol. 12:355–374.

56.   Kranz, J. 1978. Comparative anatomy of epidemics. Pages 33–62 in: Plant Disease: An Advanced Treatise, Vol. II: How Disease Develops in Populations. J.G. Horsfall and E.B. Cowling, eds. Academic Press, New York, NY.

57.   Kranz, J. 1988. The methodology of comparative epidemiology. Pages 279–289 in: Experimental Techniques in Plant Disease Epidemiology. J. Kranz and J. Rotem, eds. Springer-Verlag, Berlin, Germany.

58.   Larkin, R.P., Ristaino, J.B., and Campbell, C.L. 1995. Detection and quantification of *Phytophthora capsici* in soil. Phytopathology 85:1057–1063.

59.   Lussenhop, J. 1992. Mechanisms of microarthropod-microbial interactions in soil. Pages 1–33 in: Advances in Ecological Research, Vol. 23. M. Begon and A.H. Fitter, eds. Academic Press, San Diego, CA.

60.   Lussenhop, J., and Fogel, R. 1993. Observing soil biota in situ. Geoderma 56:25–36.

61.   Lussenhop, J., Fogel, R., and Pregitzer, K. 1991. A new dawn for soil biology: video analysis of root-soil-microbial-faunal interactions. Agric. Ecosyst. Environ. 34:235–249.

62.   MacDonald, J.D., Stites, J., and Kabashima, J. 1990. Comparison of serological and culture plate methods for detecting species of *Phytophthora*, *Pythium*, and *Rhizoctonia* in ornamental plants. Plant Dis. 74:655–659.

63.   MacGuidwin, A.E., and Stanger, B.A. 1991. Changes in vertical distribution of *Pratylenchus scribneri* under potato and corn. J. Nematol. 23:73–81.

64.   Madden, L.V. 1980. Quantification of disease progression. Prot. Ecol. 2:159–176.

65.   Menzies, J.D. 1963. The direct assay of plant pathogen populations in soil. Annu. Rev. Phytopathol. 1:127–142.

66. Mihail, J.D., and Alcorn, S.M. 1982. Quantitative recovery of *Macrophomina phaseolina* sclerotia from soil. Plant Dis. 66:662–663.

67. Miller, S.A., Bhat, R.G., and Schmitthenner, A.F. 1994. Detection of *Phytophthora capsici* in pepper and cucurbit crops in Ohio with two commercial immunoassay kits. Plant Dis. 78:1042–1046.

68. Mitchell, D.J., Kannwischer-Mitchell, M.E., and Zentmyer, G.A. 1986. Isolating, identifying, and producing inoculum of *Phytophthora* spp. Pages 63–66 in: Methods for Evaluating Pesticides for Control of Plant Pathogens. K.D. Hickey, ed. APS Press, St. Paul, MN.

69. Moore, J.C., and de Ruiter, P.C. 1991. Temporal and spatial heterogeneity of trophic interactions within below-ground food webs. Agric. Ecosyst. Environ. 34:371–397.

70. Moore, J.C., Walter, D.E., and Hunt, H.W. 1988. Arthropod regulation of micro- and mesobiota in below-ground detrital food webs. Annu. Rev. Entomol. 33:419–439.

71. Morrall, R.A.A., and Verma, P.R. 1981. Disease progress curves, linear transformations and common root rot of cereals. Can. J. Plant Pathol. 3:182–183.

72. Nakai, T., and Ui, T. 1977. Population and distribution of sclerotia of *Rhizoctonia solani* Kühn in sugar beet field soil. Soil Biol. Biochem. 9:377–381.

73. Neher, D.A. 1990. Inoculum density, furrow irrigation and soil temperature effects on the epidemiology of Phytophthora root rot of processing tomato. Ph.D. thesis, University of California, Davis, CA.

74. Neher, D.A. 1992. Ecological sustainability in agricultural systems: definition and measurement. J. Sust. Ag. 2(3):51–61.

75. Neher, D.A. 1996. Biological diversity in soils of agricultural and natural ecosystems. Pages 55–72 in: Exploring the Role of Diversity in Sustainable Agriculture. R.K. Olson, C.A. Francis, and S. Kaffka, eds. ASA Press, Madison, WI.

76. Neher, D.A., and Campbell, C.L. 1994. Nematode communities and microbial biomass in soils with annual and perennial crops. Appl. Soil Ecol. 1:17–28.

77. Neher, D., and Duniway, J.M. 1991. Relationship between amount of *Phytophthora parasitica* added to field soil and the development of root rot in processing tomatoes. Phytopathology 81:1124–1129.

78. Neher, D.A., McKeen, C.D., and Duniway, J.D. 1993. Relationships among Phytophthora root rot development, *Phytophthora parasitica* populations in soil, and yield of tomatoes under commercial field conditions. Plant Dis. 77:1106-1111.

79. Neher, D.A., Peck, S.L., Rawlings, J.O., and Campbell, C.L. 1995. Measures of nematode community structure and sources of variability among and within agricultural fields. Plant Soil 170:167-181.

80. Nicot, P.C., and Rouse, D.I. 1987. Precision and bias of three quantitative soil assays for *Verticillium dahliae*. Phytopathology 77:875-881.

81. Noe, J.P. and Barker, K.R. 1985. Relation of within-field spatial variation of plant-parasitic nematode population densities and edaphic factors. Phytopathology 75:247-252.

82. Old, K.M. 1967. Effects of natural soil on survival of *Cochliobolus sativus*. Trans. Br. Mycol. Soc. 50:615-624.

83. O'Neill, R.V., DeAngelis, D.L., Waide, J.B., and Allen, T.F.H. 1986. A Hierarchical Concept of Ecosystems. Princeton University Press, Princeton, NJ. 253 pp.

84. Ott, P., Hansen, S., and Vogtmann, V. 1983. Nitrates in relation to composting and use of farmyard manures. Pages 145-154 in: Environmentally Sound Agriculture. W. Lockeretz, ed. Praeger, New York, NY.

85. Papavizas, G.C., and Davey, C.B. 1959. Isolation of *Rhizoctonia solani* Kuehn from naturally infested and artificially inoculated soils. Plant Dis. Rep. 43:404-410.

86. Papavizas, G.C., and Klag, N.G. 1975. Isolation and quantitative determination of *Macrophomina phaseolina* from soil. Phytopathology 65:182-187.

87. Paul, E.A., and Clark, F.E. 1989. Soil Microbiology and Biochemistry. Academic Press, San Diego, CA. 273 pp.

88. Pfender, W.F. 1982. Monocyclic and polycyclic root diseases: distinguishing between the nature of the disease cycle and the shape of the disease progress curve. Phytopathology 72:31-32.

89. Pfender, W.F., Rouse, D.I., and Hagedorn, D.J. 1981. A "most probable number" method for estimating inoculum density of *Aphanomyces euteiches* in naturally infested soil. Phytopathology 71:1169-1172.

90. Punja, Z.K., Smith, V.L., Campbell, C.L., and Jenkins, S.F. 1985. Sampling and extraction procedures to estimate numbers, spatial pattern, and temporal distribution of sclerotia of *Sclerotium rolfsii* in soil. Plant Dis. 69:469–474.

91. Reynolds, K.M., Gold, H.J., Bruck, R.I., Benson, D.M., and Campbell, C.L. 1986. Simulation of the spread of *Phytophthora cinnamomi* causing a root rot of Fraser fir in nursery beds. Phytopathology 76:1190–1201.

92. Ristaino, J.B., and Averre, C.W. 1992. Effects of irrigation, sulfur, and fumigation on Streptomyces soil rot and yield components in sweetpotato. Phytopathology 82:670–676.

93. Ristaino, J.B., Duniway, J.M., and Marois, J.J. 1989. Phytophthora root rot and irrigation schedule influence growth and phenology of processing tomatoes. J. Am. Soc. Hort. Sci. 114:556–561.

94. Ristaino, J.B., Larkin, R.P., and Campbell, C.L. 1993. Spatial and temporal dynamics of *Phytophthora* epidemics in commercial bell pepper fields. Phytopathology 83:1312–1320.

95. Ristaino, J.B., Larkin, R.P., and Campbell, C.L. 1994. Spatial dynamics of disease symptom expression during *Phytophthora* epidemics in bell pepper. Phytopathology 84:1015–1024.

96. Schmitthenner, A.F. 1988. ELISA detection of *Phytophthora* from soil. Phytopathology 78:1576. (Abstr.).

97. Shannon, C.E., and Weaver, W. 1949. The Mathematical Theory of Communication. University of Illinois Press, Urbana, IL. 117 pp.

98. Singleton, L.L., Mihail, J.D., and Rush, C.M. 1992. Methods for Research on Soilborne Phytopathogenic Fungi. APS Press, St. Paul, MN. 265 pp.

99. Sneh, B., Katan, J., Henis, Y., and Wahl, I. 1966. Methods for evaluating inoculum density of *Rhizoctonia* in naturally infested soil. Phytopathology 56:74–78.

100. Thornton, C.R., Dewey, F.M., and Gilligan, C.A. 1994. Development of monoclonal antibody-based immunological assays for the detection of live propagules of *Rhizoctonia solani* in soil. Pages 29–35 in: Modern Assays for Plant Pathogenic Fungi: Identification, Detection and Quantification. A. Schots, F.M. Dewey, and R.P. Oliver, eds. CAB International, Wallingford, UK.

101. Timmer, L.W., Menge, J.A., Zitko, S.E., Pond, E., Miller, S.A., and Johnson, E.L.V. 1993. Comparison of ELISA techniques and standard isolation methods for *Phytophthora* detection in citrus orchards in Florida and California. Plant Dis. 77:791–796.

102. Tsao, P.H. 1983. Factors affecting isolation and quantitation of *Phytophthora* from soil. Pages 219–236 in: *Phytophthora*: Its Biology, Taxonomy, Ecology, and Pathology. D.C. Erwin, S. Bartnicki-Garcia, and P.H. Tsao, eds. APS Press, St. Paul, MN.

103. Upchurch, D.R., and Ritchie, J.T. 1983. Root observations using a video recording system in mini-rhizotrons. Agron. J. 75:1009–1015.

104. Upchurch, D.R., and Ritchie, J.T. 1984. Battery-operated color video camera for root observations in mini-rhizotrons. Agron. J. 76:1015–1017.

105. Van Bruggen, A.H.C., and Grünwald, N.J. 1994. The need for a dual hierarchical approach to study plant disease suppression. Appl. Soil Ecol. 1:91–95.

106. Vanderplank, J.E. 1963. Plant Diseases: Epidemics and Control. Academic Press, New York, NY. 349 pp.

107. Waage, J.K. 1991. Biodiversity as a resource for biological control. Pages 149–163 in: The Biodiversity of Microorganisms and Invertebrates: Its Role in Sustainable Agriculture. D.L. Hawksworth, ed. CASAFA Report Series No. 4. CAB International, Wallingford, UK.

108. Wiens, J.A. 1989. Spatial scaling in ecology. Funct. Ecol. 3:385–397.

109. Woods, L.E., Cole, C.V., Elliott, E.T., Anderson, R.V., and Coleman, D.C. 1982. Nitrogen transformations in soil as affected by bacterial-microfaunal interactions. Soil Biol. Biochem. 14:93–98.

110. Workneh, F., and Van Bruggen, A.H.C. 1994. Suppression of corky root of tomatoes in soils from organic farms associated with soil microbial activity and nitrogen status of soil and tomato tissue. Phytopathology 84:688–694.

111. Yeates, G.W., Bamforth, S.S., Ross, D.J., Tate, K.R., and Sparling, G.P. 1991. Recolonization of methyl bromide sterilized soils under four different field conditions. Biol. Fertil. Soils 11:181–189.

Chapter 3

THEORETICAL BASIS FOR MICROBIAL INTERACTIONS LEADING TO BIOLOGICAL CONTROL OF SOILBORNE PLANT PATHOGENS

R. Baker and T.C. Paulitz

Robust science requires mind-stretching theory. Theoretical notions lie at the heart of the amalgamation of knowledge gained in the past. Future directions of research are largely governed by theory development. Perhaps it is opportune, therefore, to attempt to develop a basis for theory in biological control of plant pathogens in soil. As such, we present some current developments based on three strategies for obtaining biological control: (i) protection of infection courts, (ii) reduction of inoculum potential in sites not necessarily associated with the infection court, and (iii) induction of host resistance.

PROTECTION OF INFECTION COURTS

Fixed Infection Courts

Seed treatment with biological control agents

Perhaps the most efficient strategy for achieving biological control is the protection of a fixed infection court, e.g. a seed. Since the infection court remains stationary and does not encounter new inoculum over time, a single application of a biological control agent (BCA) applied inundatively may protect the site. For this reason, a wide variety of BCAs have been effective in control of seed decay and seedling damping-off. BCAs most often reported in the literature are *Gliocladium virens* (83), *Trichoderma* spp. (22), *Bacillus subtilis* (79) and *Pythium oligandrum* (77). Reviews are available on the subject (43,48,85,90).

Bacillus subtilis produces the antibiotic iturin A which is assumed to be the antagonistic principle responsible for biological control (64). The BCA is available as a seed treatment under the label of Quantum-4000. Howell and Stipanovic (54) identified the antibiotic gliovirin as responsible for biological control of *Pythium ultimum* by *G. virens*. Howell (53) also demonstrated that *G. virens* produced viridin and gliotoxin and still provided biological control of damping-off of cotton seedlings, induced by *Rhizoctonia solani*, when mutated to eliminate mycoparasitism. Therefore, the important mechanism of antagonism in *G. virens* appears to be antibiosis.

Mycoparasitism, however, appears to be the chief mechanism responsible for antagonism by *Trichoderma* spp. against *R. solani*. Evidence is summarized by Baker (13). Repeated replants of radish induced suppressiveness to damping-off with proportionate decreases in inoculum density of *R. solani*. There was no change in conduciveness when only radish was planted or when the pathogen was added alone. Therefore, the radish was necessary for *R. solani* to convert from dormant sclerotia resistant to mycoparasitism to vegetative thalli susceptible to parasitism by *Trichoderma* spp. Again, during radish monoculture in the presence of *R. solani*, the population density of *Trichoderma harzianum* increased from nondetectable levels to almost 10^6 colony-forming units (CFU) g^{-1} of soil. Furthermore, suppressiveness developed whether or not organic matter and radish residues were screened out between successive crops. Another evidence for the ability of *Trichoderma* spp. to increase biomass principally by mycoparasitism is that increases in population density of the antagonist were observed when an avirulent strain of *R. solani* was added to soil in which radishes were cropped but no increase occurred when the strain was not added. How can these phenomena be related to seed treatment with *Trichoderma* spp.? Only those units of inoculum of *R. solani* in the spermosphere have the potential to encounter mycoparasitism. At low inoculum densities typical of those observed in soil (58), few propagules would touch the seed and potential infection hyphae arriving later (94) would either be subjected to mycoparasitism or encounter maturing resistant host tissue.

Possible mechanisms involved in damping-off of pea by *P. ultimum* were explored by Lifshitz et al. (68). Systematic elimination of various mechanisms can be achieved by a study of the interaction of host, pathogen and BCA. The pathogen germinates in a matter of hours and colonizes the seed coat within 10 hours. In contrast, conidia of *Trichoderma* spp. require 12 to 14 hours to germinate. This factor alone decreases the probability of competition for nutrients on the spermoplane.

Indeed, addition of potentially limiting nutrients essential for germination of *P. ultimum* did not nullify suppressiveness in the presence of conidia of the BCA on the seed coat. Mycoparasitism in vitro by *Trichoderma* spp. did not occur until 24 hours after contact with the pathogen and then in low frequency (25). Thus, mycoparasitism was not initiated soon enough to protect the "window of susceptibility" of the host. The only mechanism that explained the phenomenon of biological control in this case was the observation of an antibiotic-like reaction induced by *Trichoderma* spp. that was named "the routing factor".

Harman (43) wrote: "the seed, because it is a senescent dry unit, is amenable to treatment and manipulation to favor microbial antagonists." Even so, most reports of successful biological control by seed treatment with antagonists were done under laboratory or greenhouse conditions (44,45). Under field conditions, results are not as satisfactory. As pointed out by Harman (43) and Deacon (31), seeds treated with BCA in soil are subject to environmental conditions of variable temperature, moisture relationships, organic matter, soil texture, plant nutrients and temperature which may not be conducive to the activity of BCAs. The seed itself is a concentrated source of nutrients for microorganisms and pathogens. Thus, for satisfactory control by use of BCAs on seeds, enhancement strategies are appropriate.

Trichoderma spp. are favored by a low soil pH (24). Harman and Taylor (47) approximately doubled the amount of biological control by combining *T. harzianum* seed treatment with HCl. Solid matrix priming, in which seeds are mixed with ground coal material at a pH conducive for development of the BCA, enhances seed vigor (47). Sufficient water is added to the seed-coal mixture to permit hydration, but not radicle emergence, of the seed. This priming decreases the "window of susceptibility" of the seed to infection. At the same time the BCA grows on the moist, exudate-coated seed surface and increases in population density by approximately 10-fold. A liquid coating technique using a suspension of binder, finely ground particulates and *T. harzianum* gave better control compared to conventional slurry or solid particulate formulations (103). The addition of polyethylene glycol to culture medium as an osmoticum enhanced the ability of *Trichoderma* conidia to survive drying and protect cucumber seeds (46). Enhancement of BCA activity by optimising formulations appears to be a promising strategy. The nature of the thalli of BCAs also may be used for enhancement. McQuilkin et al. (78) reported that oospore seed coatings with *P. oligandrum*, derived from liquid fermentation, gave significant control of Pythium damping-off in cress.

The great attention given by scientists to biological control of fixed infection courts alone would provide evidence of the potential of this strategy for success in biological control. Theory, elaborated above, provides evidence for this optimism. Unfortunately, pathogens involved in preemergence damping-off can induce infection in the seedling stage and the BCA may not be as motile as the moving infection courts. We shall treat this problem later.

Competition: Rhizoctonia preemergence damping-off

Inoculum of *R. solani* can infect a susceptible seed when present in the spermosphere (10); however, if the effect could be reduced to a spermoplane, where only those units of inoculum that touch the surface of the seed infect (16), fewer units of inoculum would be successful in infection. In other words, if there was a spermosphere effect, an inoculum density versus infections curve would have a slope value of one infection per unit of inoculum on a log-log scale. If there was a spermoplane effect, the slope value would be 0.67 (14).

Rouse and Baker (94) tested this hypothesis by adding cellulose to spermosphere soil. Such treatment with an organic amendment with an infinitely large carbon:nitrogen ratio immobilizes nitrogen which is essential for germination and infectivity of many soilborne pathogens (76). In this type of competition, inoculum is not killed, but only held in a dormant condition (17). Indeed, when a cellulose amendment was added to soil infested with *R. solani* at different inoculum levels, the slope value of the inoculum density versus infection curve (log-log) was reduced from 1 (in raw soil) to 0.67.

In contrast to the effect of cellulose on pathogen propagules, chitin added to soil is thought to break down into fungitoxic compounds that kill propagules (99). In this case the efficiency of inoculum throughout the spermosphere is decreased (14). Therefore, inoculum density–infection curves should be parallel when chitin is added, all with a slope value of 1, as demonstrated by Rouse and Baker (94).

The relative efficiency of each of the above treatments in biological control can be compared by calculating the areas under the inoculum density–disease curves for each by use of the following equation (94):

$$A = \int_{x=I_0}^{x=I} (b_1 - b_1') x + (b_0 - b_0') \, dx$$

where A equals the change of the area under the curves due to control, b_0

and b_1 are regression coefficients for the raw soil curve, b'_0 and b'_1 are the regression coefficients for the second treatment curve and I and I_0 are the inoculum density limits for integration. By this analysis, cellulose amendment was more efficient than chitin amendment in biological control.

Motile-Fixed Infection Courts

As the plant grows, various organs are formed that may remain fixed in space and are susceptible to infection by soilborne pathogens. In some instances, the infection court is formed, inoculum is present but infection does not occur unless environmental conditions are favorable within a relatively narrow range of conditions. In this case, placement of the BCA in the infection court may be difficult during the mobile phase of host growth, and various strategies must be developed to counteract such difficulties. Alternatively, if environmental conditions are critical for infection by the pathogen, these may be manipulated to achieve biological control.

Infection courts invaded when a factor in the environment is conducive

Stanghellini and associates described an interesting example of biological control in which inoculum of *Pythium aphanidermatum* is encountered by a sugar beet tap root, which grows vertically as well as horizontally, but infection occurs only when environmental conditions are favorable on infection courts "fixed" in time (100).

Penetration and infection of the tap root of mature sugar beets by *P. aphanidermatum* coincides with the onset of specific soil temperatures ($>27°C$ for 12 consecutive hours per day at the 10-cm soil depth) 8 to 9 months after planting (107). Inoculum is not limiting since high population densities of exogenously dormant oospores occur in the rhizosphere soil of naturally infested fields throughout the growing season (101,102). Again, oospores germinated and hyphal growth occurred over a temperature range of 10 to 43°C. Therefore, some factors, in addition to soil temperature, influence the timing of the onset of root infection.

Bacterial antagonism was important in suppressing activity of *P. aphanidermatum* (104). These workers conclude that oospore germination and root infection occur primarily within a short period of time (1 to 2 hours immediately following irrigation). Up to this point in the relatively dry soil, pathogen and bacterial activity would be minimal. When irrigation occurs, however, soluble nutrients become available and *P. aphanidermatum* would have a competitive advantage over resident bacteria by virtue of rapid oospore germination, if temperatures are favorable. Therefore, the pathogen plays a waiting game, prepared for an opportunity to exploit the brief period of time when the pathogen has a competitive advantage over rhizosphere bacteria.

This is an outstanding example of how a study of basic mechanisms leads to formulation of strategies for biological control of a plant pathogen. Obviously, irrigation practices may be managed so that bacterial antagonism is sustained to induce suppressiveness. M.E. Stanghellini (personal communication) believes that this strategy for control of opportunistic pathogens can be extended to many other systems.

Rhizoctonia postemergence damping-off

As pointed out in the last section, fixed infection courts are relatively easy to protect. The primary objective in agricultural plant production, however, is growth ultimately resulting in yield. Therefore, the movement of the infection court complicates the appropriate placement of a BCA. What strategies can be employed to provide extended protection against a pathogen such as *R. solani* (AG4) that attacks a tender seedling at the crown/soil surface? Howell (52) planted seeds of cotton in peat infested with *R. solani* and distributed *G. virens* inoculum as an in-furrow treatment. There was no significant control of postemergence damping-off. However, Lumsden and Locke (72) obtained 24% control of Rhizoctonia postemergence damping-off of zinnia by use of the formulation G20 alginate-grain mixed in soilless potting mix. Obviously, mixing a formulation of *G. virens* has its application in high income, intensive agriculture operations but is likely to be less efficient in large-scale field conditions. An isolate of *G. virens* has been registered by the Environmental Protection Agency, is in the process of commercial testing, and will be marketed in the near future by W. R. Grace and Co., a producer of greenhouse substrates (110).

Trichoderma spp. were screened for biological control of Phytophthora root and crown rot of apple seedlings (93,97). Similarly, *Bacillus cereus*, selected from 700 bacterial isolates, was effective in control of Phytophthora damping-off of alfalfa seedlings (42). In these cases, again, the BCAs were distributed about the infection court in artificial substrates.

Competition: Fusarium root rot of bean

Competition may be a mechanism for the protection of a fixed or motile-fixed infection court. Amendments with high carbon:nitrogen ratios reduced the severity of hypocotyl infection of bean in soils infested with *Fusarium solani* f. sp. *phaseoli* (76), even though the population density of the pathogen was not reduced (17). The chlamydospores of this pathogen require nitrogen to germinate in the soil (27) or at high spore densities (40). The suppression was often eliminated by addition of fertilizer or crop debris containing high levels of nitrogen. All these studies suggest that nitrogen competition can be important in disease suppression, and that microbial immobilization of nitrogen can induce this suppression. This is a case where cultural management may be more effective than inundative inoculation in inducing biological control, and should be examined in relation to minimum tillage practices. But very little recent work has focused on this mechanism in biological control.

Motile Infection Courts

Vascular wilt diseases

Protection of infection courts in the development of vascular wilt diseases is a most difficult objective. For example, chlamydospores of *Fusarium oxysporum* f. sp. *cucumerinum* germinate on the rhizoplane of a cucumber root and penetrate about 40 hours later (75,98). Even though efficiency is only 4%, approximately three infections per root tip occur each day in raw soil at 25°C at an inoculum density of 1,000 chlamydospores g^{-1} soil (R. Baker and T.C. Paulitz, unpublished), and there are many root tips encountering chlamydospores in a typical field or experimental situation. It is obvious that efficiencies would have to be drastically reduced to obtain control of systemic infections in a typical susceptible host. Even so, control of Fusarium wilt diseases by competition for available iron (in soils of high pH) by use of fluorescent pseudomonads producing siderophores was accomplished (15). These results were obtained, however, while testing the hypothesis under experimental conditions, that siderophores produced by pseudomonads were instrumental in competition for available iron. Field experiments failed and control was only temporary. Thus, competition for iron, carbon (35) and nitrogen (41) does not reduce infection efficiency of inoculum enough to provide substantial biological control in systems involving moving infection courts.

Take-all

Take-all induced by *Gaeumannomyces graminis* var. *tritici* (Ggt) is the most important wheat disease world wide (6). Upon infection, roots become black and yellowing occurs on the foliage. Lesions on the tiller bases induce premature blighting of the plants during or shortly after heading. The pathogen survives between host crops as a saprophyte in fragments of previous host crops but the inoculum potential of the fungus is reduced as food reserves decline or the pathogen is displaced by other microorganisms. These characteristics ensure great attention in the literature for development of biological control strategies for suppression of take-all.

Disease potential relationships involved in host-parasite reactions are probably the most complex of any of the systems treated here. Deacon and Henry (33) observed that severity of infection by Ggt was greatest on root tip tissue, intermediate on 5-day-old root tissue, and least on 15-day-old tissue. After initial infections, runner hyphae are produced inducing secondary infections.

This complex system becomes even more complicated by early root cortex death (RCD) in the wheat seminal and root axes. Deacon (30) reviewed the effects of RCD as it affects the pathogen and potential BCAs. Even though roots of seedlings may appear white and healthy, RCD occurs even without involvement of microorganisms. This could imply that RCD can benefit Ggt through reduced host resistance in the cortex and the release of nutrients supporting infection; however, this may be countered by increased resistance of the endodermis in the maturing stele. Deacon interprets observations that inoculum can be present in the soil at low inoculum density yet take-all lesions are seen in the new crop later in the growing season (51) as indicating that "Ggt undergoes a prolonged 'feeding stage' on senescing cortices and eventually accumulates a nutritional base from which it can infect later-formed roots".

These relationships suggested that the senescing cortex tissues of the root could be colonized by low grade parasites that compete for infection sites with Ggt, notably *Phialophora graminicola* and other low-grade parasites (30); however, as with all biological control systems, success is usually achieved only when there are low inoculum levels of Ggt.

Phialophora graminicola is found in relatively low amounts on cereal roots and activity must be enhanced by a rotation with grass pasture, since population densities of the BCA do not increase in cereals (29). Therefore, a rotation with grass pastures (30) is obviously an option developed from theory.

It is also possible to suppress Ggt in the lesions formed on roots and tiller bases. In take-all-suppressive soils, just as many lesions form as in conducive soils but they do not expand (115). This is explained by the observation that *Pseudomonas* spp. suppressing Ggt from roots were more inhibitory in take-all decline (TAD) than in nonsuppressive soil (28). Inhibition of Ggt in TAD is attributed to a phenazine antibiotic produced by the florescent pseudomonads (109).

REDUCTION OF INOCULUM POTENTIAL IN SITES NOT NECESSARILY ASSOCIATED WITH THE INFECTION COURT

In these systems, an intimate interaction of the BCA with a pathogen in the plant rhizosphere may or may not occur. Usually, there is uniform distribution of suppressiveness wherever biological control is operating. In many ways, this strategy presents a much greater challenge to the practitioner of biological control than protection of the infection court. The assumption seems sound—since most soilborne pathogens survive in the soil as dormant propagules or resting structures, why not attack these propagules with BCAs such as mycoparasites, thus reducing the inoculum potential? But these pathogen propagules are distributed in a large volume of soil, and the population densities may range from 0.1 to 2 propagules g^{-1} soil for sclerotial pathogens such as *R. solani*, *Verticillium dahliae*, or *Sclerotinia minor*, from 10 to 300 propagules g^{-1} for *Pythium* spp, or from 100 to 1,000 propagules g^{-1} for Fusarium wilt pathogens. For a BCA to be effective, it must be close to the pathogen propagule (within the sporosphere), or have the ability to grow saprophytically from a food base toward the propagule. As a result, high application rates of the BCA to the soil are needed to reduce the inoculum potential effectively. When inoculum densities of BCAs from laboratory or greenhouse experiments are extrapolated to the field, thousands of kilograms of inoculum per hectare would have to be applied in many cases, leaving only greenhouse crops as an economically feasible market. We call this "the barrier of inoculum application rates".

Adams (2) compared the relative efficiencies of various BCAs, expressed as the ratio of the number of propagules of the mycoparasite required to obtain disease control to the typical inoculum density of the pathogen. These numbers ranged from 1.6 for *Sporidesmium sclerotivorum-S. minor*, through 15 for *Pythium nunn-P. ultimum*, to 5 × 10^4 for *Trichoderma* spp.-*R. solani*. However, other factors need to be considered, such as the cost of inoculum production, the ability of the

BCA to reproduce on the pathogen propagule or crop debris, the ability to spread to other inoculum, and the dispersal within the soil by cultural methods. Furthermore, biological control practitioners rarely study the distribution of the pathogen in the soil or take into account that the distribution is nonrandom and usually clustered or aggregated (20). The aggregated distribution may make the job easier, since instead of having to target many individual propagules, it may only be necessary to target fewer larger foci or clusters of propagules.

In the remaining section, we would like to explore a few well-studied examples of this strategy, emphasizing some theoretical principles that can be gleaned from this work. We will not cover all the possible examples in each category. This subject, more specifically mycoparasitism, has been adequately reviewed by others in the past few years (2,13,23,82,112,113).

Addition Of Specific Antagonists To Reduce Survival And Inoculum Potential Of Pathogen (Parasitism Of Propagules)

Specific sclerotial mycoparasites

Coniothyrium minitans. Recently, Whipps and Gerlagh (112) comprehensively reviewed the potential for use of *C. minitans* as a BCA. It is an effective agent in the control of several sclerotium-forming fungi, but mainly species of *Sclerotinia* and *Sclerotium*. In nature, it is found specifically associated with sclerotia, although it can be easily cultured in vitro. While mycoparasitic hyphal interactions occur, the most favorable aspect of the fungus as a BCA is its ability to colonize and destroy sclerotia at any location in soil.

Biological control efficiencies can be remarkably high. Huang and Kzub (56) reported an 80% decline in incidence of Sclerotinia wilt associated with monocropping of sunflower. This was related to indigenous mycoparasites, especially *C. minitans*. Whipps and coworkers (112) concentrated research on glasshouse crops. They achieved 50% control of *Sclerotinia sclerotiorum* in lettuce if inoculum of *C. minitans* was applied when 50 to 90% of lettuce plants in control plots were infected, and obtained increases in yield of up to 100% (19). For maximum efficiency in biological control, the agent had to be applied prior to planting and treatments had to be repeated (112). Enhancement strategies have not been explored extensively; however, integration with compatible fungicides and/or other BCAs could be considered.

Sporidesmium sclerotivorum. The story of this biological control system is a classic example of where an understanding of the ecology and biology of the BCA has been used to overcome one of the primary constraints of the commercial application of mycoparasites—the dilemma of unrealistic inoculum application rates. Since its discovery in the late 1970s, this dematiaceous Hyphomycete has been isolated from soils all over the world, where it behaves as an obligate parasite of sclerotia of *S. sclerotiorum, S. minor, Sclerotium cepivorum,* and *Botrytis cinerea* (7). Although inoculum can be produced on infected sclerotia, the fungus grows very slowly and is difficult to culture, but a vermiculite-based medium was developed (8). In early trials, the mycoparasite caused a 75 to 95% reduction in sclerotia of *S. minor*, and reduced disease over a 2-year period when 2,300 kg inoculum ha^{-1} was applied (3). But subsequent research showed that sclerotia were distributed in clusters (1), that the mycoparasite could grow up to 3 cm from an infected sclerotium to infect a healthy sclerotium, and that 15,000 macroconidia of the mycoparasite were produced on each infected sclerotium. Based on this knowledge, it was hypothesized that a smaller amount of inoculum could be applied, and the secondary spread of the mycoparasite from the food base of the pathogen sclerotia would be sufficient to build up suppressiveness. This hypothesis was confirmed by Adams and Fravel (4), who induced suppressiveness to Sclerotinia lettuce drop in lettuce fields with the application of 0.2, 2, and 20 kg inoculum ha^{-1}. In addition, the mycoparasite even spread into the nontreated plots and reduced disease. This is a landmark paper, one that should be emulated by all biological control workers for two reasons: (i) by application of ecological principles, practical obstacles were overcome, and (ii) the authors did an economic analysis to show the potential profitability of using the BCA. This BCA made the transition from the pampered conditions of the laboratory to the harsh reality of the field, by the application of ecological and biological principles. It is currently in the process of commercialization (110).

Most of the research with sclerotial mycoparasites has focused on their effect on the viability and myceliogenic germination of sclerotia. For many pathogens like *S. minor,* sclerotia are the most important inoculum in the field (84). But for *S. sclerotiorum*, ascospores ejected from apothecia are more important, and it may be more difficult to control disease caused by carpogenically-germinating sclerotia, since one apothecium can form thousands of ascospores. However, very little research has focused on the effect of mycoparasites on apothecial formation by sclerotia. Sclerotia germinating carpogenically in the spring may be more susceptible to mycoparasitism than dormant sclerotia (55). It may be possible to reduce the inoculum potential of the sclerotia enough to prevent carpogenic germination, even though the sclerotia may still be viable. Cold-adapted mycoparasites may colonize sclerotia during the fall and early spring in temperate climates and affect subsequent apothecial formation in the summer. Because of the difficulty of working with the sexual stage in the laboratory, this area has been virtually ignored by biological control workers, with a few exceptions (81).

Hyphal mycoparasites: *Trichoderma* and *Gliocladium*

Gliocladium and *Trichoderma* have already been mentioned in the context of protecting infection courts, and can also degrade or parasitize sclerotia (105,111). However, they can also attack vegetative hyphae, although the exact nutritional relationship between the parasite and host in the strict sense of mycoparasitism is controversial (32). These necrotrophic mycoparasites produce cell wall-degrading enzymes (23,34), antibiotics such as gliovirin (52) and gliotoxin (92), and other unidentified lytic substances (65). For this strategy, actively growing hyphae may be more effective than quiescent conidia or chlamydospores (62), and the incorporation of a food base into the inoculum may be desirable. Biomass of *Trichoderma* and *Gliocladium*, mixed with clay carriers and a food base, has been successfully formulated into alginate pellets (63). Use of these two BCAs has focused on horticultural greenhouse crops, since the organisms can be applied to the entire growing medium, an unrealistic strategy in field crops.

Addition of specific antagonists with organic matter

The effect of crop residues or organic matter in the soil should also be considered as an integral part of the interaction between the BCA, the pathogen, and the plant. For many BCAs, such as *Trichoderma*, the addition of a food base with the inoculum has improved the performance of the BCA (9). However, it may be possible to use crop residues or organic matter to enhance the level of suppressiveness via a BCA. The story of *P. nunn* provides a classic example of this. This mycoparasite was isolated from a grassland soil suppressive to *P. ultimum* (67). After bean leaves were added to the soil, the population of *P. ultimum* increased, then declined again to very low levels. At the same time, the population density of a slow-growing *Pythium* sp. (*P. nunn*) increased dramatically (66). *Pythium nunn* was shown to be a mycoparasite of

Pythium spp, *Phytophthora* spp., and other plant pathogenic fungi (66). The mycoparasite produced cellulase, chitinase and β-1,3-glucanase when grown with the cell walls of pathogenic fungi. However, unlike other mycoparasites such as *Trichoderma* that used the fungal host as a food base (71), the activity of *P. nunn* against *P. ultimum* was dependent on organic substrates that contained high levels of labile carbon substrates, such as green leaves. Disease suppression against *P. ultimum* on cucumber operated only if organic amendments like bean leaves or rolled oats were also added to the soil (0.3%, w/w), along with a low inoculum density (300 CFU g^{-1}) of *P. nunn* (86). *Pythium nunn* did not increase saprophytically in raw soil, but remained at low levels. The addition of these substrates to soil with *P. ultimum* alone would result in high levels of disease, but when *P. nunn* was added, the level of disease would decline over several replants (87). This suggested that *P. nunn* requires organic substrates rather than a fungal host for suppression to operate, and that the two *Pythium* spp. compete for the same substrates and occupy overlapping niches. This hypothesis was further substantiated by the observation that *P. nunn* stimulated the rapid formation of secondary sporangia by *P. ultimum* (88). These secondary sporangia are formed in response to nutrient deprivation, and may have a reduced inoculum potential, compared to primary sporangia. Studies of the colonization of bean leaves by *P. ultimum* and *P. nunn* showed that both could act as primary colonizers. If leaves were precolonized by *P. ultimum*, *P. nunn* could colonize the substrate and displace *P. ultimum* (89). However, if the substrate was colonized by *P. nunn* first, *P. ultimum* could not establish on the substrate. In this case, mycoparasitism may be a mechanism operating in resource capture, and the benefits arising from the utilization of the fungal host for nutrients may be of secondary importance to the mycoparasite (73).

Pythium nunn still continues to elicit the curiosity of biological control researchers, and the theory is still evolving. The fungus was capable of inducing lysis of sporangia of *P. ultimum* (66), and Elad et al. (36) isolated a factor from cultures of *P. nunn* that inhibited the growth of *P. ultimum*, two observations indicative of an antibiotic. Laing and Deacon (60), in a fascinating paper using real-time video microscopy, found that *P. nunn* caused an early stoppage of growth of host hyphal tips, branching at the point of contact, lysis, and cytoplasmic disruption, all within 10 to 30 minutes after contact with the host. They did not observe any directed growth of *P. nunn* to the host. This rapid occurrence of events is too fast for mycoparasite-derived lytic enzymes to be produced, and suggests that preexisting host-derived wall-degrading enzymes may be liberated by the mycoparasite (60). There is still some controversy about the host range of *P. nunn*. In the same study, Laing and Deacon reported that *Pythium* spp. tested were not hosts of *P. nunn*, although they did not test *P. ultimum*. They believe that hyphal coiling indicates a resistant rather than susceptible host. However, it is possible that a range of susceptibility exists. Some species of fungi may be susceptible to rapid disruption by the host within minutes of contact, before hyphal coiling occurs. With other hosts, *P. nunn* may require a longer period of time to produce cell wall-degrading enzymes and penetrate the host. The substrate may also affect the outcome of mycoparasitism, e.g. *P. nunn* may use organic substrates as a food base for a more vigorous attack than that observed by Laing and Deacon, who grew the fungus on water agar.

ACTIVATION OF RESISTANCE GENES IN THE HOST: INDUCED RESISTANCE

The problems involved in protecting moving infection courts against penetration by soilborne pathogens, like the formae speciales of *Fusarium oxysporum*, could be solved by inhibiting the parasite after colonization. This is accomplished in biological control by induced resistance, defined as the initiation of structural or chemical changes brought about by a BCA that occur only after a plant is challenged by a potential pathogen (95). The BCA may be a fungus (75) or bacterium (106,109). Much of the focus of research on induced resistance has been on foliar pathogens, such as anthracnose of cucumber caused by *Colletotrichum lagenarium* (59). Fusarium wilt diseases, caused by formae speciales of *F. oxysporum*, have been the main target of induced resistance research in soil systems. Preinoculation of plants with nonpathogenic formae speciales of *F. oxysporum* has reduced the severity of wilt disease (18,39,116). A thorough study of the mechanisms of protection against Fusarium wilt of cucumber by nonpathogenic isolates of *F. oxysporum* will serve to illustrate some of the basic principles involved in induced resistance of plants to root pathogens (75).

Inoculation of roots of cucumber with microconidia or chlamydospores of a nonpathogenic isolate of *F. oxysporum* (C14), induced resistance to *C. lagenarium*, which was inoculated on the leaves. C14 applied to the roots also reduced the size of stem lesions caused by *F. oxysporum* f. sp. *cucumerinum* and germination of microconidia in wounded stem tissue. Using a split root design, which assures separation of the inducer from the challenger, chlamydospores of *F. oxysporum* f. sp. *cucumerinum* were applied to one half of the root system and C14 or water were applied to the other half. In another experiment, chlamydospores of the pathogenic isolate were mixed with soil and added to the bottom of a planting tube, with inoculum of C14 added to the top of the tube. In this way, the root of a cucumber seedling would first grow through inoculum of the nonpathogen, before encountering the pathogenic fungus. In both experiments, the progress of Fusarium wilt was reduced when part of the root system encountered C14.

Recent work also suggests that induced resistance may operate against root-rotting pathogens such as *P. aphanidermatum* (117). The inducers were *Pseudomonas* spp. isolated from the rhizosphere of cucumber and previously shown to be antagonistic to *P. aphanidermatum* on cucumber (91). In a split root system with cucumber grown in rockwool, *Pseudomonas* spp. were applied to one half of the root system, along with the simultaneous inoculation of the other half with zoospores of *P. aphanidermatum*. The bacterial treatment reduced disease incidence and severity, with an increase in plant shoot and root growth, compared to treatments protected with water. This is the first example of *Pseudomonas* spp. inducing resistance to *Pythium* spp. in the root system, when both the BCA and pathogen are applied to the root. Kloepper and coworkers have also demonstrated that PGPR (plant growth-promoting rhizobacteria) can induce resistance to *Colletotrichum orbiculare* (108) and cucumber mosaic virus (70).

One of the principles of induced resistance is that the inducer must come in contact with the living tissue of the plant in order to initiate a defense reaction. Both of these examples illustrate this point, which has been confirmed by more recent work. A nonpathogenic mutant of *Colletotrichum magna* which could grow in watermelon seedlings as an endophyte, was capable of inducing resistance to *F. oxysporum* f. sp. *niveum* (38). Bacterial endophytes, including *Pseudomonas* spp., have been shown to colonize the root cortex (57,106), presumably from wounds in the epidermal tissue from abrasion or lateral root formation. Nonpathogenic fungi, such as C14, have the advantage of developing well-formed penetration mechanisms that breach living intact epidermal tissue (75). This interaction between living cells is responsible for generating the signal that leads to systemic resistance.

This mechanism has several advantages over the other strategies of protecting the infection court or reducing inoculum potential outside the infection court. Organisms that induce resistance can trigger protection in tissue far removed from the site of interaction between the BCA and the plant, so the BCA and pathogen do not have to be in contact either spatially or temporally. *Pseudomonas* spp. and nonpathogenic *F.*

oxysporum can grow and multiply within the plant without causing disease, and can continually induce a reaction in the plant, overcoming one of problems with single inducing treatments, the diminution of protection over time (116). In addition, these BCAs can continually penetrate a moving infection court. If the induction signal can be expressed in newly formed tissue, then the perplexing task of protecting moving infection courts such as root tips may be accomplished.

Sequeira (95) adequately treated the various mechanisms involved in induced resistance and these will only be summarized here. When a BCA—usually an avirulent strain of the pathogen or a nonpathogen—is introduced into a host, various resistance reactions are initiated. All plants, whether resistant or susceptible, have the capacity to respond in this fashion; however, the rapidity and extent of host response establishes the difference between a resistant and a susceptible interaction. Of most interest, however, is the initiation of a signal that is communicated systemically in the induced plant that "alerts" the host to a possible attack by a pathogen.

Effective inducers of resistance include injury, heat shock, and numerous chemicals such as ethylene and degradation products of host and pathogen cell walls. Therefore, it is difficult to separate inducers from signals. Recently, however, evidence was presented that an increase in endogenous salicylic acid paralleled the rise in resistance upon inoculation of induced hosts with tobacco mosaic virus (74). No exogenous application of salicylic acid was required, suggesting that this compound functions as the transduction signal. There is also an association between the increase in salicylic acid and the induction of PR (pathogenesis-related) genes (21). Metraux et al. (80) observed that the concentration of salicylic acid in the phloem sap of cucumber plants increased transiently just prior to the establishment of systemic induced resistance by inoculation with either tobacco necrosis virus or the fungal pathogen *C. lagenarium*. Even so, Wildon et al. (114) suggest that a propagative electrical signal, rather than a mobile chemical, may be the messenger that transmits the signal in a wound response. In any case, we are getting closer to solving the molecular puzzle of induced resistance, especially the unknown black box between the induction of a signal and the final resistance gene product.

After a BCA (usually a nonpathogen) is introduced into a plant and the systemic signal is initiated, a challenge inoculation by a pathogen induces resistant responses at the point of attack. These responses were comprehensively reviewed by Sequeira (95). Resistance reactions are triggered by biosynthesis in the fungal cell wall of glucans and other elicitors. Indeed, glucans have high biological activity and can, for example, induce production of widely different phytoalexins in diverse plants. Moreover, elicitors induce activation of genes other than those involved in phytoalexin synthesis, such as hydroxyproline-rich glycoproteins that may function in cell wall strengthening and accumulation of suberin, lignin and cellulose. In addition, the host may produce enzymes, such as chitinase, β-1,3-glucanase and peroxidase, that are inhibitory to pathogens.

DEVELOPING THEORETICAL FOUNDATIONS

The foregoing analyses show that many factors influence the success or failure of BCAs in biological control. It was not our intent to analyze exhaustively the many systems reported in the literature. Rather, selection was based upon the existence of extensive information or detailed knowledge of mechanisms.

In theory, the efficiency with which BCAs protect the infection court is largely dependent on whether those courts are stationary or motile. Obviously, if the infection court is stationary, the BCA does not have to move. Thus, seed treatments have been used successfully in control of preemergence damping-off. Even in this case, however, success has been achieved in controlled environments where temperatures are especially conducive for activity of BCAs.

At least two general strategies have been employed for protection of moving infection courts. In seedling diseases (72), the BCA can be mixed into a soilless medium. Alternatively, fungal BCAs could be mutated so that they are rhizosphere competent (5); that is, able to "follow" the infection court (root) as it develops. Thus, population densities and germination of sporangia of *P. ultimum* along a developing root were reduced by treatment of cucumber seed with rhizosphere-competent *T. harzianum* (5).

There are two systems that are special cases where infection courts are motile, fixed, or both. In the system of Tedla and Stanghellini (104), oospore germination about a mature sugar beet root is inhibited by manipulation of irrigation so as to increase the activity of bacterial antagonists. Again, early root cortex death affords the opportunity to preempt infection courts in the take-all disease by occupation of the court with a BCA (30). In most instances, however, protection of moving infection courts is not feasible, especially in the case of vascular wilt diseases. The strategy of employing competition to control disease in systems involving moving infection courts is particularly inefficient.

Competition for elements essential to the germination or penetration stages leads to fungistasis only in the sense that inoculum densities are the same before and after instigation of biological control (17).

Three chemical elements may be necessary in the infection process for a soilborne pathogen: carbon, nitrogen and, for small-spored species, iron (15). It is possible to decrease the availability of these elements to limiting levels; however, competition is not an efficient means to initiate significant biological control. For example, Guy and Baker (41) added cellulose to soil, a procedure which is known to immobilize nitrogen (76). Even though biological control was achieved in the fixed infection court of Fusarium hypocotyl rot of bean by use of cellulose, no control was noticed in the motile infection courts involved with Fusarium wilt of pea (41). Further, in spite of the extensive attention given to siderophores as competitors for iron, control of Fusarium wilt diseases with siderophores is observed only under carefully controlled conditions (11). The iron activity has to be less than 10^{-19} M in order to suppress germination of small (approximately 10 μm diameter) chlamydospores of *F. oxysporum* f. sp. *cucumerinum* (96). Leong and Expert (61) suggest that genetic engineering research could produce a *Pseudomonas* strain that is a "super-producer" of high affinity siderophores or of multiple siderophores active over a wide range of environmental conditions. Even so, such a super strain would only be effective in fixed infection courts for small-spored fungal species or in lesions (37).

For these reasons, induced resistance appears to be the only potentially efficient system for biological control involving motile infection courts.

Early on in biological control studies, researchers hypothesized that the addition of organic matter to soils would stimulate the activity of antagonists deleterious to pathogens. For the most part, this was a vain hope. For example, analyses were performed on six different systems when organic matter was mixed into soil and survival of pathogens studied (12). In all cases with one exception, death rate of "weak" propagules was increased in comparison with nonamended controls. Death rate of resistant propagules important in survival was not affected. The one exception was Clark's (26) classic investigation involving digestion of sclerotia of *Phymatotrichum omnivorum* in which survival slope values over time decreased when manure was added for all types of propagules.

Proper management of organic matter can lead to suppression of certain plant diseases in specific situations, such as use of compost-amended substrates (49,50). For example, compost high in cellulose suppresses hyperparasitism by *Trichoderma hamatum* and is conducive to aggressiveness of *R. solani*. In contrast, mature compost colonized by this BCA is suppressive and biological control occurs. This illustrates the principle that the nutritional status of organic matter conditions activity of *T. hamatum*. Again, high levels of microbial biomass restrict germination of *Pythium* and *Phytophthora* spp. and their entry into roots, resulting in disease suppression. Besides the biological control in such systems, allelopathic toxins, the available carbon:nitrogen ratio, the concentration of soluble salts, and perhaps chloride ion concentration in composts are the principal chemical mediators of disease control.

Obviously, composting is only applicable in relatively intensive agriculture such as the ornamental industry. Lin et al. (69), however, have applied the so-called "S-H mixture" to field situations and improved root health and increased plant stands. Mechanisms associated with control may be increase of pH and calcium content of soil, urea fungitoxic to plant pathogens, chitin contained in oystershell powder, the calcium, aluminum, and silicon elements in siliceous slag inhibiting chlamydospore germination in *Fusarium* spp., and the influence of plant nutrients on the general health of plants. This may be one of the few examples in which various mechanisms and control measures have been field tested. The example of *P. nunn* cited earlier also illustrates how manipulation of organic matter can affect the outcome of biological control.

The theoretical foundations treated above can form the basis of strategies for control of soilborne plant diseases with BCAs. They also explain failures in disease control, and reduced efficiency of BCAs. The nature of infection courts, the characteristics of the pathogen and environmental considerations are keys to efficiency.

ACKNOWLEDGEMENT

Ralph "Tex" Baker died on 11 April 1994 after a long battle with cancer. He was a pioneer in the fields of ecology and biological control of soilborne plant pathogens, starting his research in the 1950s. He was an innovative thinker, developing seminal ideas and theories on inoculum potential, mathematical modeling of inoculum density relationships in the rhizosphere and soil fungistasis. His search to understand the mechanisms of biological control led him to investigate antibiosis, mycoparasitism, iron competition and induced resistance. He studied biological control agents such as *Trichoderma, Pseudomonas, Pythium nunn*, and nonpathogenic *Fusarium oxysporum* against the pathogens *Rhizoctonia, Pythium*, and *Fusarium*. But his greatest contribution may be the infectious enthusiasm he instilled in his many students and visitors from all over the world. He was an outstanding individual on a personal level outside of science with his family, friends, students, colleagues and church. As Tex would have said in his imaginative literary way, the quest to unlock the tantalizing secrets of the plant-soil-microbe universe beneath our feet will continue. But his presence in this adventure will be greatly missed.

LITERATURE CITED

1. Adams, P.B. 1986. Production of sclerotia of *Sclerotinia minor* on lettuce in the field and their distribution in soil after disking. Plant Dis. 70:1043–1046.

2. Adams, P.B. 1990. The potential of mycoparasites for biological control of plant diseases. Annu. Rev. Phytopathol. 28:59–72.

3. Adams, P.B., and Ayers, W.A. 1982. Biological control of Sclerotinia lettuce drop in the field by *Sporidesmium sclerotivorum*. Phytopathology 72:485–488.

4. Adams, P.B., and Fravel, D.R. 1990. Economical biological control of Sclerotinia lettuce drop by *Sporidesmium sclerotivorum*. Phytopathology 80:1120–1124.

5. Ahmad, J.S., and Baker, R. 1988. Implications of rhizosphere competence of *Trichoderma harzianum*. Can. J. Microbiol. 34:229–234.

6. Asher, M.J.C., and Shipton, P.J., eds. 1981. Biology and Control of Take-All. Academic Press, New York, NY. 538 pp.

7. Ayers, W.A., and Adams, P.B. 1979. Mycoparasitism of sclerotia of *Sclerotinia* and *Sclerotium* species by *Sporidesmium sclerotivorum*. Can. J. Microbiol. 25:17–23.

8. Ayers, W.A., and Adams, P.B. 1983. Improved media for growth and sporulation of *Sporidesmium sclerotivorum*. Can. J. Microbiol. 29:325–330.

9. Backman, P.A., and Rodriguez-Kabana, R. 1975. A system for the growth and delivery of biological control agents to the soil. Phytopathology 65:819–821.

10. Baker, R. 1971. Analyses involving inoculum density of soil-borne plant pathogens in epidemiology. Phytopathology 61:1280–1292.

11. Baker, R. 1978. Inoculum potential. Pages 137–157 in: Plant Disease: An Advanced Treatise, Vol. II: How Disease Develops in Populations. J.G. Horsfall and E.B. Cowling, eds. Academic Press, New York, NY.

12.   Baker, R. 1981. Biological control: Eradication of plant pathogens by adding organic amendments to soil. Pages 317–327 in: CRC Handbook of Pest Management in Agriculture, Vol. II. D. Pimentel, ed. CRC Press, Boca Raton, FL.

13.   Baker, R. 1987. Mycoparasitism: ecology and physiology. Can. J. Plant Pathol. 9:370–379.

14.   Baker, R., and Drury R. 1981. Inoculum potential and soilborne pathogens: The essence of every model is within the frame. Phytopathology 71:363–372.

15.   Baker R., Elad, Y., and Sneh, B. 1986. Physical, biological and host factors in iron competition in soils. Pages 77–84 in: Iron, Siderophores and Plant Diseases. T.R. Swinburne, ed. Plenum Press, New York, NY.

16.   Baker, R., Maurer, C.L., and Maurer, R.A. 1967. Ecology of plant pathogens in soil. VII. Mathematical models and inoculum density. Phytopathology 57:662–666.

17.   Baker, R., and Nash, S.M. 1965. Ecology of plant pathogens in soil. VI. Inoculum density of *Fusarium solani* f. sp. *phaseoli* in bean rhizosphere as affected by cellulose and supplemental nitrogen. Phytopathology 55:1381–1382.

18.   Biles, C.L, and Martyn, R.D. 1989. Local and systemic resistance induced in watermelons by formae speciales of *Fusarium oxysporum*. Phytopathology 79:856–860.

19.   Budge, S.P., and Whipps, J.M. 1991. Glasshouse trials of *Coniothyrium minitans* and *Trichoderma* species for the biological control of *Sclerotinia sclerotiorum* in celery and lettuce. Plant Pathol. 40:59–66.

20.   Campbell, C.L., and Noe, J.P. 1985. The spatial analysis of soilborne pathogens and root diseases. Annu. Rev. Phytopathol. 23:129–148.

21.   Chen, Z., and Klessig, D.F. 1991. Identification of a soluble salicylic acid-binding protein that may function in signal transduction in the plant disease-resistance response. Proc. Natl. Acad. Sci. USA 88:8179–8183.

22.   Chet, I. 1987. *Trichoderma*—application, mode of action and potential as a biocontrol agent of soil-borne plant pathogenic fungi. Pages 137–160 in: Innovative Approaches to Plant Disease Control. I. Chet, ed. John Wiley & Sons, New York, NY.

23. Chet, I. 1990. Mycoparasitism: recognition, physiology, and ecology. Pages 725-735 in: New Directions in Biological Control: Alternatives for Suppressing Agricultural Pests and Diseases. R.R. Baker and P.E. Dunn, eds. Alan R. Liss Inc., New York, NY.

24. Chet, I., and Baker, R. 1980. Induction of suppressiveness to *Rhizoctonia solani* in soil. Phytopathology 70:994-998.

25. Chet, I., Harman, G.E., and Baker, R. 1981. *Trichoderma hamatum*: Its hyphal interactions with *Rhizoctonia solani* and *Pythium* spp. Microb. Ecol. 7:29-38.

26. Clark, F.E. 1942. Experiments toward the control of the take-all disease of wheat and the Phymatotrichum root rot of cotton. USDA Tech. Bull. 835. 27 pp.

27. Cook, R.J., and Schroth, M.N. 1965. Carbon and nitrogen compounds and germination of chlamydospores of *Fusarium solani* f. *phaseoli*. Phytopathology 55:254-256.

28. Cook, R.J., and Weller, D.M. 1987. Management of take-all in consecutive crops of wheat or barley. Pages 41-76 in: Innovative Approaches to Plant Disease Control. I. Chet, ed. John Wiley & Sons, New York, NY.

29. Deacon, J.W. 1981. Ecological relationships with other fungi: Competitors and hyperparasites. Pages 75-101 in: Biology and Control of Take-All. M.J.C. Asher and P.J. Shipton, eds. Academic Press, London, UK.

30. Deacon, J.W. 1987. Programmed cortical senescence: a basis for understanding root infection. Pages 285-297 in: Fungal Infection of Plants. G.F. Pegg and P.G. Ayres, eds. Cambridge University Press, Cambridge, UK.

31. Deacon, J.W. 1991. Significance of ecology in the development of biocontrol agents against soil-borne plant pathogens. Biocontrol Sci. Technol. 1:5-20.

32. Deacon, J.W., and Berry, L.A. 1992. Modes of action of mycoparasites in relation to biocontrol of soilborne pathogens. Pages 157-174 in: Biological Control of Plant Diseases, Progress and Challenges for the Future. E.C. Tjamos, G.C. Papavizas, and R.J. Cook, eds. Plenum Press, New York, NY.

33. Deacon, J.W., and Henry, C.M. 1980. Age of wheat and barley roots and infection by *Gaeumannomyces graminis* var. *tritici*. Soil Biol. Biochem. 12:113-118.

34.   Elad, Y., Chet, I., and Henis, Y. 1982. Degradation of plant pathogenic fungi by *Trichoderma harzianum*. Can. J. Microbiol. 28:719–725.

35.   Elad, Y., and Baker, R. 1985. The role of competition for iron and carbon in suppression of chlamydospore germination of *Fusarium* spp. by *Pseudomonas* spp. Phytopathology 75:1053–1059.

36.   Elad, Y., Lifshitz, R., and Baker, R. 1985. Enzymatic activity of the mycoparasite *Pythium nunn* during interaction with host and non-host fungi. Physiol. Plant Pathol. 27:131–148.

37.   Enard, C., Diolez, A., and Expert, D. 1988. Systemic virulence of *Erwinia chrysanthemi* 3937 requires a functional iron assimilation system. J. Bacteriol. 170:2419–2426.

38.   Freeman, S., and Rodriguez, R.J. 1993. Genetic conversion of a fungal plant pathogen to a nonpathogenic, endophytic mutualist. Science 260:75–78.

39.   Gessler, C., and Kuć, J. 1982. Induction of resistance to Fusarium wilt in cucumber by root and foliar pathogens. Phytopathology 72:1439–1441.

40.   Griffin, G.J. 1970. Exogenous carbon and nitrogen requirements for chlamydospore germination by *Fusarium solani*: dependence on spore density. Can. J. Microbiol. 16:1366–1368.

41.   Guy, S.O., and Baker, R. 1977. Inoculum potential in relation to biological control of Fusarium wilt of peas. Phytopathology 67:72–78.

42.   Handelsman, J., Raffel, S., Mester, E.H., Wunderlich, L., and Grau, C.R. 1990. Biological control of damping-off of alfalfa seedlings with *Bacillus cereus* UW85. Appl. Environ. Microbiol. 56:713–718.

43.   Harman, G.E. 1990. Deployment tactics for biocontrol agents in plant pathology. Pages 729–792 in: New Directions in Biological Control: Alternatives for Suppressing Agricultural Pests and Diseases. R.R. Baker and P.E. Dunn, eds. Alan R. Liss Inc., New York, NY.

44.   Harman, G.E., Chet, I. and Baker, R. 1980. *Trichoderma hamatum* effects on seed and seedling disease induced in radish and pea by *Pythium* spp. or *Rhizoctonia solani*. Phytopathology 70:1167–1172.

45.   Harman, G.E., Chet, I. and Baker, R. 1981. Factors affecting *Trichoderma hamatum* applied to seeds as a biocontrol agent. Phytopathology 71:569–572.

46. Harman, G.E., Jin, X., Stasz, T.E., Peruzzotti, G., Leopold, A.C., and Taylor, A.G. 1991. Production of conidial biomass of *Trichoderma harzianum* for biological control. Biol. Control 1:23-28.

47. Harman, G.E., and Taylor, A.G. 1988. Improved seedling performance by integration of biological control agents at favorable pH levels with solid matrix priming. Phytopathology 78:520-525.

48. Harman, G.E., and Taylor, A.G. 1990. Development of an effective biological seed treatment system. Pages 415-426 in: Biological Control of Soil-borne Plant Pathogens. D. Hornby, ed. CAB International, Wallingford, UK.

49. Hoitink, H.A.J., and Fahy, P.C. 1986. Basis for the control of soilborne plant pathogens with composts. Annu. Rev. Phytopathol. 24:93-114.

50. Hoitink, H.A.J., Boehm, M.J., and Hadar, Y. 1993. Mechanisms of suppression of soil-borne plant pathogens in compost-amended substrates. Pages 601-621 in: Science and Engineering of Composting. H.A.J. Hoitink and H.M. Keener, eds. Renaissance Publications, Worthington, OH.

51. Hornby, D. 1975. Inoculum of the take-all fungus: Nature, measurement, distribution and survival. EPPO Bull. 5:319-333.

52. Howell, C.R. 1982. Effect of *Gliocladium virens* on *Pythium ultimum*, *Rhizoctonia solani*, and damping-off of cotton seedlings. Phytopathology 72:496-498.

53. Howell, C.R. 1987. Relevance of mycoparasitism in the biological control of *Rhizoctonia solani* by *Gliocladium virens*. Phytopathology 77:992-994.

54. Howell, C.R., and Stipanovic, R.D. 1983. Gliovirin, a new antibiotic from *Gliocladium virens*, and its role in the biological control of *Pythium ultimum*. Can. J. Microbiol. 29:321-324.

55. Huang, H.C. 1992. Ecological basis of biological control of soilborne plant pathogens. Can. J. Plant Pathol. 14:86-91.

56. Huang, H.C., and Kozub, G.C. 1991. Monocropping to sunflower and decline of sclerotinia wilt. Bot. Bull. Acad. Sin. 32:163-170.

57. Kloepper, J.W., Schippers, B., and Bakker, P.A.H.M. 1992. Proposed elimination of the term *endorhizosphere*. Phytopathology 82:726-727.

58.   Ko, W., and Hora, F.K. 1971. A selective medium for the quantitative determination of *Rhizoctonia solani* in soil. Phytopathology 61:707–710.

59.   Kuć, J. 1987. Plant immunization and its applicability for disease control. Pages 255–275 in: Innovative Approaches to Plant Disease Control. I. Chet, ed. John Wiley & Sons, New York, NY.

60.   Laing, S.A.K., and Deacon, J.W. 1991. Video microscopical comparison of mycoparasitism by *Pythium oligandrum, P. nunn* and an unnamed *Pythium* species. Mycol. Res. 95:469–479.

61.   Leong, S.A., and Expert, D. 1989. Siderophores in plant-pathogen interactions. Pages 62–83 in: Plant-Microbe Interactions: Molecular and Genetic Perspectives, Vol. 3. T. Kosuge and E.W. Nester, eds. McGraw-Hill, New York, NY. 511 pp.

62.   Lewis, J.A., and Papavizas, G.C. 1984. A new approach to stimulate population proliferation of *Trichoderma* species and other potential biocontrol fungi introduced into natural soil. Phytopathology 74:1240–1244.

63.   Lewis, J.A., and Papavizas, G.C. 1991. Biocontrol of plant diseases: the approach for tomorrow. Crop Prot. 10:95–105.

64.   Leyns, F., Lambert, B., Juos, H., and Swings, I. 1990. Antifungal bacteria from different crops. Pages 437–444 in: Biological Control of Soil-borne Plant Pathogens. D. Hornby, ed. CAB International, Wallingford, UK.

65.   Lifshitz, R., Dupler, M., Elad, Y., and Baker, R. 1984. Hyphal interactions between *Pythium nunn*, a mycoparasite, and several soil fungi. Can. J. Microbiol. 30:1482–1487.

66.   Lifshitz, R., Sneh, B., and Baker, R. 1984. Soil suppressiveness to a plant pathogenic *Pythium* species. Phytopathology 74:1054–1061.

67.   Lifshitz, R., Stanghellini, M.E., and Baker, R. 1984. A new species of *Pythium* isolated from soil in Colorado. Mycotaxon 20:373–379.

68.   Lifshitz, R., Windham, M.T., and Baker, R. 1986. Mechanism of biological control of preemergence damping-off of pea by seed treatment with *Trichoderma* spp. Phytopathology 76:720–725.

69.   Lin, Y.S., Sun, S.K., Hsu, S.T., and Hsieh, W.H. 1990. Mechanisms involved in the control of soil-borne plant pathogens by S-H mixture. Pages 249–259 in: Biological Control of Soil-borne Plant Pathogens. D. Hornby, ed. CAB International, Wallingford, UK.

70. Liu, L., Kloepper, J.W., and Tuzun, S. 1992. Induction of systemic resistance against cucumber mosaic virus by seed inoculation with select rhizobacterial strains. Phytopathology 82:1108. (Abstr.).

71. Liu, S., and Baker, R. 1980. Mechanism of biological control in soil suppressive to *Rhizoctonia solani*. Phytopathology 70:404–412.

72. Lumsden, R.D., and Locke, J.C. 1989. Biological control of damping-off caused by *Pythium ultimum* and *Rhizoctonia solani* with *Gliocladium virens* in soilless mix. Phytopathology 79:361–366.

73. Lutchmeah, R.S., and Cooke, R.C. 1984. Aspects of antagonism by the mycoparasite *Pythium oligandrum*. Trans. Br. Mycol. Soc. 83:696–700.

74. Malamy, J., Carr, J.P., Klessig, D.F, and Raskin, I. 1990. Salicylic acid: a likely endogenous signal in the resistance response of tobacco to viral infection. Science 250:1002–1004.

75. Mandeel, Q., and Baker, R. 1991. Mechanisms involved in biological control of Fusarium wilt of cucumber with strains of nonpathogenic *Fusarium oxysporum*. Phytopathology 81:462–469.

76. Maurer, C.L., and Baker, R. 1965. Ecology of plant pathogens in soil. II. Influence of glucose, cellulose, and inorganic nitrogen amendments on development of bean root rot. Phytopathology 55:69–72.

77. McQuilken, M.P., Whipps, J.M., and Cooke, R.C. 1990. Control of damping-off in cress and sugar-beet by commercial seed-coating with *Pythium oligandrum*. Plant Pathol. 39:452–462.

78. McQuilken, M.P., Whipps, J.M., and Cooke, R.C. 1992. Use of oospore formulations of *Pythium oligandrum* for biological control of *Pythium* damping-off in cress. J. Phytopathol. 135:125–134.

79. Merriman, P.R., Price, R.D., Baker, K.F., Kollmorgan, J.F., Piggott, T., and Ridge, E.H. 1975. Effect of *Bacillus* and *Streptomyces* spp. applied to seed. Pages 130–133 in: Biology and Control of Soil-Borne Plant Pathogens. G.W. Bruehl, ed. The American Phytopathological Society, St. Paul, MN.

80. Métraux, J.P., Signer, H., Ryals, J., Ward, E., Wyss-Benz, M., Gaudin, J., Raschdorf, K., Schmid, E., Blum, W., and Inverardi, B. 1990. Increase in salicylic acid at the onset of systemic acquired resistance in cucumber. Science 250:1004–1006.

81. Mueller, J.D., Cline, M.N., Sinclair, J.B., and Jacobsen, B.J. 1985. An in vitro test for evaluating efficacy of mycoparasites on sclerotia of *Sclerotinia sclerotiorum*. Plant Dis. 69:584–587.

82. Papavizas, G.C., and Lumsden, R.D. 1980. Biological control of soilborne fungal propagules. Annu. Rev. Phytopathol. 18:389–413.

83. Papavizas, G.C. 1985. *Trichoderma* and *Gliocladium*: Biology, ecology, and potential for biocontrol. Annu. Rev. Phytopathol. 23:23–54.

84. Patterson, C.L., and Grogan, R.G. 1985. Differences in epidemiology and control of lettuce drop caused by *Sclerotinia minor* and *S. sclerotiorum*. Plant Dis. 69:766–770.

85. Paulitz, T.C. 1992. Biological control of damping-off diseases with seed treatments. Pages 145–156 in: Biological Control of Plant Diseases. E.C. Tjamos, G.C. Papavizas, and R.J. Cook, eds. Plenum Press, New York, NY.

86. Paulitz, T.C., and Baker, R. 1987. Biological control of Pythium damping-off of cucumbers with *Pythium nunn*: Population dynamics and disease suppression. Phytopathology 77:335–340.

87. Paulitz, T.C., and Baker, R. 1987. Biological control of Pythium damping-off of cucumbers with *Pythium nunn*: Influence of soil environment and organic amendments. Phytopathology 77:341–346.

88. Paulitz, T.C., and Baker, R. 1988. The formation of secondary sporangia by *Pythium ultimum*: The influence of organic amendments and *Pythium nunn*. Soil Biol. Biochem. 20:151–156.

89. Paulitz, T.C., and Baker, R. 1988. Interactions between *Pythium nunn* and *Pythium ultimum* on bean leaves. Can. J. Microbiol. 34:947–951.

90. Paulitz, T.C., and Fernando, W.G.D. 1995. Biological seed treatments for the control of soil-borne plant pathogens. Pages 185–217 in: Management of Soil Borne Diseases. R. Utkhede and V. K. Gupta, eds. Kalyani Publishers, New Delhi.

91. Paulitz, T.C., Zhou, T., and Rankin, L. 1992. Selection of rhizosphere bacteria for biological control of *Pythium aphanidermatum* on hydroponically grown cucumber. Biol. Control 2:226–237.

92. Roberts, D.P., and Lumsden, R.D. 1990. Effect of extracellular metabolites from *Gliocladium virens* on germination of sporangia and mycelial growth of *Pythium ultimum*. Phytopathology 80:461–465.

93. Roiger, D.J., and Jeffers, S.N. 1991. Evaluation of *Trichoderma* spp. for biological control of Phytophthora crown and root rot of apple seedlings. Phytopathology 81:910–917.

94. Rouse, D.I., and Baker, R. 1978. Modeling and quantitative analysis of biological control mechanisms. Phytopathology 68:1297–1302.

95. Sequeira, L. 1990. Induced resistance: Physiology and biochemistry. Pages 663–678 in: New Directions in Biological Control: Alternatives for Suppressing Agricultural Pests and Diseases. R.R. Baker and P.E. Dunn, eds. Alan R. Liss Inc., New York, NY.

96. Simeoni, L.A., Lindsay, W.L., and Baker, R. 1987. Critical iron level associated with biological control of Fusarium wilt. Phytopathology 77:1057–1061.

97. Smith, V.L., Wilcox, W.F., and Harman, G.E. 1990. Potential for biological control of *Phytophthora* root and crown rots of apple by *Trichoderma* and *Gliocladium* spp. Phytopathology 80:880–885.

98. Sneh, B., Dupler, M., Elad, Y., and Baker, R. 1984. Chlamydospore germination of *Fusarium oxysporum* f. sp. *cucumerinum* as affected by fluorescent and lytic bacteria from a Fusarium-suppressive soil. Phytopathology 74:1115–1124.

99. Sneh, B., Katan, J., and Henis, Y. 1971. Mode of inhibition of *Rhizoctonia solani* in chitin-amended soil. Phytopathology 61:1113–1117.

100. Stanghellini, M.E., and Nigh Jr., E.L. 1972. Occurrence and survival of *Pythium aphanidermatum* under arid soil conditions in Arizona. Plant Dis. Rep. 56:507–510.

101. Stanghellini, M.E., Stowell, L.J., Kronland, W.C., and von Bretzel, P. 1983. Distribution of *Pythium aphanidermatum* in rhizosphere soil and factors affecting expression of the absolute inoculum potential. Phytopathology 73:1463–1466.

102. Stanghellini, M.E., von Bretzel, P., Kronland, W.C., and Jenkins, A.D. 1982. Inoculum densities of *Pythium aphanidermatum* in soils of irrigated sugar beet fields in Arizona. Phytopathology 72:1481–1485.

103. Taylor, A.G., Min, T.-G., Harman, G.E., and Jin, X. 1991. Liquid coating formulation for the application of biological seed treatments of *Trichoderma harzianum*. Biol. Control 1:16–22.

104. Tedla, T., and Stanghellini, M.E. 1992. Bacterial population dynamics and interactions with *Pythium aphanidermatum* in intact rhizosphere soil. Phytopathology 82:652–656.

105. Trutmann, P., and Keane, P.J. 1990. *Trichoderma koningii* as a biological control agent for *Sclerotinia sclerotiorum* in Southern Australia. Soil Biol. Biochem. 22:43–50.

106. Van Peer, R., Neimann, G.J., and Schippers, B. 1991. Induced resistance and phytoalexin accumulation in biological control of Fusarium wilt of carnation by *Pseudomonas* sp. strain WCS417r. Phytopathology 81:728–734.

107. von Bretzel, P., Stanghellini, M.E., and Kronland, W.C. 1988. Epidemiology of Pythium root rot of mature sugar beets. Plant Dis. 72:707–709.

108. Wei, G., Kloepper, J.W., and Tuzun, S. 1991. Induction of systemic resistance of cucumber to *Colletotrichum orbiculare* by select strains of plant growth-promoting rhizobacteria. Phytopathology 81:1508–1512.

109. Weller, D.M., and Thomashow, L.S. 1990. Antibiotics: Evidence for their production and sites where they are produced. Pages 703–711 in: New Directions in Biological Control: Alternatives for Suppressing Agricultural Pests and Diseases. R.R. Baker and P.E. Dunn, eds. Alan R. Liss Inc., New York, NY.

110. Whipps, J.M. 1992. Status of biological disease control in horticulture. Biocontrol Sci. Technol. 2:3–24.

111. Whipps, J.M., and Budge, S.P. 1990. Screening for sclerotial mycoparasites of *Sclerotinia sclerotiorum*. Mycol. Res. 94:607–612.

112. Whipps, J.M., and Gerlagh, M. 1992. Biology of *Coniothyrium minitans* and its potential for use in disease control. Mycol. Res. 96:897–907.

113. Whipps, J.M., Lewis, K., and Cooke, R.C. 1988. Mycoparasitism and plant disease control. Pages 161–187 in: Fungi in Biological Control Systems. M.N. Burge, ed. Manchester University Press, Manchester, UK.

114. Wildon, D.C., Thain, J.F., Minchin, P.E.H., Gubb, I.R., Reilly, A.J., Skipper, Y.D., Doherty, H.M., O'Donnell, P.J., and Bowles, D.J. 1992. Electrical signalling and systemic proteinase inhibitor induction in the wounded plant. Nature (Lond.) 360:62–65.

115. Wilkinson, H.T., Cook, R.J., and Alldredge, J.R. 1985. Relation of inoculum size and concentration to infection of wheat roots by *Gaeumannomyces graminis* var. *tritici*. Phytopathology 75:98–103.
116. Wymore, L.A., and Baker, R. 1982. Factors affecting cross-protection in control of Fusarium wilt of tomato. Plant Dis. 66:908–910.
117. Zhou, T., and Paulitz, T.C. 1994. Induced resistance in the biocontrol of *Pythium aphanidermatum* by *Pseudomonas* spp. on cucumber. J. Phytopathol. 142:51–63.

Chapter 4

MOLECULAR BASIS OF PATHOGEN SUPPRESSION BY ANTIBIOSIS IN THE RHIZOSPHERE

L.S. Thomashow and D.M. Weller

Bacteria introduced on seeds or other planting material for the purpose of suppressing soilborne plant diseases may interact with pathogens directly through one or more mechanisms of antagonism including competition, parasitism, predation, and antibiosis, or they may function indirectly to limit the initiation or spread of disease by triggering systemic defense responses in the host plant. It has become increasingly clear over the past decade that antibiosis, "the inhibition or destruction of one organism by a metabolic product of another" (4), has a dominant role in the suppression of several important fungal root and seed pathogens by bacterial biological control agents, and especially by fluorescent *Pseudomonas* species (34,40,67). In this chapter we review the various approaches currently used to evaluate the contribution of antibiotics to the control of soilborne plant diseases. Emphasis is on the use of molecular techniques to provide new insights into the genetics and regulation of antibiotic production by *Pseudomonas*, one of comparatively few bacterial genera that synthesize an array of compounds with broad-spectrum antibiotic activity. We also discuss ways in which our current understanding of the role and production of antibiotics by plant-associated microorganisms can be applied to help make biological control technology more reliable and effective.

TRADITIONAL, MOLECULAR, AND BIOANALYTIC ASSESSMENT OF ANTIBIOSIS IN PLANT DISEASE SUPPRESSION

Antibiotics include low molecular weight organic compounds that are produced by microorganisms and are deleterious to the growth or other metabolic activities of other microorganisms (10). This broad definition encompasses numerous bacterial metabolites that have been studied in relation to biological control including phenazines, 2,4-diacetylphloroglucinol, pyrrolnitrin, pyoluteorin, oomycin A, agrocin 84, and herbicolin A; fungal metabolites such as chaetomin, gliovirin, gliotoxin, viridin, and iturin A; volatiles such as hydrogen cyanide, ammonia and ethanol; and the siderophores, pyoverdine and pyochelin. Many of these compounds are produced late in growth or during stationary phase, but the definition includes products of both so-called primary and secondary metabolism.   On the other hand, enzymes including chitinases, cellulases, glucanases, and proteases are excluded by the definition because of their molecular weight.

A variety of experimental approaches have been, and continue to be, used to assess the importance of antibiotics in biological control systems. Each strategy has merits and limitations; strategies are not mutually exclusive, and are in fact more informative when used together. Antibiosis traditionally has been implicated by establishing a correlation between the ability of an agent to suppress a particular plant disease in situ and to inhibit growth of the causal pathogen in vitro. Traditional studies often have provided the first indication, albeit circumstantial, that antibiotics are involved in a biological control system, and in some cases the approach has been strengthened by establishing a more specific or quantitative relationship between biological control activity in situ and the production of a particular compound in vitro. For example, the beneficial effect of viable cells has been duplicated by cell-free filtrates, crude extracts, or antibiotics purified from cultures of the biological control agent, and the level of disease suppression has been correlated with the amount of antibiotic produced (66,67). Alternatively, certain antibiotic-nonproducing mutants have been shown to be less suppressive than the parental strains.   Even with these refinements, however, conclusions derived via traditional approaches alone remain uncertain because it cannot be assumed that the in vitro environment has adequately simulated the conditions in which the antagonist and the pathogen interact in situ.

The physical and biological constraints to antibiotic production, activity and detection in soil environments have been reviewed (66,69,70) and include instability or microbial degradation of antibiotics, adsorption to soil colloids, spatially or temporally restricted production, insufficient sensitivity of detection, and insufficient nutrient availability. With current recognition that nutrients in plant-associated microhabitats are at least transiently adequate to support microbial growth and antibiotic production, and through the use of more powerful analytic techniques, reports of antibiotic production associated with plant disease suppression increasingly are becoming more common. Thus, suppression of take-all of wheat, caused by *Gaeumannomyces graminis* var. *tritici*, was correlated respectively with production in situ of phenazine-1-carboxylic acid (PCA) by *Pseudomonas fluorescens* 2-79 and *Pseudomonas chlororaphis* (formerly *P. aureofaciens*) 30-84 (60) or with production of 2,4-diacetylphloroglucinol (Phl) by *P. fluorescens* CHA0 (26). The cyclic peptide antibiotic herbicolin A was detected in crown and root tissues of wheat seedlings colonized by *Erwinia herbicola* B247, which suppresses diseases caused by *Fusarium culmorum* and *Puccinia recondita* f. sp. *tritici* (28,29). Pyrrolnitrin was extracted from the rhizosphere soil of sugar beet seedlings (24) and cotton, and from the spermosphere of barley bacterized with *Pseudomonas* spp. (30). Control of Pythium damping-off of zinnia was correlated with the concentration of gliotoxin produced in situ by *Gliocladium virens* GL-21 (35).

These studies establish that antibiotic production occurs not only in vitro, but also in association with plants, and that such production can be of significance to plant protection. However, the direct analytical approach requires that the active metabolite be known or at least characterized biochemically, that it be reasonably stable, and that sensitive and specific assay procedures be available for its detection. For many substances it may not be possible to meet these criteria. The need for specialized analytical equipment and the time and expense involved in handling large numbers of replicated samples are further impediments to routine use of the direct analytical approach. Collectively, these obstacles have provided incentive to apply methods based on molecular genetic technology to biological control systems.

The power of the molecular approach as applied to studies of biological control mechanisms is in being able to evaluate the role of a specific genetic or phenotypic determinant both in vitro and in situ. The strategy, which has been likened to Koch's postulates, consists of: (i) mutagenesis to inactivate the trait of interest; (ii) screening of mutants and characterization of the mutant phenotype; (iii) complementation of the mutant with DNA from a wild-type genomic library to genetically reconstitute the trait; and (iv) comparison of the mutant, wild-type and restored phenotypes. A fifth step, in which the complementing gene is mutated, introduced into the wild-type strain, and shown to confer the mutant phenotype, is desirable, but is not always completed. Transposons, such as Tn5, are preferred for mutagenesis because it can easily be shown that single insertions (i.e. single-site mutations) have occurred, and target genes can be recovered because of the introduction of a selectable marker. It is critical that assays be highly specific for the trait in question. Many biological control strains produce more than one antibiotic, and assays and mutant characterization in vitro must eliminate or compensate for the effects of metabolites other than the one of interest. Finally, strain evaluation must include studies in situ because phenotypic changes that impact on competitiveness or biological control activity may not be detected in vitro, and many mutations resulting in antibiotic deficiency are now known to have occurred in global regulatory loci that control phenotypes other than antibiotic production.

GENETICS AND ROLE OF ANTIBIOTIC PRODUCTION IN PATHOGEN SUPPRESSION BY SELECTED BACTERIAL BIOLOGICAL CONTROL AGENTS

A survey of emerging model systems for the genetics and regulation of antibiotic production by *Pseudomonas* spp. that suppress plant diseases reveals certain important similarities. First, the biosynthetic genes themselves are clustered, and the biosynthetic loci contain linked regulatory elements. Secondly, expression of the biosynthetic locus is highly responsive to conditions in the environment and is modulated by a complex network of additional, unlinked regulatory genes that function in environmental sensing, signal transduction, and the global coordination of cellular metabolism and homeostasis. Mutations in such global regulatory genes typically give rise to pleiotropic phenotypes because they also control the expression of target genes other than those required for antibiotic production. The first indications of these organizational and regulatory aspects of antibiotic gene expression came from studies of the production of oomycin A.

Pseudomonas fluorescens Hv37a: Production Of Oomycin A

Approximately 50% of the increase in cotton seedling emergence and 70% of the reduction in root infection provided by *P. fluorescens* Hv37a against *Pythium ultimum* have been attributed to oomycin A, a heat-stable antibiotic of 700 to 800 Da (17,25). Production of oomycin A was induced by glucose and certain other sugars, and inhibited by a combination of amino acids (16,17). Glucose sensing required the activity of glucose dehydrogenase and expression of the regulatory operon *afuAB*. However, neither *afuAB* nor another regulatory gene, *afuP*, was induced by glucose, and mutations in any of these genes gave rise to pleiotropic phenotypes. The product of *afuP* activated expression of yet a fourth regulatory gene, *afuR*, thought to encode a positive transcriptional activator of the adjacent oomycin A biosynthetic operon. In the absence of glucose, the product of a fifth regulatory gene, *cin*, prevented transcription of the genes within the biosynthetic locus, apparently through interference with the *afuP* gene product (16).

The oomycin A biosynthetic gene cluster encompasses some 15 kb and includes *afuH* (of unknown function), *afuR*, and the divergently transcribed operon *afuDEFG*, encoding proteins respectively of 48 kDa, 31 kDa, 37 kDa and 145 kDa. Direct participation of *afuE* in oomycin A biosynthesis was inferred from results showing 250-fold induction by glucose of *afuE::lacZ* gene fusions under conditions supportive of a 200-fold increase in accumulation of the antibiotic (16). Transcription of the *afuE* operon required activation by the product of the divergently transcribed gene *afuR*. Because expression of *afuR* depends in turn on the product of *afuP* as well as the concentration of a diffusible coinducer, possibly oomycin A itself, synthesis of this antibiotic is both positively autoregulated and dependent on cell density (16). When the population is small or substrate is limiting, the pathway remains uninduced, conserving available resources for growth or maintenance but preventing accumulation of the autoinducer to the threshold required for induction. Only with adequate nutrients will the concentration of inducer reach the level required for induction, triggering diversion of carbon and energy to antibiotic synthesis. This positive autoregulatory mechanism is ideally suited to the "feast or famine" environment of the rhizosphere, where nutrients typically are transiently available and the metabolic priorities of the resident microbial population must fluctuate accordingly.

Derivatives of strain Hv37a containing transcriptional fusions between *afuE* and β-galactosidase (*lacZ*) or bioluminescence (*lux*) reporter genes have proven useful to demonstrate the dependence of control of *P. ultimum* on expression of the oomycin biosynthetic locus in the spermosphere and rhizosphere of cotton. On seeds, expression of *afuE* in reporter strains was detected about 10 hours after planting, several hours after the onset of bacterial growth and approximately 4 hours after infection of the seed by *Pythium* already had begun. Expression varied greatly among individual seedlings; under optimal conditions the highest levels in situ were only a few percent of those observed in vitro, and levels were reduced further when moisture and temperature conditions were suboptimal (16,17,25). These results indicate that variable expression of genes involved in antibiotic production is likely to contribute to the inconsistent field performance of biological control agents that depend largely on antibiosis to suppress disease.

Phenazine Antibiotics

Phenazine antibiotics are produced by *Pseudomonas* and *Streptomyces* spp. and compose a family of more than 50 pigmented, nitrogen-containing heterocyclic compounds derived from chorismic acid, a key branchpoint metabolite of the shikimic acid pathway (61). Many phenazines exhibit broad-spectrum toxicity to bacteria and fungi as well as to some higher plants and animals (reviewed in 11). The most intensively studied in relation to plant disease suppression is phenazine-1-carboxylic acid (PCA), a major determinant in the control by *P. fluorescens* 2-79 and *P. chlororaphis* 30-84 of take-all, an important root and crown rot of wheat and barley caused by *G. graminis* var. *tritici* (48,59). Both strains were isolated from the roots of wheat grown in soils naturally suppressive of take-all (65; W. Bockus, personal communication), and both produce PCA, although strain 30-84 also produces lesser amounts of the PCA derivatives 2-hydroxyphenazine-1-carboxylic acid and 2-hydroxyphenazine (2,48). Other strains with biological control activity include *P. chlororaphis* (formerly *P. aureofaciens*) PGS12, which produces the same phenazines as strain 30-84 and has broad-spectrum activity in vitro against plant pathogenic fungi (15); *Pseudomonas aeruginosa* LEC1, for which pyocyanine production contributes to the suppression of Septoria tritici blotch of wheat caused by *Septoria tritici* (9); and *P. aeruginosa* In-b-109 and In-b-784, which produce PCA and pyocyanine and are effective against *R. solani* AG 1, the rice sheath blight pathogen (49).

Phenazine-deficient (Phz$^-$) Tn5 mutants of strains 2-79 and 30-84 did not inhibit *G. graminis* var. *tritici* in vitro and were 60% to 90% less suppressive of take-all on seedlings of wheat than were the wild-type strains (19,43,48,59). A pyoverdine siderophore and anthranilic acid made only minor contributions to the suppressiveness of strain 2-79 (19). When phenazine production was restored to Phz$^-$ mutants by complementation, they also regained activity against the take-all pathogen both in vitro and in situ (48,59). PCA subsequently was isolated from the roots and rhizosphere of wheat treated with strains 2-79 or 30-84 and grown in raw soil, providing the first direct evidence for the specific role of the antibiotic in disease suppression in situ (60). No PCA was recovered from control roots or roots colonized by Phz⁻ mutants, and take-all was reduced only on roots from which PCA was isolated. In another study, an inverse linear relationship was demonstrated between the population size of strain 2-79 or its PCA-producing derivatives applied to the seed and the number of primary lesions due to take-all on the roots of wheat grown in *G. graminis* var. *tritici*-infested soil. In contrast, a Phz⁻ mutant failed to reduce lesion number at any dose (3). PCA-deficient mutants did not exhibit reduced ecological competence in these studies, which were no longer than 3 weeks in duration. Over longer periods up to 100 days, however, populations of Phz⁻ mutants of 2-79 and 30-84 declined more rapidly than did those of PCA-producing parental strains or Phz⁺ complemented mutants in raw rhizosphere and bulk soil but not in steam-pasteurized soil, suggesting that the antibiotic contributes to competitiveness against indigenous microorganisms (37).

Environmental effects on biological control and phenazine gene expression

Strain 2-79 and its Phz⁺ mutant derivatives suppressed take-all across a range of soil pH from 4.9 to 8.0 (43). In ten different soils, suppression was positively correlated with chemical and physical factors including sulfate-sulfur, % sand, pH, sodium, zinc and ammonium-nitrogen, and negatively correlated with cation exchange capacity, exchangeable acidity, manganese, iron, % silt, % clay, % organic matter, and total carbon (41,42). These properties may affect the production or activity of PCA, and variation among soils from different locations may help to explain differences in the field performance of strain 2-79 from one site to another.

The availability of carbon profoundly affected the expression in vitro of ice nucleation activity from a Tn*3*-spice fusion in the phenazine biosynthetic locus of *P. chlororaphis* PGS12. Ice nucleation activity was several orders of magnitude greater in complex than in defined media, and expression in defined media supplemented with combinations of a carbohydrate or organic acid and an amino acid generally was greater than when the medium contained only a single carbon source (14). In situ, seeds from seven plant species (presumably differing in the quantity and quality of their exudates) supported different levels of expression of the PGS12 phenazine reporter in the spermosphere, but expression was only slightly affected by soil type and not at all by soil matric potential. On seeds of wheat, expression was first detected at 12 hours and peaked by 40 hours; expression levels were only a fraction of those in vitro, and varied lognormally among individual seeds (15).

Genetics and regulation of phenazine production

Phenazine biosynthetic loci have been cloned from *P. fluorescens* 2-79 (58), *P. chlororaphis* 30-84 (48), and *P. chlororaphis* PGS12 (14). The cosmid pLSP259 from strain 30-84 restored production of PCA, 2-hydroxyphenazine-1-carboxylic acid and 2-hydroxyphenazine to ten Phz⁻ mutants, each containing a single Tn*5* insertion in one of at least five different *Eco*RI restriction fragments. Subclones of pLSP259 containing a 9.2 kb *Eco*RI fragment downstream of a *lac* promoter were expressed in *Escherichia coli*, enabling synthesis of the three phenazine antibiotics produced by strain 30-84 (48). Genes designated *phzB* and *phzC*, encoding 55-kDa and 19-kDa proteins, were required respectively for production of PCA and the 2-hydroxyphenazines (48,58).

The phenazine biosynthetic locus from *P. fluorescens* 2-79 is contained within a 12-kb fragment in the clone pPHZ108A, which complemented Phz⁻ mutants of 2-79, hybridized to subcloned fragments from pLSP259, and transferred PCA biosynthetic capability to all of 27 recipient strains of *Pseudomonas* spp. (20,58). Mutagenesis of pPHZ108A with the transposon Tn*3*HoHo1 revealed divergently transcribed units of about 5.0 kb and 0.75 kb that were required for PCA production and were strongly and weakly expressed, respectively (L.S. Thomashow and D.K. Fujimoto, unpublished). DNA sequence analysis within the 5-kb transcriptional unit has identified five open reading frames, three of which encode products similar in sequence to 3-deoxy-D-*arabino*-heptulosonate-7-phosphate synthase, isochorismatase, and the large and small subunits of *p*-aminobenzoate and anthranilate synthases (Genbank accession number L48616). Such enzymatic activities are consistent with participation directly in PCA biosynthesis.

Production of phenazines by *P. chlororaphis* 30-84, and presumably also by *P. fluorescens* 2-79, resembles production of oomycin A by *P. fluorescens* Hv37a (see above) in that synthesis is both positively autoregulated and cell-density dependent. Autoinduction enables bacteria to monitor their own population density, and was recognized over 20 years ago as the means by which bioluminescent marine vibrios regulate light production (reviewed in 12,38,68). Actively growing cultures of *Vibrio fischeri* accumulate a diffusible acylated homoserine lactone (HSL) derivative, the product of the *luxI* gene. At high cell density this autoinducer reaches a critical concentration and interacts with LuxR, the product of the transcriptional activator gene *luxR*, thus triggering expression of the remaining bioluminescence genes. HSL derivatives are now known to function as inducers in diverse phenomena including bioluminescence, Ti plasmid transfer, and the synthesis of antibiotics and exoenzymes (12,38,68). In each case, gene regulation coincides with transition to a host plant-associated or animal-associated state, leading to the idea that autoinduction, signaled by HSL derivatives, may represent a general mechanism by which bacteria modulate gene expression under conditions conducive to increased population density.

In *P. chlororaphis* 30-84, the regulatory genes *phzR* (47) and *phzI* (72) appear to correspond to *luxR* and *luxI*, respectively, and are located upstream of the biosynthetic genes *phzB* and *phzC*. Gene *phzR* is transcribed divergently from *phzBC*, can activate phenazine gene expression *in trans*, and encodes a predicted product of 27 kDa with homology to proteins including *LasR* from *P. aeruginosa*, *LuxR* from *V. fischeri*, and *TraR* from *Agrobacterium tumefaciens* (47). Synthesis of the 30-84 autoinducer requires expression of *phzI*, which is located only 30 bp downstream of *phzR*, is oriented in the opposite direction, and encodes a predicted product with sequence similarity to other proteins required for synthesis of HSL autoinducer molecules (72; L.S. Pierson III, personal communication). Cell-free supernatants of *E. coli* strains expressing the cloned *phzI* gene induced phenazine gene expression at lower cell densities than did controls lacking *phzI* (72). Furthermore, supernatants from cultures of a number of other bacterial isolates from the rhizosphere of wheat also induced phenazine gene expression, suggesting that heterologous strains can produce signal molecules that may influence the expression of the phenazine biosynthetic locus in situ (73).

2,4-Diacetylphloroglucinol

Phloroglucinol compounds are phenolic bacterial and plant metabolites with antiviral, antibacterial, antifungal, antihelminthic, and phytotoxic properties (reviewed in 11). The antibiotic 2,4-diacetylphloroglucinol (Phl) is produced by soil-associated or plant-associated fluorescent pseudomonads of diverse geographic origin including isolates from England, Ireland, France, Switzerland, Ukraine, and the United States. Recent interest has focused on four strains with biological control activity. The first, *P. fluorescens* CHA0, was isolated from a Swiss soil naturally suppressive to black root rot of tobacco caused by *Thielaviopsis basicola* and is active against take-all of wheat and a number of other soilborne diseases (6,26,27,36,63). *Pseudomonas fluorescens* Q2-87 also suppresses take-all, and was isolated from roots of wheat grown in a take-all-suppressive soil near Quincy, WA, in the Pacific Northwest of the United States (21,47,62). More than 20% of fluorescent pseudomonads isolated from wheat grown at the Quincy site produced Phl (21; A. Rovira and D.M. Weller, unpublished; C. Keel, D.M. Weller, and L.S. Thomashow, unpublished), leading to speculation that such strains are highly enriched in certain take-all-suppressive soils. The third strain, *Pseudomonas* sp. F113, was isolated in Ireland from the root hairs of a mature sugar beet plant and protects against preemergence damping-off caused by *P. ultimum* (8). Finally, *P. fluorescens* Pf-5, isolated in the southwestern United States and studied initially for its ability to produce pyoluteorin and pyrrolnitrin, effective against damping-off diseases caused by *P. ultimum* and *R. solani*, respectively, recently was shown to produce a third antibiotic identified as Phl (39).

Pseudomonas fluorescens CHA0 produces a complex repertoire of secondary metabolites including not only Phl, but also monoacetylphloroglucinol, pyoluteorin, salicylic acid, indoleacetic acid, hydrogen cyanide and a pyoverdine siderophore (reviewed in 63). Genetic analyses and plant assay systems have been indispensable in defining the role of each metabolite. Phl is the major determinant in the control of take-all by strain CHA0 and, with hydrogen cyanide, it contributes to control of black root rot of tobacco (6,26,27,64), but not to protection of cucumber against *P. ultimum*, which was mediated mainly by pyoluteorin (36). The Phl⁻ Tn5 mutant CHA625 was less inhibitory in vitro to *G. graminis* var. *tritici* and *T. basicola*, and less suppressive of take-all and black root rot, than was CHA0. Complementation of CHA625 with pME3128, containing an 11-kb fragment from a CHA0 genomic library, restored Phl production and antagonistic activity in vitro and on plants (26,27). In contrast, mutants deficient in production of the pyoverdine siderophore or of indoleacetic acid remained fully suppressive on both wheat and tobacco, indicating that these metabolites do not contribute to biological control (6,18).

Phl was isolated from the roots of wheat colonized by CHA0 in a gnotobiotic system but not from roots colonized by CHA625 or control roots (26). The amounts of Phl and pyoluteorin (Plt) produced in situ by wild-type CHA0 in the absence of pathogens were not sufficient to cause visible symptoms of phytotoxicity on tobacco, wheat (26,27), cucumber, sweet corn, or cress (36). However, CHA0(pME3090), containing a 22-kb genomic fragment, overproduced Phl and Plt and reduced the growth of cress and sweet corn as compared to the parental strain. Both Phl and

Plt probably contributed to the overall herbicidal effect (36). Among pathogens, *G. graminis* var. *tritici* was comparatively sensitive to Phl (50% inhibition of growth in vitro at 16 to 32 μg ml^{-1} of Phl), and among plants, wheat was the most resistant species tested (50% inhibition of growth and seed germination at 32 to 64 and >1,024 μg ml^{-1}, respectively) (26). Phl generally is more toxic to dicotyledons than monocotyledons; activity resembles that of 2,4-dichlorophenoxyacetate, with the two acyl and three hydroxyl groups on the phenolic nucleus all contributing to inhibition of photosystem II activity (reviewed in 11).

Phl$^-$ Tn5 mutants of *P. fluorescens* F113 were significantly less suppressive of disease than the respective wild-type strains, and activity was partially restored upon complementation with a 6-kb fragment of wild-type DNA in the plasmid pCU203 (8). Cell-free extracts from strain F113 have monoacetylphloroglucinol (MAPG) acetylase activity, which is encoded on pCU203 (7) and catalyzes the conversion of MAPG to Phl in the final enzymatic step of the Phl biosynthetic pathway proposed by Shanahan et al. (54). Strain F113 produced Phl optimally at 12°C and in cultures with a high surface to volume ratio; the latter effect appeared less due to oxygen tension than to need for a minimum amount of surface contact. Fructose, sucrose and mannitol supported high yields of Phl, with ammonium ion as the preferred nitrogen source (54,55). Glucose reduced Phl production by strain F113, but enhanced production by strain Pf-5 (39).

Genetics and regulation of 2,4-diacetylphloroglucinol biosynthesis

A Phl biosynthetic locus from *P. fluorescens* Q2-87. Overlapping cosmid clones of 25 or 37 kb containing the putative Phl biosynthetic locus from Q2-87 complemented the Phl$^-$ mutant Q2-87::Tn5 and were expressed in *Pseudomonas* strains 2-79 and 5097, neither of which themselves produced Phl (62). A subclone of approximately 6.5 kb in the plasmid pMON5122 conferred Phl biosynthesis to all and improved the biological control activity against take-all of some of 12 other strains of *Pseudomonas* from the rhizosphere of wheat (20). At least six open reading frames (ORFs) spanning approximately 5 kb of DNA are required for the transfer of Phl biosynthetic capability by pMON5122 (1; M.G. Bangera and L.S. Thomashow, unpublished). The predicted products of ORFs 1 and 2 showed a high degree of protein sequence similarity to thiolase, and to chalcone and stilbene synthase enzymes, respectively. These enzymes function in acyl condensation and cyclization reactions of precisely the sort expected in the synthesis of monoacetylphloroglucinol, the immediate Phl precursor in the pathway proposed by Shanahan et al. (54). PS2, predicted by ORF 2, may be the first prokaryotic enzyme with significant similarity to the chalcone and stilbene synthases, which are unique to plants and have key roles in the production of phytoalexins and stilbenes important in plant defense (53).

ORFs 1 and 2, as well as ORFs 5 and 6, reside upstream of ORF 3, with a product PS3 thought to belong to a large superfamily of transmembrane solute facilitator pumps specific for simple sugars, oligosaccharides, organic acids, organophosphate esters and drugs (51). PS3 exhibited structural features of these integral membrane permeases including a central hydrophilic loop bordered on either side by strongly hydrophobic α-helical domains. The expression of such proteins typically is negatively regulated, with the transport and repressor genes divergently transcribed from tandem or overlapping promoters. ORF 4 in pMON5122 is divergently transcribed from ORF 3, and the predicted protein PS4 had similarity with other repressors including those that regulate tetracycline resistance. Thus, PS3 and PS4 may represent an export locus that has been modified by the introduction of a DNA segment including ORFs 1 and 2, encoding predicted Phl biosynthetic enzymes, and ORFs 5 and 6, for which functions are yet to be proposed, between the putative transporter and repressor genes.

Unlinked regulatory loci: _gacA_, _lemA_, and _rpoS_. Three loci that regulate the production of Phl have been characterized. Of these, two are thought to interact as a two-component regulatory system that activates other genes required for the synthesis of products of secondary metabolism. Two-component regulators are sensory transduction systems in which a membrane-bound sensor receives and transmits signals from the environment to a cytoplasmic response regulator that binds to specific DNA sequences to mediate changes in gene expression (44). The _gacA_ (_g_lobal _a_ntibiotic and _c_yanide control) response regulator was identified initially in _P. fluorescens_ CHA0, where it is required for production of Phl, pyoluteorin, HCN (33), and an extracellular protease and phospholipase (50). The homologous gene also has been cloned from _P. fluorescens_ BL915, in which it activates production of pyrrolnitrin, chitinase and HCN, and is required for protection of cotton seedlings from damping-off caused by _R. solani_ (13).

Spontaneous mutations in _gacA_ account for some of the losses in antagonistic activity that have occurred after biological control agents were maintained in culture. For example, the original _gacA_ mutation in strain CHA0 arose by deletion (33), and pleiotropic _gacA_ mutants of

BL915 were recovered after only 1 week on rich media (13). Mutations in *gacA* also occur in nature. Transfer of the *gacA* locus from BL915 to two heterologous strains of *P. fluorescens* activated the expression of latent genes for the synthesis of cyanide, chitinase, and pyrrolnitrin, and conferred biological control activity in a cotton-*Rhizoctonia* assay system (13). Other pleiotropic mutants were not complemented, indicating that additional regulatory elements are involved in the production of secondary metabolites. One of these, the sensor hypothesized to function in conjunction with *gacA*, has been identified in the biological control agents BL915 (31,32) and Pf-5 (5) as a homologue of *lemA*. This gene is extensively conserved among *Pseudomonas* spp. including the phytopathogen *P. syringae* pv. *syringae* (71). Mutations in the *lemA* homologues of BL915 and Pf-5 gave rise to complex phenotypic changes similar to those of *gacA* mutants; typically included were altered colony morphology on some media, loss of ability to produce secondary metabolites, and reduction or failure to inhibit fungal pathogens in vitro and to control diseases in situ.

The third locus known to influence Phl production was identified in a Tn*5* mutant of *P. fluorescens* Pf-5 that overproduced pyoluteorin and Phl, produced no pyrrolnitrin, and was less inhibitory to *Pyrenophora tritici-repentis* than was the wild type (45). Sequence analysis of the DNA flanking the site of Tn*5* insertion revealed strong similarity to *rpoS* of *E. coli*, and stationary phase cells of the Pf-5 mutant exhibited hypersensitivity to osmotic and oxidative stresses typical of *rpoS* mutants (52). The *rpoS* gene in *E. coli* controls the expression of over 30 other genes, many of which are regulatory and function in the transition to stationary phase and survival under adverse environmental circumstances (22,23). It is especially interesting that a mutation in the *rpoS* gene of strain Pf-5 differentially affected the production of pyrrolnitrin, pyoluteorin and Phl, indicating that antibiotic biosynthetic pathways need not be expressed coordinately.

MECHANISMS, GENETICS, AND IMPROVED BIOLOGICAL CONTROL

Knowledge of the mechanisms and the determinants critical to biological control systems can: (i) help to optimize the performance and reliability of existing agents; (ii) provide a rational basis for the development of more effective strain combinations; and (iii) facilitate the selection of new strains with desirable attributes from native populations adapted to different host plants or environments where biological control may be needed. The tools of molecular genetics also can be used directly to alter the timing or level of expression of biological control determinants in existing agents and to combine desirable traits within a single organism.

Optimizing The Selection And Use Of Naturally Occurring Agents

When conditions unfavorable to the expression or activity of a particular biological control mechanism have been defined, sites in which those conditions prevail can be modified or avoided, lessening the failure rate and the site-to-site variation in field performance typical of biological control agents. A regression model ($R^2 = 0.96$) including six variables (cation exchange capacity, ammonium-nitrogen, soil pH, iron, zinc, and % silt) was developed to help predict the effectiveness of strain 2-79 in soils of known composition (41,42). These soil factors were correlated with suppression of take-all by 2-79 and its PCA-producing derivatives, but not by Phz$^-$ mutants, suggesting that they influence phenazine production or activity in the rhizosphere. For example, the negative correlation between percent silt and suppression of take-all suggests that PCA is bound or inactivated on the surface of soil particles. Similarly, the positive relationship between zinc and disease suppression is consistent with the report that zinc enhanced PCA production without increasing biomass accumulation in culture (56) and suggests that zinc may directly influence PCA biosynthesis. Furthermore, suppression of take-all was greater on wheat grown in a Woodburn silt loam (naturally low in zinc) amended with 50 μg of zinc (as Zn-EDTA) g^{-1} soil than on wheat in the nonamended soil (B.H. Ownley and D.M. Weller, unpublished).

Strain combinations offer a second approach to improve the performance of existing biological control agents. A mixture of strains Q2-87 + Q1c-80 + Q8d-80 + Q65c-80 increased the yield of spring wheat in a field plot at Pullman by 20% over the control, whereas each strain applied individually increased the yield by no more than 5% (46). Mixtures of strains that performed well individually did not necessarily yield effective combinations, and strains from the same soil functioned better together than did strains from different locations. The empirical development of superior strain mixtures is both time-consuming and expensive, and the process might be facilitated by combining agents known to differ in their antagonistic mechanisms, in the antibiotics they produce, or in possible synergistic interactions (72).

Finally, biosynthetic genes can serve as probes to rapidly identify additional potential antibiotic producers among members of populations adapted to a particular host plant or agroecosystem. Isolates with one or more genes known to confer activity against the target pathogen would be candidates for further evaluation in situ. Even when biosynthetic genes have not yet been cloned, specific assays based on their expression in vitro sometimes can expedite the screening process. Simple screens based on in vitro inhibition of target pathogens often are not specific enough for this purpose, and "positives" detected by the in vitro screen also must be subjected to chemical analysis to avoid the potentially confounding effects of other biologically active compounds. In a recent study using this approach, 11% of 4,307 isolates of *Bacillus cereus* and *Bacillus thuringiensis* from diverse sources were shown to produce either zwittermycin A or antibiotic B; strains producing one or both antibiotics were more suppressive of damping-off of alfalfa caused by *Phytophthora medicaginis* than those that produced neither. This approach appears well suited to identify producers of zwittermycin A that will function in locations where *B. cereus* UW85 performs poorly (57).

Genetic Manipulation To Enhance Performance

Earlier and higher levels of antibiotic production have been achieved by manipulation of both gene copy number and regulation (16,17,36,47), and the few studies conducted to date indicate that such modifications can enhance biological control activity. For example, production of oomycin A in the spermosphere of cotton lagged behind the onset of seed infection by *P. ultimum* (16,25). Replacement of the autoinducible and cell density-dependent *afuE* native promoter with the constitutive *tac* promoter from *E. coli* permitted earlier antibiotic production to levels estimated to be 1,000-fold or 50- to 100-fold higher, respectively, than those of the parental strain under noninducing conditions in vitro or under optimal conditions in the rhizosphere. The modified strain was 20 to 25% more effective in improving seedling emergence and controlling infections caused by *Pythium* than was Hv37a (16). In another study, a derivative of *P. fluorescens* CHA0 containing an uncharacterized locus on the plasmid pME3090 produced increased amounts of pyoluteorin and Phl in vitro and was twofold more effective than strain CHA0 against damping-off of cucumber caused by *P. ultimum* in a gnotobiotic assay system (36). The impact of enhanced or constitutive antibiotic production on the ecological fitness of the modified strains has not yet been evaluated but is likely to be substantial, particularly when mutations have been introduced into global regulatory loci that control genes in addition to those required for antibiotic production.

Biological control activity also has been enhanced through the transfer and expression of genetic loci for the synthesis of antifungal metabolites to nonproducing strains. For example, introduction of the *hcn* biosynthetic locus from CHA0 into *P. fluorescens* P3, a noncyanogenic strain with weak activity against *T. basicola*, rendered the recipient Hcn[+] and more suppressive of black root rot of tobacco than the parent (64). A 6-kb fragment from *Pseudomonas* strain F113 in pCU203 partially restored Phl production to the Phl[−] Tn*5* mutant F113G22 and transferred Phl biosynthetic capability to M114, one of eight nonproducer strains into which it was introduced. The Phl[+] plasmid-bearing derivatives were more inhibitory to *P. ultimum* in vitro and increased sugar beet seedling emergence in soil as compared to the parental strains (8). In another study, plasmid-borne PCA and Phl biosynthetic genes cloned from *P. fluorescens* 2-79 and Q2-87, respectively, were expressed in all of 27 and 12 strains of fluorescent pseudomonads into which they were introduced. In growth chamber experiments, PCA-producing and Phl-producing derivatives of strain Q69c-80 more effectively reduced the incidence and the severity, respectively, of take-all of wheat than did unmodified Q69c-80 (20).

As presented in this chapter, our current understanding of the role of antibiosis in biological control and of the genetic basis of antibiotic production has been emerging rapidly over the past 15 years. Much of our recent progress can be attributed to the application of modern molecular biology and analytical techniques to biological control research, and the findings will be useful to facilitate the selection of better agents, to reduce impediments to consistent performance and to genetically engineer superior agents. The development of superior agents is perhaps the most exciting outcome of contemporary research on biological control antibiotics, but social and regulatory concerns about the safety of engineered strains must be addressed if the technology is to be implemented. This will require that our base of knowledge about the ecology of bacteria that produce biological control antibiotics in agroecosystems be expanded. For example, it is now known that Phl-producing strains are abundant in certain suppressive soils. Evidence that naturally occurring antibiotic-producing strains already contribute to the suppression of soilborne pathogens should help to alleviate concerns about the introduction of existing or engineered antibiotic-producing strains.

LITERATURE CITED

1. Bangera, M.G., Weller, D.M., and Thomashow, L.S. 1994. Genetic analysis of the 2,4-diacetylphloroglucinol biosynthetic locus from *Pseudomonas fluorescens* Q2-87. Pages 383-386 in: Advances in Molecular Genetics of Plant-Microbe Interactions, Vol. 3. M.J. Daniels, J.A. Downie, and A.E. Osbourn, eds. Kluwer Academic Publishers, Dordrecht, The Netherlands.

2. Brisbane, P.G., Janik, L.J., Tate, M.E., and Warren, R.F.O. 1987. Revised structure for the phenazine antibiotic from *Pseudomonas fluorescens* 2-79 (NRRL B-15132). Antimicrob. Agents Chemother. 31:1967-1971.

3. Bull, C.T., Weller, D.M., and Thomashow, L.S. 1991. Relationship between root colonization and suppression of *Gaeumannomyces graminis* var. *tritici* by *Pseudomonas fluorescens* strain 2-79. Phytopathology 81:954-959.

4. Cook, R.J., and Baker, K.F. 1983. The Nature and Practice of Biological Control of Plant Pathogens. The American Phytopathological Society, St. Paul, MN. 539 pp.

5. Corbell, N.A., Kraus, J., and Loper, J.E. 1994. Global regulation of secondary metabolism in *Pseudomonas fluorescens* Pf-5. Mol. Ecol. 3:608.

6. Défago, G., Berling, C.H., Burger, U., Haas, D., Kahr, G., Keel, C., Voisard, C., Wirthner, P., and Wüthrich, B. 1990. Suppression of black root rot of tobacco and other root diseases by strains of *Pseudomonas fluorescens*: potential applications and mechanisms. Pages 93-108 in: Biological Control of Soil-borne Plant Pathogens. D. Hornby, ed. CAB International, Wallingford, UK.

7. Fenton, A.M., Crowley, J.J., Shanahan, P., and O'Gara, F. 1993. Cloning and hererologous expression of genes involved in 2,4-diacetylphloroglucinol biosynthesis in *Pseudomonas* strains. Page 49 in: Abstracts, Fourth International Symposium on *Pseudomonas*: Biotechnology and Molecular Biology. R.E.W. Hancock, ed. Vancouver, Canada.

8. Fenton, A.M., Stephens, P.M., Crowley, J., O'Callaghan, M., and O'Gara, F. 1992. Exploitation of gene(s) involved in 2,4-diacetylphloroglucinol biosynthesis to confer a new biocontrol capability to a *Pseudomonas* strain. Appl. Environ. Microbiol. 58:3873-3878.

9. Flaishman, M., Eyal, Z., Voisard, C., and Haas, D. 1990. Suppression of *Septoria tritici* by phenazine- or siderophore-deficient mutants of *Pseudomonas*. Curr. Microbiol. 20:121–124.

10. Fravel, D.R. 1988. Role of antibiosis in the biocontrol of plant diseases. Annu. Rev. Phytopathol. 26:75–91.

11. Fujimoto, D.K., Weller, D.M., and Thomashow, L.S. 1994. The role of secondary metabolites in disease suppression. Pages 330–347 in: Allelopathy: Organisms, Processes and Applications. Inderjit, K.M.M. Dakshini, and F.A. Einhellig, eds. American Chemical Society, Washington, DC.

12. Fuqua, W.C., Winans, S.C., and Greenberg, E.P. 1994. Quorum sensing in bacteria: the LuxR-LuxI family of cell density-responsive transcriptional regulators. J. Bacteriol. 176:269–275.

13. Gaffney, T.D., Lam, S.T., Ligon, J., Gates, K., Frazelle, A., Di Maio, J., Hill, S., Goodwin, S., Torkewitz, N., Allshouse, A.M., Kempf, H.-J., and Becker, J.O. 1994. Global regulation of expression of antifungal factors by a *Pseudomonas fluorescens* biological control strain. Mol. Plant-Microbe Interact. 7:455–463.

14. Georgakopoulos, D.G., Hendson, M., Panopoulos, N.J., and Schroth, M.N. 1994. Cloning of a phenazine biosynthetic locus of *Pseudomonas aureofaciens* PGS12 and analysis of its expression in vitro with the ice nucleation reporter gene. Appl. Environ. Microbiol. 60:2931–2938.

15. Georgakopoulos, D.G., Hendson, M., Panopoulos, N.J., and Schroth, M.N. 1994. Analysis of expression of a phenazine biosynthesis locus of *Pseudomonas aureofaciens* PGS12 on seeds with a mutant carrying a phenazine biosynthesis locus-ice nucleation reporter gene fusion. Appl. Environ. Microbiol. 60:4573–4579.

16. Gutterson, N. 1990. Microbial fungicides: recent approaches to elucidating mechanisms. Crit. Rev. Biotechnol. 10:69–91.

17. Gutterson, N., Howie, W., and Suslow, T. 1990. Enhancing efficiencies of biocontrol agents by the use of biotechnology. Pages 749–765 in: New Directions in Biological Control: Alternatives for Suppressing Agricultural Pests and Diseases. R.R. Baker and P.E. Dunn, eds. Alan R. Liss Inc., New York, NY.

18.   Haas, D., Keel, C., Laville, J., Maurhofer, M., Oberhänsli, T., Schnider, U., Voisard, C., Wüthrich, B. and Défago, G. 1991. Secondary metabolites of *Pseudomonas fluorescens* strain CHA0 involved in the suppression of root diseases. Pages 450–456 in: Advances in Molecular Genetics of Plant-Microbe Interactions, Vol. 1. H. Hennecke and D.P.S. Verma, eds. Kluwer Academic Publishers, Dordrecht, The Netherlands.

19.   Hamdan, H., Weller, D.M., and Thomashow, L.S. 1991. Relative importance of fluorescent siderophores and other factors in biological control of *Gaeumannomyces graminis* var. *tritici* by *Pseudomonas fluorescens* 2-79 and M4-80R. Appl. Environ. Microbiol. 57:3270–3277.

20.   Hara, H., Bangera, M., Kim, D.-S., Weller, D.M., and Thomashow, L.S. 1994. Effect of transfer and expression of antibiotic biosynthesis genes on biological control activity of fluorescent pseudomonads. Pages 247–249 in: Improving Plant Productivity with Rhizobacteria. M.H. Ryder, P.M. Stephens, and G.D. Bowen, eds. CSIRO Division of Soils, Adelaide, Australia.

21.   Harrison, L.A., Letendre, L., Kovacevich, P., Pierson, E., and Weller, D. 1993. Purification of an antibiotic effective against *Gaeumannomyces graminis* var. *tritici* produced by a biocontrol agent, *Pseudomonas aureofaciens*. Soil Biol. Biochem. 25:215–221.

22.   Hengge-Aronis, R. 1993. Survival of hunger and stress: the role of *rpoS* in early stationary phase gene regulation in *E. coli*. Cell 72:165–168.

23.   Hengge-Aronis, R. 1993. The role of *rpoS* in early stationary-phase gene regulation in *Escherichia coli* K12. Pages 171–200 in: Starvation in Bacteria. S. Kjellberg, ed. Plenum Press, New York, NY.

24.   Homma, Y. 1994. Mechanisms in biological control—focused on the antibiotic pyrrolnitrin. Pages 100–103 in: Improving Plant Productivity with Rhizobacteria. M.H. Ryder, P.M. Stephens, and G.D. Bowen, eds. CSIRO Division of Soils, Adelaide, Australia.

25.   Howie, W.J., and Suslow, T.V. 1991. Role of antibiotic biosynthesis in the inhibition of *Pythium ultimum* in the cotton spermosphere and rhizosphere by *Pseudomonas fluorescens*. Mol. Plant-Microbe Interact. 4:393–399.

26.   Keel, C., Schnider, U., Maurhofer, M., Voisard, C., Laville, J., Burger, U., Wirthner, P., Haas, D., and Défago, G. 1992. Suppression of root diseases by *Pseudomonas fluorescens* CHA0: importance of the bacterial secondary metabolite 2,4-diacetylphloroglucinol. Mol. Plant-Microbe Interact. 5:4–13.

27.   Keel, C., Wirthner, P.H., Oberhänsli, T.H., Voisard, C., Berger, U., Haas, D., and Défago, G. 1990. Pseudomonads as antagonists of plant pathogens in the rhizosphere: role of the antibiotic 2,4-diacetylphloroglucinol in the suppression of black root rot of tobacco. Symbiosis 9:327–341.

28.   Kempf, H.-J., and Wolf, G. 1989. *Erwinia herbicola* as a biocontrol agent of *Fusarium culmorum* and *Puccinia recondita* f. sp. *tritici* on wheat. Phytopathology 79:990–994.

29.   Kempf, H.-J., Bauer, P.H., and Schroth, M.N. 1993. Herbicolin A associated with crown and roots of wheat after seed treatment with *Erwinia herbicola* B247. Phytopathology 83:213–216.

30.   Kempf, H.J., Sinterhauf, S., Muller, M., Becker, J.O., and Pachlatko, P. 1993. Production of pyrrolnitrin by a biocontrol bacterium in the rhizosphere of cotton and the spermosphere of barley. Page 266 in: Abstracts, 6th International Congress of Plant Pathology, Montreal, Canada. National Research Council of Canada, Ottawa, Canada.

31.   Lam, S.T., Gaffney, T.D.,Frazelle, R.A., Gates, K., DiMaio, J., Torkewitz, N., Ligon, J., Hill, S., Goodwin, S., and Kempf, H.-J. 1993. Two genes which regulate the coordinated expression of antifungal activities in *Pseudomonas fluorescens*. Page 209 in: Abstracts, Fourth International Symposium on *Pseudomonas*: Biotechnology and Molecular Biology. R.E.W. Hancock, ed., Vancouver, Canada.

32.   Lam, S.T., Gaffney, T.D.,Frazelle, R.A., Gates, K., DiMaio, J., Torkewitz, N., Ligon, J., Hill, S., Goodwin, S., and Kempf, H.-J. 1994. *LemA* and *GacA* regulate the coordinated expression of antifungal activities in *Pseudomonas fluorescens*. Mol. Ecol. 3:620.

33.   Laville, J., Voisard, C., Keel, C., Maurhofer, M., Défago, G., and Haas, D. 1992. Global control in *Pseudomonas fluorescens* mediating antibiotic synthesis and suppression of black root rot of tobacco. Proc. Natl. Acad. Sci. USA 89:1562–1566.

34.   Lugtenberg, B.J.J., de Weger, L.A., and Bennett, J.W. 1991. Microbial stimulation of plant growth and protection from disease. Curr. Opin. Biotechnol. 2:457–464.

35. Lumsden, R.D., Locke, J.C., Adkins, S.T., Walter, J.F., and Rideout, C.J. 1992. Isolation and localization of the antibiotic gliotoxin produced by *Gliocladium virens* from alginate prill in soil and soilless media. Phytopathology 82:230-235.

36. Maurhofer, M., Keel, C., Schnider, U., Voisard, C., Haas, D., and Défago, G. 1992. Influence of enhanced antibiotic production in *Pseudomonas fluorescens* strain CHA0 on its disease suppressive capacity. Phytopathology 82:190-195.

37. Mazzola, M., Cook, R.J., Thomashow, L.S., Weller, D.M., and Pierson III, L.S. 1992. Contribution of phenazine antibiotic biosynthesis to the ecological competence of fluorescent pseudomonads in soil habitats. Appl. Environ. Microbiol. 58:2616-2624.

38. Meighen, E.A. 1991. Molecular biology of bacterial bioluminescence. Microbiol. Rev. 55:123-142.

39. Nowak-Thompson, B., Gould, S.J., Kraus, J., and Loper, J.E. 1994. Production of 2,4-diacetylphloroglucinol by the biocontrol agent *Pseudomonas fluorescens* Pf-5. Can. J. Microbiol. 40:1064-1066.

40. O'Sullivan, D.J., and O'Gara, F. 1992. Traits of fluorescent *Pseudomonas* spp. involved in suppression of plant root pathogens. Microbiol. Rev. 56:662-676.

41. Ownley, B.H., Weller, D.M., and Alldredge, J.R. 1990. Influence of soil edaphic factors on suppression of take-all by *Pseudomonas fluorescens* 2-79. Phytopathology 80:995. (Abstr.).

42. Ownley, B.H., Weller, D.M., and Alldredge, J.R. 1991. Relation of soil chemical and physical factors with suppression of take-all by *Pseudomonas fluorescens* 2-79. Pages 299-301 in: Plant Growth-Promoting Rhizobacteria: Progress and Prospects. C. Keel, B. Koller, and G. Défago, eds. West Palaeartic Regional Section Bulletin 1991/XIV/8, Interlaken, Switzerland.

43. Ownley, B.H., Weller, D.M., and Thomashow, L.S. 1992. Influence of in situ and in vitro pH on suppression of *Gaeumannomyces graminis* var. *tritici* by *Pseudomonas fluorescens* 2-79. Phytopathology 82:178-184.

44. Parkinson, J.S., and Kofoid, E.C. 1992. Communication modules in bacterial signaling proteins. Annu. Rev. Genet. 26:71-112.

45. Pfender, W.F., Kraus, J., and Loper, J.E. 1993. A genomic region from *Pseudomonas fluorescens* Pf-5 required for pyrrolnitrin production and inhibition of *Pyrenophora tritici-repentis* in wheat straw. Phytopathology 83:1223-1228.

46. Pierson, E.A., and Weller, D.M. 1994. Use of mixtures of fluorescent pseudomonads to suppress take-all and improve the growth of wheat. Phytopathology 84:940–947.

47. Pierson III, L.S., Keppenne, V.D., and Wood, D.W. 1994. Phenazine antibiotic biosynthesis in *Pseudomonas aureofaciens* 30-84 is regulated by PhzR in response to cell density. J. Bacteriol. 176:3966–3974.

48. Pierson III, L.S., and Thomashow, L.S. 1992. Cloning and heterologous expression of the phenazine biosynthetic locus from *Pseudomonas aureofaciens* 30-84. Mol. Plant-Microbe Interact. 5:330–339.

49. Rosales, A., Thomashow, L. Cook, R.J., and Mew, T.W. 1993. Isolation and identification of antifungal metabolites produced by rice-associated antagonistic *Pseudomonas* spp. Page 268 in: Abstracts, 6th International Congress of Plant Pathology, Montreal, Canada. National Research Council of Canada, Ottawa, Canada.

50. Sacherer, P., Défago, G., and Haas, D. 1994. Extracellular protease and phospholipase C are controlled by the global regulatory gene *gacA* in the biocontrol strain *Pseudomonas fluorescens* CHA0. FEMS Microbiol. Lett. 116:155–160.

51. Saier Jr., M.H. 1994. Computer-aided analysis of transport protein sequences: gleaning evidence concerning function, structure, biogenesis, and evolution. Microbiol. Rev. 58:71–93.

52. Sarniguet, A., Kraus, J., and Loper, J.E. 1994. An *rpoS* homologue affects antibiotic production, ecological fitness, and biological control activity of *Pseudomonas fluorescens* Pf-5. Mol. Ecol. 3:607.

53. Schröder, J., and Schröder, G. 1990. Stilbene and chalcone synthases: related enzymes with key functions in plant-specific pathways. Z. Naturforsch. 45c:1–8.

54. Shanahan, P., Glennon, J.D., Crowley, J.J., Donnelly, D.F., and O'Gara, F. 1993. Liquid chromatographic assay of microbially derived phloroglucinol antibiotics for establishing the biosynthetic route to production, and the factors affecting their regulation. Anal. Chem. Acta 272:271–277.

55. Shanahan, P., O'Sullivan, D.J., Simpson, P., Glennon, J.D., and O'Gara, F. 1992. Isolation of 2,4-diacetylphloroglucinol from a fluorescent pseudomonad and investigation of physiological parameters influencing its production. Appl. Environ. Microbiol. 58:353–358.

56.   Slininger, P.J., and Jackson, M.A. 1992. Nutritional factors
        regulating growth and accumulation of phenazine-1-carboxylic
        acid by *Pseudomonas fluorescens* 2-79. Appl. Microbiol.
        Biotechnol. 37:388–392.

57.   Stabb, E.V., Jacobson, L.M., and Handelsman, J. 1994.
        Zwittermycin A-producing strains of *Bacillus cereus* from diverse
        soils. Appl. Environ. Microbiol. 60:4404–4412.

58.   Thomashow, L.S., Essar, D.W., Fujimoto, D.K., Pierson III,
        L.S., Thrane, C., and Weller, D.M. 1993. Genetic and
        biochemical determinants of phenazine antibiotic production in
        fluorescent pseudomonads that suppress take-all disease of wheat.
        Pages 535–541 in: Advances in Molecular Genetics of Plant-
        Microbe Interactions, Vol. 2. E.W. Nester and D.P.S. Verma,
        eds. Kluwer Academic Publishers, Dordrecht, The Netherlands.

59.   Thomashow, L.S., and Weller, D.M. 1988. Role of a phenazine
        antibiotic from *Pseudomonas fluorescens* in biological control of
        Gaeumannomyces graminis var. *tritici*. J. Bacteriol.
        170:3499–3508.

60.   Thomashow, L.S., Weller, D.M., Bonsall, R.F., and Pierson III.,
        L.S. 1990. Production of the antibiotic phenazine-1-carboxylic
        acid by fluorescent *Pseudomonas* species in the rhizosphere of
        wheat. Appl. Environ. Microbiol. 56:908–912.

61.   Turner, J.M., and Messenger, A.J. 1986. Occurrence,
        biochemistry and physiology of phenazine pigment production.
        Adv. Microbial Physiol. 27:211–275.

62.   Vincent, M.N., Harrison, L.A., Brackin, J.M., Kovacevich,
        P.A., Mukerji, P., Weller, D.M., and Pierson, E.A. 1991.
        Genetic analysis of the antifungal activity of a soilborne
        Pseudomonas aureofaciens strain. Appl. Environ. Microbiol.
        57:2928–2934.

63.   Voisard, C., Bull, C.T., Keel, C., Laville, J., Maurhofer, M.,
        Schnider, U., Défago, G., and Haas, D. 1994. Biocontrol of root
        diseases by *Pseudomonas fluorescens* CHA0: current concepts and
        experimental approaches. Pages 67–89 in: Molecular Ecology of
        Rhizosphere Microorganisms: Biotechnology and the Release of
        GMOs. F. O'Gara, D.N. Dowling, and B. Boesten, eds. VCH,
        Weinheim, Germany.

64.   Voisard, C., Keel, C., Haas, D., and Défago. G. 1989. Cyanide
        production by *Pseudomonas fluorescens* helps suppress black root
        rot of tobacco under gnotobiotic conditions. EMBO J. 8:351–358.

65.   Weller, D.M., and Cook, R.J. 1983. Suppression of take-all of wheat by seed treatments with fluorescent pseudomonads. Phytopathology 73:463–469.

66.   Weller, D.M., and Thomashow, L.S. 1990. Antibiotics: evidence for their production and sites where they are produced. Pages 703–711 in: New Directions in Biological Control: Alternatives for Suppressing Agricultural Pests and Diseases. R.R. Baker and P.E. Dunn, eds. Alan R. Liss Inc., New York, NY.

67.   Weller, D.M., and Thomashow, L.S. 1993. Microbial metabolites with biological activity against plant pathogens. Pages 172–180 in: Pest Management: Biologically Based Technologies. R.D. Lumsden and J.L. Vaughn, eds. American Chemical Society, Washington, DC.

68.   Williams, P., Bainton, N.J., Swift, S., Chhabra, S.R., Winson, M.K., Stewart, G.S.A.B., Salmond, G.P.C., and Bycroft, B.W. 1992. Small molecule-mediated density-dependent control of gene expression in prokaryotes: bioluminescence and the biosynthesis of carbapenem antibiotics. FEMS Microbiol. Lett. 100:161–168.

69.   Williams, S.T. 1982. Are antibiotics produced in soil? Pedobiologia 23:427–435.

70.   Williams, S.T., and Vickers, J.C. 1986. The ecology of antibiotic production. Microb. Ecol. 12:43–52.

71.   Willis, D.K., Rich, J.J., Kinscherf, T.G., and Kitten, T. 1994. Genetic regulation in plant pathogenic pseudomonads. Pages 167–193 in: Genetic Engineering: Principles and Methods, Vol. 15. J.T. Setlow, ed. Plenum Press, New York, NY.

72.   Wood, D.W., and Pierson III., L.S. 1994. A diffusible signal molecule regulates phenazine expression in *Pseudomonas aureofaciens* 30–84. Phytopathology 84:1082. (Abstr.).

73.   Wood, D.W., and Pierson III., L.S. 1994. Cell to cell interactions among rhizosphere bacteria influence the expression of phenazine antibiotics in *Pseudomonas aureofaciens* 30–84. Phytopathology 84:1134. (Abstr.).

Chapter 5

METHODS FOR QUANTITATIVE AND IN SITU DETECTION OF LOW LEVELS OF VIABLE SOILBORNE BACTERIA AND FUNGI.

J.W.L. Van Vuurde and J. Postma

There is an increasing demand for more sensitive and reliable detection methods for soilborne microorganisms such as antagonists for biological control and genetically modified organisms in risk assessment studies. The aim of this paper is to highlight some procedures that have been successful in quantitative and in situ studies of viable target organisms. Several of these procedures combine traditional isolation with serology or gene technology to improve sensitivity and reliability. Examples of the application of these procedures will be presented for the study of population dynamics in soil or for in situ detection on roots.

FUNDAMENTAL PRINCIPLES AND METHODS

The Three Basic Detection Principles

The characteristics of the three basic methods for detection of bacteria and fungi—isolation, serology and gene probes—are presented in Table 1. As the principles for the detection of both groups are the same, our unspecified remarks on a detection method generally refer to its application for both fungi and bacteria.

Isolation-based methods are suitable only for viable and active (culturable) organisms. However, when pure cultures are obtained from dilution plating series, results for a positive sample are 100% reliable after proper identification. It is often overlooked that the theoretical detection level of the plating technique per ml of sample extract is still 10 to 100 times more sensitive than that of the later developed serological and

Table 1. Characteristics of the three basic methods for detecting bacteria and fungi.

Method	Target	Reactant	Examples
Isolation	Colony-forming unit (CFU)		Media (selective)
Serology	Antigen epitope	Antibody	ELISA Immunofluorescence
Gene probes	DNA or RNA	Probe	Colony hybridization
		Primer set	PCR

gene probe techniques. One colony-forming unit (CFU) per ml can be detected by plating on agar media in the absence, or at low levels, of background organisms. For immunofluorescence microscopy (IF) and polymerase chain reaction (PCR), the detection level per test sample is also one target cell, but the detection level per ml of sample extract without previous concentration is about 100 times higher due to the small size of the test sample (about 10 μl). The advantage of the two more recent techniques is that they are relatively insensitive to microbial interference. This often results in practice in better detection levels than with isolation on media, even when these media were designed to be selective for the target organism.

Serological techniques can be used for target organisms that produce a typical antigenic compound. For bacteria, this is a useful technique at the species and subspecies levels and sometimes even at the strain level. The variation in sufficiently available specific cell wall antigens is less for fungi than for bacteria. Serology is useful for fungi at the genus level and for several important fungi at the species level (9).

Gene probe-based techniques have the advantage over serological techniques that the probes can use differences in nucleic acid sequences without the need for gene expression. This enormously increases the number of potentially specific target sequences compared with antigens.

Procedures Developed From The Basic Principles

Different formats of detection procedures, based on the basic principles in Table 1 are chronologically presented in Table 2. All techniques listed in Table 2 are still being used for detection of microorganisms. The characteristics of the various methods can be found in handbooks and review articles on various techniques for bacteria (11,21), fungi (35), and fungi and bacteria (1), and on serology for bacteria (15), serology for fungi (9), gene probes for bacteria (24), gene probes for fungi (10), and gene probes for fungi and bacteria (16).

Selection Of The Most Suitable Method

The choice of the method is determined by the characteristics of the various methods in regard to the aim of the experiment. The most important questions in regard to the data to collect are:
1. Is a method needed to detect active cells, culturable cells or the total population of the target?
2. Is in situ detection on a root or a soil particle needed or will the detection be done in a sample extract?
3. Should the target organism be determined quantitatively or is only qualitative information needed?
4. Should the detection be directed towards an introduced "marked" genotype, or towards the variety of indigenous genotypes of the target organism?
5. What is the desired detection level?
6. Should a positive result be made 100% reliable by additional confirmation or can a certain risk of false-positive results be accepted?
7. What are the risks of a false-negative reaction, e.g. due to target strain variation or to interference of sample components with the test?
8. What is the number of samples to deal with?

The most suitable method can be chosen based on the answers to these questions. A combination of techniques will often be essential, especially when low levels of target organisms should reliably be detected.

For introduced strains, there are detection methods in addition to those used for natural populations, because such strains can be selected or genetically modified to present strain-specific characteristics suitable for detection.

In this chapter, we focus on methods that can be used to detect and quantify viable target organisms, since these methods are usually the most

Table 2. Chronology of the development of detection techniques and procedures.

Year	Serology	Isolation	Gene probes
1900 – 1950	Agglutination	Solid medium Selective medium	
1951 – 1975	Immuno-fluorescence	Antibiotics in selective medium	
1976 – 1980	ELISA	Selected mutants	
1981 – 1985		Genetically engineered mutants	DNA-dot hybridization
		DNA-colony hybridization	
1986 – 1995	Immunofluorescence colony-staining Immunomagnetic isolation		Polymerase chain reaction Fluorescent oligonucleotide probes

important as tools for phytopathological research. Special attention is also paid to methods that can be used for in situ detection. Selected techniques will be discussed based on examples of a specific application. Examples are presented for bacteria and fungi regarding the study of: (i) population dynamics in complex substrates at low target densities, and (ii) in situ colonization of roots.

QUANTITATIVE DETECTION OF VIABLE TARGET ORGANISMS

The detection of an introduced marked strain in a model experiment has possibilities that are not available for the detection of wild-type strains in natural substrates. In model experiments, the strain can be selected or genetically modified for a specific and detectable characteristic not present in the background microflora. Under these conditions, the risk of a false-negative or false-positive result will be limited. On the other hand, one should know if the selected or modified organisms are representative of the natural population. Mutants may be less fit (4), but mutants with improved antagonistic properties have been described (28). The process of preparing and selecting a marked strain, and comparing important ecological properties of this strain with those of the wild type, is laborious.

For the study of undefined natural populations of the target, extensive basic research on the specificity of an assay is therefore needed; strain variation may easily result in false negatives if the method is too specific. The increasing risk of false-positive reactions at decreasing detection levels of the assay will make additional verification of positive samples often necessary.

Fungicide-resistant Mutant On Mutant-specific Medium

Although several antibiotic-resistance markers are available for bacteria, there are only few reliable markers present for fungi. Only benomyl resistance is commonly used as a selective marker for different fungi. Valuable information about benomyl and several other markers and how to deal with them is given by Andrews (3).

Selected mutants, irrespective of the origin of the mutation (spontaneous, UV or chemically induced, or genetically transformed), must always be tested thoroughly for their similarity with the wild type and for stability of the marker. It has not been established which screening methods are the most convenient and which yield relevant information about the in vivo ecological properties. For example, a benomyl-resistant mutant of *Chaetomium globosum*, which was as antagonistic as its wild type in growth chamber experiments, did not survive as well under field conditions (7). In general, methods to study the competition between mutant and wild-type strains are more sensitive in determining differences in ecological fitness than in evaluating wild type and mutant separately. An example of several tests of increasing complexity is given in Table 3. Postma and Luttikholt (30) tested UV-induced benomyl-resistant mutants of an antagonistic *Fusarium oxysporum* strain by this procedure and selected a mutant that was reliable for biological control studies. Population dynamics of the antagonistic *F. oxysporum*

Table 3. Tests with benomyl-resistant mutants of an antagonistic *Fusarium oxysporum* strain.

Fitness character	Marker stability determined
Growth rate on agar Competition with wild type	On agar without benomyl
Saprophytic ability in sterile soil	After growth in sterile soil
Antagonism in natural environment	After growth in natural environment

From Postma and Luttikholt (30).

could be studied in soil and within plants in the vicinity of a strain of *F. oxysporum* pathogenic on carnation. In another study, the population dynamics and biological control effect of this mutant were investigated in several soil samples naturally infested with *Fusarium solani* pathogenic on pea (27). The mutant was as antagonistic as the wild-type strain and as competitive as the wild type on agar media as well as in soil. Furthermore, the marker appeared to be stable under different conditions.

The isolation method in which resistant mutants grow on a mutant-specific medium only detects culturable propagules. The method is quantitative with a detection level between 1 and 100 CFU g^{-1} soil, depending on the dilution factors during the extraction procedure. To detect very low levels of organisms (1 CFU g^{-1} soil), it is necessary to mix the soil directly with the agar medium (pour-plate technique). The agar medium has to be sufficiently selective to minimize growth of nontarget organisms. Colonies resembling the target can be confirmed by pure culturing and standard identification procedures.

Wild-type Strains By Immunofluorescence Colony-staining (IFC)

The detection of culturable wild-type strains in natural substrates is illustrated for *Pseudomonas* sp. in soil and for *Erwinia* spp. on potato tubers.

IFC is based on pour plating of the sample combined with immunofluorescence detection of the target colonies in the medium

(43,44). A compound microscope with low objective magnification (e.g. 4 to 10 ×) or a stereo microscope adapted for specific fluorescence can be used to count the positive (fluorescent) colonies against a dark background. The method is miniaturized for routine handling of hundreds of samples per day per person by using wells of tissue culture plates and automatic pipettes. Comparative experiments with *Erwinia* spp. showed that even a miniaturized routine format of IFC is 10 to 100 times more sensitive than traditional dilution plating on (s)elective media (19,23,46).

Important characteristics are that the method is quantitative and can be used in the miniaturized routine format at a detection level of 10 to 100 CFU g^{-1} soil. The possibilities for direct verification of the IFC-positive colonies, either by sampling cells from that colony directly for PCR (32,42) or by sampling for pure culturing followed by proper identification procedures (46,48), are shown in Figure 1. For *Erwinia* spp., fatty-acid profiling of pure cultures from IFC-positive colonies proved sufficiently specific to differentiate the target from cross-reacting species. Two features confer additional specificity of the IFC test compared to other serological assays, such as IF and ELISA. First, only bacteria that can form visible colonies in the selected agar medium can interfere with the test. This excludes the many nonculturable species from soil which often interfere with the specificity of IF. Secondly, cross-reacting nontarget colonies in IFC can often be morphologically recognized as such and can be isolated for verification directly from the positive colony. After isolation of the cross-reacting bacterium, the specificity of the IFC system can often be improved, either by making the medium unsuitable for its growth or by cross-absorption of the antiserum. The testing of cells from IFC-positive colonies by PCR proved to be an important tool for the routine confirmation of *Erwinia* spp. No cross-reactions were found between the PCR primers for *Erwinia chrysanthemi* and the five known types of cross-reacting *Pseudomonas* spp. that react with the anti-*E. chrysanthemi* serum (41).

The IFC technique proved especially successful for studies of low numbers of target cells in complex substrates such as soil (23), cattle manure (46), tubers or roots (19,20) and stems (26). Isolation from IFC-positive colonies, especially from those with a different colony morphology or fluorescence pattern, proved an effective strategy to obtain pure cultures of cross-reacting bacteria for further research on assay improvement (47).

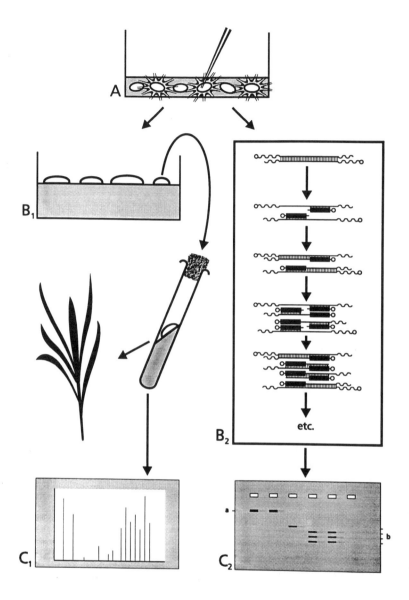

Figure 1. Diagram showing the two possible routes for the confirmation of IFC-positive colonies by (1) isolation of pure cultures and (2) by PCR. (A) Sampling from IFC-positive colonies in pour plates. (B₁) Isolation of a pure culture for testing of ecological behavior or characterization by fatty-acid profiling (C₁), or by PCR (B₂) prior to gel electrophoresis (C₂) and identification by band patterns (a and b). Reproduced from IPO-DLO Annual Report 1992.

Detection Of Metabolically Active Cells With Labeled 16S rRNA-targeted Oligonucleotide Probes

RNA-directed probes can be used for organisms that are nonculturable, but that are active or can be activated in the substrate under investigation. Fluorochrome-labeled oligonucleotide probes, directed against specific sequences of ribosomal RNA, can penetrate into the cell (2,8). Sufficient fluorescent probe will be bound in the cell to be seen with a fluorescence microscope (14) only if sufficient target copies per cell are available, indicating recent activity of the cell. Inspection of the microscope preparation for positive cells is similar to immunofluorescence cell-staining. The sensitivity of the detection can be improved by using a laser scanning confocal fluorescence microscope linked to an imaging system. With this equipment, nonspecific background fluorescence can be suppressed, and a three-dimensional picture can be made from specific signals not clear in traditional fluorescence microscopy.

An example of the use of fluorescent oligoprobes is the detection of *Pseudomonas cepacia* in three different soil types (12). Introduced target bacteria were fixed directly in soil that was applied to microscope slides after removal of large particles. After treatment to improve the permeability of the cell wall, the preparation was stained with fluorescent oligonucleotide probes and numbers of fluorescent cells were counted. The intensity of the fluorescence was related to the metabolic state of the cell. Naturally present target bacteria could be detected only after addition of nutrients to activate the cells.

The major aspect of this assay is the possibility for quantitative detection of active nonculturable microorganisms. Hahn et al. (13) used the method to study strains of *Frankia* sp. directly in nodule homogenate from *Alnus* roots. Compared with serological probes, oligoprobes also allow the detection of functional or phylogenetic groups (2). The detection level for active cells is comparable to that of immunofluorescence cell-staining at about 10^3 cells ml^{-1} sample extract. In principle, the method can be combined with immunofluorescence by using a different fluorochrome for the gene probe (active target cells) than for the antibody probe (total numbers of target cells).

IN SITU DETECTION OF TARGET ORGANISMS

In Situ Detection Of *Lux*-marked Bacterial Mutant

The detection of luminescence-marked microorganisms is based on the introduction of *lux* genes in the target organism. When the bacterium is metabolically active, luminescent light is produced which can be monitored with optical systems and sensitive photographic film. The *lux* marking system can be used in model studies to follow the process of colonization of plant parts in time without the need to disturb the experimental model. For example, the external bacterial colonization of roots from target-inoculated seeds germinating on wet filter paper can be followed day after day for the same roots. The use of *lux* genes for bacteria is reviewed by Stewart and Williams (38) and, together with other molecular markers, by Prosser (31).

The following examples illustrate the potential of *lux* genes for the in situ detection of marked target bacteria.

Mahaffee et al. (25) used an operon from *Vibrio fisheri* encoding for bacterial luciferase as a marker for bacteria of several genera. A plant growth-promoting *lux*-marked strain of *Pseudomonas putida* was inoculated on the seed of cotton and seeds were germinated on wet filter paper and in soil. A charge-coupled device (CCD) camera coupled to an image analyzing computer was used to record the colonization pattern of the cotton root on wet filter paper with *lux*-positive bacteria. For seedlings on filter paper, the spatial distribution was visualized showing the highest density of *lux*-positive cells in the root hair zone. Attempts to detect *P. putida* in field soil failed.

Rattray et al. (33) used a CCD camera linked to a microscope to study *lux*-marked cells on roots. In situ spatial organization of microcolonies on the roots could be observed for all the *lux*-positive strains. For some constructs the luminescence was sufficient to detect individual *lux*-positive cells. *Enterobacter cloacae* was used to study root colonization at a cellular level in the presence of the natural background microflora. The in situ detection enabled the assessment of the different regions of the root system that were colonized and of the regions with the highest bacterial activity.

The major characteristic of this technique is the possibility of obtaining information on the colonization pattern of plant parts or soil particles with metabolically active *lux*-positive bacteria. Detection is possible at a cellular level for constructs with high luminescence, but for less luminescent constructs microcolonies of about 1,000 cells are needed.

In Situ Detection Of GUS-marked Fungal Mutant

The GUS gene fusion system (*Escherichia coli* β-D-glucuronidase gene) is a powerful tool for the assessment of gene activity in transgenic plants, but is also used to mark and monitor populations of microorganisms (17).

Couteaudier et al. (6) successfully transformed the plant-pathogenic fungus *F. oxysporum* with this system. No loss of pathogenicity occurred. After addition of a histochemical substrate (X-Gluc, 5-bromo-4-chloro-3-indolyl glucuronide) a blue precipitate indicates the location of glucuronidase active sites. By this method active sites in hyphae of the marked fungus on roots were visualized. Mycelium of *F. oxysporum* f. sp. *lini* colonized almost the whole root surface, but was metabolically active only where exudates are produced, close to the root tip and at the point of emergence of secondary roots (39). In addition to in situ detection, the GUS activity level was also used to quantify the active root-colonizing biomass of the pathogen in the presence of antagonistic *F. oxysporum* strains (6).

With this technique, only active cells or active sites within cells are detected. The technique is semiquantitative and can be used for efficient indexing of whole root systems for colonized sites. The detection level is one hypha per site.

In Situ Detection Of Wild-type Strains

For wild-type bacteria, introduced or present as a natural population, IFC can be used for in situ detection of the target on roots or other substrates. The procedure is based on in situ enrichment of the target bacteria after embedding the root in a suitable agar medium. After staining with fluorescent antibodies, fluorescent microcolonies can be observed on the roots using a fluorescence microscope at 4 to 10 × objective magnification.

The method is illustrated for the detection of *Erwinia* spp. on potato roots (40,45). Potato roots grown for 5 days in natural soil inoculated with *E. chrysanthemi* were washed to remove superficial soil particles and embedded in agar medium. After 1 or 2 days incubation, the agar preparations were stained with an antiserum against *E. chrysanthemi* conjugated with FITC. Green fluorescent microcolonies were visible in the rhizoplane against a weak red autofluorescence of the root. IFC-positive colonies dominated in the zone of the root tip and at the point of side root formation.

The method was also used to study the colonization of a root after simultaneous inoculation with two different target bacteria. Both targets could be observed at the same time by the use of a special filter block and antibodies with a red fluorochrome (Texas red) for one target and a green fluorochrome (FITC) for the second target (48).

An important characteristic of this technique is the possibility of searching large root systems efficiently for the locations of colony-forming units of wild-type target bacteria. In survival or risk assessment studies, a single target colony-forming unit on the root can be found. Identification can be confirmed by sampling from the IFC-positive colony followed by PCR or by pure culture procedures (48).

CONCLUDING REMARKS

The examples presented describe methods for quantitative research on active target organisms for studies of population dynamics based on target incidence in the sample extract and methods for in situ detection of the target on roots. All methods described in the examples can be used in principle for both bacteria and fungi.

For studies on population dynamics, culturable organisms can be detected by isolation. Detection level and reliability can be improved by using antibiotic- or fungicide-resistant mutant strains, as was illustrated for a *Fusarium* sp. in carnation. Marked mutants are very valuable tools in model experiments in which the marker is absent in the background microflora. The ecological behavior of the mutant should be carefully checked with the wild type under the experimental conditions of the model experiment as illustrated below. Furthermore, tests should be done for possible loss or suppression of the marker gene.

For wild-type bacteria, pour plating combined with immunofluorescence colony-staining (IFC) demonstrated an improved detectability of target *Erwinia* spp. Studies on wild-type target organisms need specific antibodies or gene probes. The specificity of selective media is usually not sufficient to eliminate the background interference. In general, an increase of the specificity of the medium will reduce the recovery of the target and increase the variation in recovery between strains (5). For this reason, pour plating media for IFC are based mainly on the use of elective compounds supporting the growth of the target, such as pectin as a carbon source for pectinolytic *Erwinia* spp. Antibiotics are used only to eliminate the growth of cross-reacting strains or of fungi or bacteria that spread in the medium.

Plating methods are indirect detection methods and have the

disadvantage that the origin of a colony is unknown. For fungi especially, it is often important to enumerate different types of propagules (e.g. mycelium fragments, spore types and resting structures) separately, because they have different ecological properties. Direct observation methods are necessary to solve this problem.

Metabolically active but nonculturable microorganisms can be detected directly with fluorochrome-labeled oligonucleotide probes, as was shown for *Pseudomonas* spp. An alternative is the incubation of the sample in the presence of a redox dye (tetrazolium chloride salt). Active cells will accumulate the salt and show red fluorescent (602 nm) crystals in the cell (34). This method is not specific for a certain target organism, but the method can be used in combination with immunofluorescence cell-staining to distinguish between active and inactive cells of the target organism (C.E. Heijnen, personal communication).

For in situ detection in and on roots the use of *lux*-marked bacteria and GUS-marked fungi enabled the study of root colonization patterns. The GUS marker was also useful for detailed studies on the localization of the GUS activity in the cell. Root colonization studies for wild-type target organisms can be performed using immunofluorescence cell- or colony-staining, as was demonstrated for *Erwinia* and *Pseudomonas* spp.

Confirmation of a positive test result is of major importance in studies where the density of the target population is low. Increasing the detection sensitivity of an assay system for the target will often also result in loss of specificity. There will be more kinds of microorganisms present at population densities of 10 to 10^2 than at densities of 10^4 to 10^5 units g^{-1} soil. This will result in an increased risk of a false-positive reaction due to a cross-reaction of a nontarget population with the antibody or gene probe used. Confirmation protocols have been worked out for direct verification of IFC-positive colonies.

PCR, the most important gene probe technique, is not yet suitable for direct quantitative detection of viable cells. However, a combination of enrichment with the use of subsamples according to the most probable number technique (29) can be used for quantitative assessment of culturable cells with PCR. The present strength of PCR is its use as a tool for identification and for qualitative assays on target bacteria or fungi independent of their viability. The theoretical detection level of PCR for target bacteria in soil is calculated at 10^3 cells g^{-1} soil (36). Extensive extraction and concentration steps can further improve the detection to 100 cells in 100 g of soil (37).

Reliable detection involves much more than the use of reliable detection methods. Factors of the assay that influence the detection of the target include: (i) the efficiency of sample preparation and extraction; (ii) the dilution factor calculated per gram of soil or root; (iii) the detection level of the detection method for the substrate under investigation; (iv) the statistical variation among replicates; (v) the risk of false-negative results due to factors interfering with the detection level (such as the variation in recovery between strains of the target, and the variation between strains in their reactivity with the probe); and (vi) the risk of false-positive results e.g. due to cross-reactions (acceptable if additional confirmation procedures are available). The whole assay system should be adapted for optimal performance based on preliminary research and literature data. Aspects such as sampling, sample homogenization, sample extraction, statistics and the selection of model test systems are reviewed by Kloepper and Beauchamp (22).

With respect to detection assays, we believe that there are still many possibilities for improvement. Best overall performance may be expected from assay systems that combine several different methods, each contributing to the specificity or sensitivity of the assay as demonstrated for the IFC-PCR system. Immunomagnetic separation, based on the use of paramagnetic beads coated with antitarget antibodies, can be used for selective extraction of the target in a detection procedure (18). For culturable targets, good prospects may be expected from the following possible combinations of pour plating with other methods, which to our knowledge are not reported in the literature.

1. Detection of *lux*-marked mutants in combination with in situ enrichment for studies of population dynamics. As for IFC, an improved routine use and sensitivity can be expected.

2. Detection of target colonies in pour plates with the use of labeled rRNA oligonucleotide probes. Using a fluorescent dye as a reporter molecule, equipment and routine procedures identical to those for IFC can be employed. Using the probes for colony staining instead of single-cell staining is expected to eliminate the risk of a false-negative result due to a too low metabolic activity in a single target cell. Preliminary experiments demonstrated a positive fluorescent staining of the target colonies (*P. cepacia*) with the fluorescent 16S rRNA oligonucleotide probes in pour plates of pure culture dilutions. In mixed suspensions, however, a high positive fluorescence was also observed for nontarget colonies (D. Hahn, J.M. Van der Wolf, and J.W.L. Van Vuurde, unpublished).

3. Detection of target fungal colonies pour plated in media that restrict mycelial growth (e.g. rose bengal medium) in combination with labeled antibody or gene probes for colony differentiation.

We conclude by noting that many good detection methods are now available as tools in the ecologist's toolbox. The selection of the right combination of tools (and often the modification of tools) is of crucial importance for the quality of the results in the study of ecology of soil microorganisms.

ACKNOWLEDGEMENTS

We would like to thank the following scientists for their support in the preparation of the presentation in Montreal: J.I. Prosser (Univ. Aberdeen), W.F. Mahaffee (Univ. Alabama), D. Hahn, (E.T.H. Zürich), C. Alabouvette (INRA Dijon), J.M. Raaijmakers (Univ. Utrecht), J.M. Van der Wolf (IPO-DLO Wageningen) and J.D. Van Elsas (IPO-DLO Wageningen). We thank H. Huttinga for critical reading of the manuscript.

LITERATURE CITED

1.   Akkermans, A.D.L., Van Elsas, J.D., and De Bruijn, F.J., eds. 1995. Molecular Microbial Ecology Manual. Kluwer Academic Publishers, Dordrecht, The Netherlands.

2.   Amann, R.I., Krumholz, L., and Stahl, D.A. 1990. Fluorescent-oligonucleotide probing of whole cells for determinative, phylogenetic, and environmental studies in microbiology. J. Bacteriol. 172:762–770.

3.   Andrews, J.H. 1986. How to track a microbe. Pages 14–34 in: Microbiology of the Phyllosphere. N.J. Fokkema and J. Van den Heuvel, eds. Cambridge University Press, Cambridge, UK.

4   Brockman, F.J., Forse, L.B., Bezdicek, D.F., and Fredrickson, J.K. 1991. Impairment of transposon-induced mutants of *Rhizobium leguminosarum*. Soil Biol. Biochem. 23:861–867.

5.   Chun, W.W.C., and Alvarez, A.M. 1983. A starch-methionine medium for isolation of *Xanthomonas campestris* pv. *campestris* from plant debris in soil. Plant Dis. 67:632–635.

6.   Couteaudier, Y., Daboussi, M.-J., Eparvier, A., Langin, T., and Orcival, J. 1993. The GUS gene fusion system (*Escherichia coli* β-D-glucuronidase gene), a useful tool in studies of root colonization by *Fusarium oxysporum*. Appl. Environ. Microbiol. 59:1767–1773.

7. Cullen, D., and Andrews, J.H. 1985. Benomyl-marked populations of *Chaetomium globosum*: survival on apple leaves with and without benomyl and antagonism to the apple scab pathogen, *Venturia inaequalis*. Can. J. Microbiol. 31:251–255.

8. DeLong, E.F., Wickham, G.S., and Pace, N.R. 1989. Phylogenetic stains: Ribosomal RNA-based probes for the identification of single cells. Science 243:1360–1363.

9. Dewey, F.M. 1992. Detection of plant-invading fungi by monoclonal antibodies. Pages 47–62 in: Techniques for the Rapid Detection of Plant Pathogens. J.M. Duncan and L. Torrance, eds. Blackwell Scientific Publications, Oxford, UK.

10. Foster, L.M., Kozak, K.R., Loftus, M.G., Stevens, J.J., and Ross, I.K. 1993. The polymerase chain reaction and its application to filamentous fungi. Mycol. Res. 97:769–781.

11. Gerhardt, P., Murray, R.G.E., Wood, W.A., and Krieg, N.R. 1993. Methods for General and Molecular Bacteriology. American Society for Microbiology, Washington, DC. 791 pp.

12. Hahn, D., Amann, R.I., Ludwig, W., Akkermans, A.D.L., and Schleifer, K.-H. 1992. Detection of micro-organisms in soil after *in situ* hybridization with rRNA-targeted, fluorescently labelled oligonucleotides. J. Gen. Microbiol. 138:879–887.

13. Hahn, D., Amann, R.I., and Zeyer, J. 1993. Whole-cell hybridization of *Frankia* strains with fluorescence- or digoxigenin-labeled, 16S rRNA-targeted oligonucleotide probes. Appl. Environ. Microbiol. 59:1709–1716.

14. Hahn, D., Amann, R.I., and Zeyer, J. 1994. Oligonucleotide probes for the detection and identification of bacteria. Pages 163–171 in: Improving Plant Productivity with Rhizosphere Bacteria. M.H. Ryder, P.M. Stephens, and G.D. Bowen, eds. CSIRO Division of Soils, Adelaide, Australia.

15 Hampton, R., Ball, E., and De Boer, S., eds. 1990. Serological Methods for Detection and Identification of Viral and Bacterial Plant Pathogens. APS Press, St. Paul, MN. 389 pp.

16. Henson, J.M., and French, R. 1993. The polymerase chain reaction and plant disease diagnosis. Annu. Rev. Phytopathol. 31:81–109.

17. Jefferson, R.A. 1989. The GUS reporter gene system. Nature (Lond.) 342:837–838.

18. Jones, J.B., and Van Vuurde, J.W.L. 1990. Magnetic immuno-isolation of *Xanthomonas campestris* pv. *pelargonii*. Pages 883–888 in: Proceedings, 7th International Conference on Plant Pathogenic Bacteria, Budapest, Hungary. Z. Klement, ed. Akadémiai Kiadó, Budapest, Hungary.

19. Jones, D.A.C., Hyman, L.J., Tumeseit, M., Smith, P., and Pérombelon, M.C.M. 1994. Blackleg potential of potato seed: determination of tuber contamination by *Erwinia carotovora* subsp. *atroseptica* by immunofluorescence colony staining and stock and tuber sampling. Ann. Appl. Biol. 124:557–568.

20. Kastelein, P., Bouman, A., Mulder, A., Schepel, E., Turkensteen, L.J., De Vries, Ph.M., and Van Vuurde, J.W.L. 1994. Green-crop-harvesting and infestation of seed potato tubers with *Erwinia* spp. and perspectives for integrated control. Pages 999–1004 in: Proceedings, 8th International Conference on Plant Pathogenic Bacteria, Versailles. INRA, Paris, France.

21. Klement, Z., Rudolph, K., and Sands, D.C., eds. 1990. Methods in Phytobacteriology. Académiai Kiadó, Budapest, Hungary. 568 pp.

22. Kloepper, J.W., and Beauchamp, C.J. 1992. A review of issues related to measuring colonization of plant roots by bacteria. Can. J. Microbiol. 38:1219–1232.

23. Leeman, M., Raaijmakers, J.M., Bakker, P.A.H.M., and Schippers, B. 1991. Immunofluorescence colony staining for monitoring pseudomonads introduced in soil. Pages 374–380 in: Biotic Interactions and Soil-borne Diseases. A.B.R. Beemster, B.J. Bollen, M. Gerlagh, M.A. Ruissen, B. Schippers, and A. Tempel, eds. Elsevier, Amsterdam, The Netherlands.

24. Macario, A.J.L., and De Macario, E.C., eds. 1990. Gene Probes for Bacteria. Academic Press, San Diego, CA. 515 pp.

25. Mahaffee, W.F., Backman, P.A., and Shaw, J.J. 1991. Visualization of root colonization by rhizobacteria using a luciferase marker. WPRS Bull. 14:248–251.

26. Mahaffee, W.F., Kloepper, J.W., Van Vuurde, J.W.L., Van der Wolf, J.M., and Van den Brink, M. 1994. Endophytic colonization of *Phaseolus vulgaris* by *Pseudomonas fluorescens* strain 89B-27 and *Enterobacter asburiae* strain JM22. Page 180 in: Improving Plant Productivity with Rhizosphere Bacteria. M.H. Ryder, P.M. Stephens, and G.D. Bowen, eds. CSIRO Division of Soils, Adelaide, Australia.

27. Oyarzun, P.J., Postma, J., Luttikholt, A.J.G., and Hoogland, A.E. 1994. Biological control of foot and root rot in pea caused by *Fusarium solani* with nonpathogenic *Fusarium oxysporum* isolates. Can. J. Bot. 72:843-852.

28. Papavizas, G.C., Lewis, J.A., and Abd-El Moity, T.H. 1982. Evaluation of new biotypes of *Trichoderma harzianum* for tolerance to benomyl and enhanced biocontrol capabilities. Phytopathology 72:126-132.

29. Picard, C., Ponsonnet, C., Paget, E., Nesme, X., and Simonet, P. 1992. Detection and enumeration of bacteria in soil by direct DNA extraction and polymerase chain reaction. Appl. Environ. Microbiol. 58:2717-2722.

30. Postma, J., and Luttikholt, A.J.G. 1993. Benomyl-resistant *Fusarium*-isolates in ecological studies on the biological control of fusarium wilt in carnation. Neth. J. Plant Pathol. 99:175-188.

31. Prosser, J.I. 1994. Molecular marker systems for detection of genetically engineered micro-organisms in the environment. Microbiology 140:5-17.

32. Raaijmakers, J.M. 1994. Accurate monitoring of wild-type *Pseudomonas putida* WCS 358 in natural environments by immunofluorescence colony-staining. Pages 43-55 in: Microbial Interactions in the Rhizosphere. Ph.D. thesis, University of Utrecht, Utrecht, The Netherlands.

33. Rattray, E.A.S., Prosser, J.I., Glover, L.A., and Killham, K. 1995. Characterization of rhizosphere colonization by a luminescent *Enterobacter cloacae* at the population and single-cell levels. Appl. Environ. Microbiol. 61:2950-2957.

34. Rodriguez, G.G., Phipps, D., Ishiguro, K., and Ridgway, H.F. 1992. Use of a fluorescent redox probe for direct visualization of actively respiring bacteria. Appl. Environ. Microbiol. 58:1801-1808.

35. Schots, A., Dewey, F.M., and Oliver, R.P., eds. 1994. Modern Assays for Plant Pathogenic Fungi: Identification, Detection and Quantification. CAB International, Wallingford, UK. 267 pp.

36. Smalla, K., Cresswell, N., Mendonca-Hagler, L.C., Wolters, A., and Van Elsas, J.D. 1993. Rapid DNA-extraction protocol from soil for polymerase chain reaction-mediated amplification. J. Appl. Bacteriol. 74:78-85.

37. Steffan, R.J., and Atlas, R.M. 1988. DNA amplification to enhance detection of genetically engineered bacteria in environmental samples. Appl. Environ. Microbiol. 54:2185-2191.

38.   Stewart, G.S.A.B., and Williams, P. 1992. *Lux* genes and the applications of bacterial bioluminescence. J. Gen. Microbiol. 138:1289–1300.

39.   Turlier, M.-F., Eparvier, A., and Alabouvette, C. 1994. Early dynamic interactions between *Fusarium oxysporum* f. sp. *lini* and the roots of *Linum usitatissimum* as revealed by transgenic GUS-marked hyphae. Can. J. Bot. 72:1605–1612.

40.   Underberg, H., and Van Vuurde, J.W.L. 1990. *In situ* detection of *Erwinia chrysanthemi* on potato roots using immunofluorescence and immunogold staining. Pages 937–942 in: Proceedings, 7th International Conference on Plant Pathogenic Bacteria, Budapest, Hungary. Z. Klement, ed. Akadémiai Kiadó, Budapest, Hungary.

41.   Van der Wolf, J.M., Van Beckhoven, J.R.C.M., De Boef, E., and Roozen, N.J.M. 1993. Serological characterization of fluorescent *Pseudomonas* strains cross-reacting with antibodies against *Erwinia chrysanthemi*. Neth. J. Plant Pathol. 99:51–60.

42.   Van der Wolf, J.M., Van Beckhoven, J.R.C.M., De Vries, Ph.M., Raaijmakers, J.M., Bertheau, Y., Bakker, P.A.H.M., and Van Vuurde, J.W.L. 1994. Polymerase chain reaction for verification of fluorescent colonies in immunofluorescence colony-staining. Pages 60–61 in: Abstracts, Conference of the European and Mediterranean Plant Protection Organization, Wageningen, The Netherlands. EPPO, Paris, France.

43.   Van Vuurde, J.W.L. 1987. New approach in detecting phytopathogenic bacteria by combined immunoisolation and immunoidentification assays. EPPO Bull. 17:139–148.

44.   Van Vuurde, J.W.L. 1990. Immunofluorescence colony staining. Pages 299–305 in: Serological Methods for Detection and Identification of Viral and Bacterial Plant Pathogens. R. Hampton, E. Ball, and S. De Boer, eds. APS Press, St. Paul, MN.

45.   Van Vuurde, J.W.L. 1991. Immunofluorescence colony-staining (IFC) and immunofluorescence cell-staining as tools for the study of rhizosphere bacteria. WPRS Bull. 14:215–222.

46.   Van Vuurde, J.W.L., and Roozen, N.J.M. 1990. Comparison of immunofluorescence colony-staining in media, selective isolation on pectate medium, ELISA and immunofluorescence cell staining for detection of *Erwinia carotovora* subsp. *atroseptica* and *E. chrysanthemi* in cattle manure slurry. Neth. J. Plant Pathol. 96:75–89.

47. Van Vuurde, J.W.L., De Vries, Ph.M., and Roozen, N.J.M. 1994. Application of immunofluorescence colony-staining (IFC) for monitoring populations of *Erwinia* spp on potato tubers, in surface water and in cattle manure slurry. Pages 741–746 in: Proceedings, 8th International Conference on Plant Pathogenic Bacteria, Versailles. INRA, Paris, France.

48. Van Vuurde, J.W.L., and Van der Wolf, J.M. 1995. Immunofluorescence colony-staining (IFC). Chapter 4.1.3, pages 1–19 in: Molecular Microbial Ecology Manual. A.D.L. Akkermans, J.D. Van Elsas, and F.J. De Bruijn, eds. Kluwer Academic Publishers, Dordrecht, The Netherlands.

Chapter 6

A THERMAL TIME BASIS FOR UNDERSTANDING PEST EPIDEMIOLOGY AND ECOLOGY

D.L. Trudgill

Organisms each have their own intrinsic potential to grow and multiply. If other factors are optimal, temperature is usually the rate-determining factor. Where this is so, an understanding of the relation between development rates and temperature provides a means of analyzing and predicting population increase and of understanding aspects of an organism's plasticity and ecology. Results with plants and nematodes will be used to demonstrate the basic principles and to explore their implications. Aspects which will be considered include whether the relation between rate and temperature is linear or sigmoidal, differences in the thermal requirements of organisms adapted to warm or cool conditions, and the ecological and practical implications of thermal time information. It will be shown how a knowledge of thermal time requirements provides a basis for comparing related organisms and understanding aspects of their ecological strategies. Our ignorance of the basic physiological processes and their inheritance will also be briefly identified.

THE BASIC RELATIONSHIP

Rates Of Development

Development rates of poikilothermic (ectothermic) organisms are determined by the interaction between their intrinsic potential and various environmental factors. Where these latter are optimal (i.e. no limitations with regard to the stimuli and resources required for development and growth) and where diapause, photoperiodic effects etc. are absent, then temperature is usually the rate-limiting factor. Experience suggests that rates of development (expressed as the reciprocal of duration; $\frac{1}{D}$) of most poikilothermic organisms increase as temperature increases, up to optimum temperature (T_o). Intuitively, many biologists would predict a sigmoidal relationship but experimentation with a range of plants and animals suggests that, for many, the relation is close to a straight line over much or all of this range.   Where this is so the intercept with the temperature axis provides an estimate of the base (T_b) or threshold temperature (Figure 1). Below T_b no development occurs. The effective temperature for development is the environment temperature (T_e) minus T_b. When the relation is linear between T_b and T_o there is a constant requirement, expressed in °C-days (or degree-days) for the developmental process being studied. This requirement, the thermal constant (K), is the reciprocal of the slope.

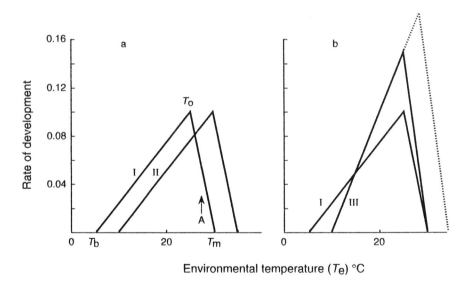

Figure 1. Hypothetical examples showing the basic thermal time relationship between temperature and rate of development and the effect (a) of increasing all the cardinal values except K by 5°C (I compared to II) and (b) of changing T_b and K inversely by the same proportion (I compared to III).

Knowledge of the value of T_b and K can be used to predict durations of development (D) from the simple equation:

$$D = \frac{K}{T_e - T_b}$$

(for values of T_e between T_b and T_o). The thermal constant K is an important ecological indicator as it reflects the amount of development to be done divided by the intrinsic rate (i.e. slope for unit amount).

Sigmoidal Or Straight Line Relationship

Whether the relation between temperature and rate of development is physiologically a straight line or is sigmoidal remains controversial. It is usually difficult to work close to T_b because of the extended duration of the experiment. Also, very accurate temperature measurements of the organism itself, which may differ from that of its surrounding environment, are required (e.g. at 1.0°C above T_b a consistent 0.1°C error in temperature measurements represents a 10% error overall!). Whether the relation appears to be a straight line or slightly sigmoidal is probably further influenced by population heterogeneity and whether temperature affects other aspects of development such as overall size (3,16). Consequently, T_b is usually determined by back extrapolation of the central, straight line part of the relationship.

There are many examples of both straight line and sigmoidal relationships in plant, nematode and insect studies. However, entomologists mostly fit shallow sigmoidal curves to their thermal time data using a logistic equation (1). But, for predictive purposes, straight line regressions are more useful (4). As estimates of T_b and K are strongly and inversely correlated their accurate determination requires precise data over as much of the temperature range between T_b and T_o as possible. Also required are clear start and end points for the process under study so that their duration can be accurately determined.

Practical Significance

The power of thermal time information is obvious as it allows one to predict the timing of certain events and plan accordingly. For example, it is used to plan sowing dates of various crops (e.g. calabrese) (11) where the maximum period of continuous harvesting is sought. Sowing dates have to be widely spaced in the coldest parts of the growing season (i.e.

when soil temperatures are just above T_b) and then need to be increasingly close together as temperatures increase. It has been used to predict dates of pest emergence (e.g. pear rust mite) (5), as part of a population dynamics model for the root-knot nematode (*Meloidogyne arenaria*) (7), and to understand the dynamics of damage by another root-knot nematode (*Meloidogyne chitwoodi*) to potato (12). It has also been used to predict the threat posed by introductions of a third root-knot nematode (*Meloidogyne hapla*) into Finland (14) and to plan the control of a cyst nematode (*Globodera* sp.) by early harvesting of potato tubers (9).

THERMAL ADAPTATION

Relations Between Cardinal Values

The relation between rate of development and temperature has been described as being wigwam shaped (Figure 1). The cardinal values are the base (T_b), optimum (T_o) and maximum (T_m) temperatures, the thermal constant (K) and the range between T_b and T_o and T_m. Some of the ways these could vary in species adapted to different temperature ranges are shown diagrammatically in Figure 1.

T_b And K Values In Temperate And Tropical Species

Tropical organisms generally have higher values of T_b than temperate ones. Hence, in field trials, the T_b for germination and 50% emergence of the tropical cereal sesame was estimated as about 16°C whereas that for a temperate cultivar of barley was near 3°C (2). A key question is why this difference occurs, especially as the effect of a higher value of T_b is to decrease the effective temperature (the difference between T_e and T_b). Assuming everything else is the same, this difference in T_b values would result in a slower rate, and hence a greater duration, of germination for sesame compared with barley at all temperatures up to and beyond T_b (see hypothetical example in Figure 1a). However, everything is not the same; the K value for germination of sesame is about 21 °C-days ($r^2 = 0.99$) compared with 79 °C-days for barley ($r^2 = 0.39$). The effect of this difference in K is to increase the slope of the regression between temperature and rate for sesame as compared to barley (as shown in the example in Figure 1b). Consequently, at lower temperatures barley will germinate before sesame, but at higher temperatures sesame will germinate first (Table 1).

Table 1. Calculated durations of germination for sesame and barley at different constant temperatures with values of T_b of 16°C and 3°C and of K of 21 °C-days and 79 °C-days respectively. It is assumed $T_o > 30$°C for both crops.

Temperature	Duration (days)	
(°C)	Barley	Sesame
10	11.3	Infinite
15	6.6	Infinite
20	4.6	5.3
25	3.6	2.3
30	2.9	0.8

A series of studies with the root-knot nematodes *M. hapla* (temperate species) and *Meloidogyne javanica* (tropical species) support the view that the values of T_b and K vary inversely for comparable processes or species. The two species are biologically similar, grow to similar sizes and were produced on the same favorable host (tomato). Based on the minimum duration for one generation, the values of T_b were estimated as 8.25 and 13.0°C and of K as 554 and 350 °C-days for *M. hapla* and *M. javanica*, respectively (8,10). As the inverse ratios of the values for T_b and K for the two species are almost identical it can be calculated (16) that the interception point of the two regressions is the sum of the two T_b values (i.e. 21.25°C). Also, the optimal base temperature is half the average environment temperature (i.e. they are adapted to T_e values of about 16.5°C and 26.0°C, respectively).

The values of T_o are similar for both species (about 27 and 29°C respectively) and T_m is about 2°C higher for *M. javanica*. However, at 31°C there was no reduction in the rate of development of *M. javanica* whereas T_m for *M. hapla* was less than 32°C (Figure 2). Overall, there appeared to be almost a 5.0°C difference in T_b values between these populations of the two species and a difference of about 3.0°C in T_m values.

The above results led to the hypothesis that there is an inverse relationship between T_b and K, potential rates of development increasing with increasing values of T_b (15,16). If this hypothesis is correct it follows that in populations heterogeneous for their thermal requirements there will be a sigmoidal relationship between temperature and the rates of development of the fastest developing individuals.

ECOLOGICAL SIGNIFICANCE OF THERMAL TIME INFORMATION

T_b

On the basis of the above hypothesis it is clear that the value of T_b reflects the thermal adaptation of an organism. Tropical species, which generally are not troubled by the threat of frost damage, have high values

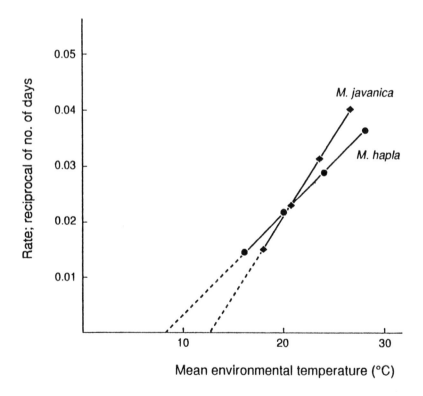

Figure 2. Relation between rates of development of *Meloidogyne hapla* and *M. javanica* and temperature. The T_b values are significantly different at $P < 0.001$.

of T_b to maximize their intrinsic potential for development (i.e. to minimize K). Those with low values are adapted to temperate environments where the priority is to maximize the growing season (at the expense of an increased K). However, T_b is usually above 0°C, presumably to avoid exposure to frost damage, and some alpine plants have relatively high T_b values, probably to prevent growth before the onset of spring. Various diapause mechanisms are frequent in temperate species to better target the onset of development. For example, before the eggs of northern populations of cereal cyst nematode (*Heterodera avenae*) will hatch they require a period of chilling followed by a period of increasing soil temperature. Populations from warmer, Mediterranean regions have the reverse behavior, hatching when soil temperatures start to decrease.

K

The thermal constant K reflects the amount of development to be done divided by the intrinsic rate. Hence, differences in K can be due to differences in the amount of development to be done, the rate, or both. We have already seen how K may be inversely related to T_b and how this relates to the thermal adaptation of species. However, this hypothesis is based on several assumptions. For example, in the interpretation of published data (2), it has been assumed that the amounts of development involved in the germination of sesame and barley are the same and that differences in K are all due to differences in rates of development associated with differences in T_b. Such an assumption would not be true when comparing the requirements for embryogenesis of nematode eggs. Most eggs hatch as first stage juveniles but those of Tylenchid nematodes (the main group of plant parasites) undergo the first molt within the egg before hatching. Inevitably, this increases their K requirement. Even so, the huge difference in K requirement for embryogenesis between species such as the bacterial feeding nematode *Caenorhabditis elegans* ($K < 20$ °C-days) and the plant parasite *Globodera rostochiensis* ($K > 120$ °C-days), both of which have a similar T_b of about 5°C, must also indicate major physiological differences (15).

The Interaction Between T_b And K

As already demonstrated (Table 1), the effect of an inverse relation between T_b and K for the same process is to give temperate species a developmental advantage in cool conditions, and tropical species the advantage in warmer conditions. Consequently, thermal requirements may be one of the key factors determining the competitive ability of species in different environments and hence their distributions. A comparison of temperate *M. hapla* and tropical *M. javanica* shows (Figure 2) that their rates of development are the same at 21°C and that each species has a greater rate than the other in the environment to which it is adapted—an arrangement which makes good ecological sense. Thermal time information indicates, therefore, not only whether an introduced organism might survive in a new environment, but also its likely competitive ability.

"r" And "K" Strategists

The comparatively small K value of *C. elegans* is associated with a short life cycle and a capacity for rapid population increase—characteristic of "r" strategy species (13). Such species exploit transient niches in contrast to "K" strategy species with tend to persist and become dominant in stable environments (6). In general, "K" strategy species grow larger and have longer life cycles and slower rates of reproduction, and hence larger K values than "r" strategy species. As K is the outcome of the amount of development divided by the intrinsic rate, differences between species in adult size will indicate differences in K, provided their rates of development are likely to be similar. Such an approach could help refine the ecological classification of the nematodes used in the Bongers (6) maturity index to classify environments.

An "r" strategy, almost by definition, is associated with a comparatively small K value. This can be achieved in several ways, some of which have already been discussed. These include small size (and hence less growth) and more rapid development associated with physiological differences (e.g. *C. elegans* compared to *G. rostochiensis*). In addition, parthenogenetic reproduction, which is extremely widespread amongst nematodes, decreases the K requirement associated with mating and some species have reduced either the number of juvenile stages or have stages which are nonfunctional, further decreasing their K requirement.

FUTURE RESEARCH/APPLICATIONS

The hypothesis that T_b and K are inversely correlated needs testing and verification, perhaps by selection within a population heterogeneous for its thermal requirements. Once contrasting populations have been selected they could be used to determine the mechanisms of inheritance of thermal requirements and their physiological basis. In particular the role of lipids in membranes and the relation between their transition (change in physical state) temperature, T_b, and membrane efficiency may repay investigation. The outcome of selection may vary depending on the intensity. Selection for the shortest generation time at low temperatures may select for smaller adults, not a lower value of T_b. The effect of selection on T_b and T_m, and whether they are linked would also be interesting to determine.

Greater application of thermal time information is desirable, especially when trying to assess the threat posed by potential introductions or the possible effects of global climate change on pest potential. However, biologically appropriate values are needed and even where thermal time studies have been done these are not always available e.g. the values available mainly relate to the most rapidly developing individuals rather than to 50% of the population. Also, not all environments are optimal and some allowance may need to be made for the effects of other environmental factors (e.g. has the host crop started to grow?).

Thermal time provides a basis, not yet fully exploited, for describing and comparing the ecology of poikilothermic organisms. It could be relevant in selecting parents in plant breeding programs and identifying appropriate environments in which to test the progeny. However, its main ecological uses are probably in assessing the risks posed by potential introductions of pathogens and as a means of classifying organisms as "r" and "K" strategists (e.g. in the Bongers (6) maturity index) on their basis of size having first taken account of likely differences in their physiological rates.

LITERATURE CITED

1.   Andrewartha, H.G., and Birch, L.C. 1954. The Distribution and Abundance of Animals. The University of Chicago Press, Chicago, IL. 782 pp.

2.   Angus, J.F., Cunningham, R.B., Moncur, M.W., and Mackenzie, D.H. 1981. Phasic development in field crops. I. Thermal response in the seedling phase in eastern Australia, ecological and physiological significance. Field Crops Res. 3:365–378.

3.   Atkinson, D. 1994. Temperature and organism size—a biological law for ectotherms. Adv. Ecol. Res. 25:1–58.

4.   Baker, C.R.B. 1980. Some problems in using meteorological data to forecast the timing of insect life cycles. EPPO Bull. 10:83–91.

5. Bergh, J.C., and Judd, G.J.R. 1993. Degree-day model for predicting emergence of pear rust mite (Acari: Eriophyidae) deutogynes from overwintering sites. Environ. Entomol. 22:1325–1332.

6. Bongers, T. 1990. The maturity index: an ecological measure of environmental disturbance based on nematode species composition. Oecologia 83:14–19.

7. Ferris, H., Du Vernay, H.S., and Small, R.H. 1978. Development of a soil-temperature data base on *Meloidogyne arenaria* for a simulation model. J. Nematol. 10:39–42.

8. Lahtinen, A.E., Trudgill, D.L., and Tiilikkala, K. 1988. Threshold temperature and minimum thermal time requirements for the complete life cycle of *Meloidogyne hapla* from northern Europe. Nematologica 34:443–451.

9. Langeslag, M., Mugniery, D., and Fayet, G. 1982. Développement embryonnaire de *Globodera rostochiensis* et *G. pallida* en fonction de la température, en conditions contrôlées et naturelles. Rev. Nématol. 5:103–109.

10. Madulu, J.D., and Trudgill, D.L. 1994. Influence of temperature on the development and survival of *Meloidogyne javanica*. Nematologica 40:230–243.

11. Marshall, B., and Thompson, R. 1987. Applications of a model to predict the time to maturity of calabrese *Brassica oleracea*. Ann. Bot. 60:521–529.

12. Pinkerton, J.N., Santo, G.S., and Mojtahedi, H. 1991. Population dynamics of *Meloidogyne chitwoodi* on Russet Burbank potatoes in relation to degree-day accumulation. J. Nematol. 23:283–290.

13. Southwood, T.R.E. 1981. Bionomic strategies and population parameters. Pages 30–52 in: Theoretical Ecology: Principles and Applications. R.M. May, ed. 2nd edition. Blackwell Scientific Publications, Oxford, UK.

14. Tiilikkala, K., Lahtinen, A., and Trudgill, D.L. 1988. The pest potential of *Meloidogyne hapla* in northern field conditions. Ann. Agric. Fenn. 27:329–388.

15. Trudgill, D.L. 1995. An assessment of the relevance of thermal time relationships to nematology. Fundam. Appl. Nematol. 18:407–417.

16. Trudgill, D.L., and Perry, J.N. 1994. Thermal time and ecological strategies—a unifying hypothesis. Ann. App. Biol. 125:521–532.

Chapter 7

NEW STRATEGIES FOR THE MANAGEMENT OF PLANT PARASITIC NEMATODES

B.R. Kerry and K. Evans

Plant parasitic nematodes are difficult to control once introduced into soil and much effort is spent on avoidance of these pests through stringent quarantine and certification schemes. The small size of nematodes and their existence in soil means that they are extremely numerous by the time they are detected causing crop damage. For example, the threshold population for damage to sugar beet caused by beet cyst nematodes is 2 eggs g^{-1} soil, which represents 4×10^9 individuals ha^{-1}; populations less than 10^5 ha^{-1} are rarely detected in soil samples. The great bulk of soil (about 2,500 t ha^{-1}) ensures that most nematode pests are relatively inaccessible and it is impossible to eradicate infestations. Hence, control measures aim to maintain populations at nondamaging densities.

For the past 40 years, soil fumigants and, more recently, granular nematicides, have provided adequate control of nematode pests in intensive and perennial cropping systems. However, the cost of most nematicidal applications has limited their use to relatively high value crops in developed farming systems. Also, the production and use of some nematicides have been associated with health and environmental hazards and several products have been withdrawn from the market. Within Europe, applications of nematicides are increasingly restricted and some countries such as The Netherlands have legislated for significant reductions in their use. By the year 2000 the widely used methyl bromide will be banned in the United States and it is unlikely to be available for use elsewhere. Hence, there is a need to establish alternative methods of nematode control which minimize the use of nematicides. However, in many situations there are few alternatives to the use of chemicals and, at present, no biological control products are available. Control of nematodes in field crops has tended to rely on the use of crop rotation integrated with the use of resistant cultivars and chemicals. Such methods have generally proved successful but may lead to the selection of virulent populations if crops with only partial resistance to the nematodes present in soil are grown too frequently.

In this brief review the need for detailed knowledge of the biology and ecology of pest nematodes in the development of sustainable methods for their management is highlighted and the integrated control of potato cyst nematodes in the UK is used as an example. We also discuss advances in the development of technologies for nematode diagnosis, of transgenic plants with novel resistance, and of biological control, all of which may lead to improved methods of nematode management. Most information concerns the cyst (*Globodera* and *Heterodera* spp.) and root-knot (*Meloidogyne* spp.) nematodes, which include the most important nematode pests in world agriculture. These genera have evolved complex relationships with their hosts and their life cycles offer a number of opportunities for perturbation and the development of novel control strategies.

INTEGRATED CONTROL OF POTATO CYST NEMATODES IN THE UK: A CASE STUDY

There are two species of potato cyst nematodes, *Globodera rostochiensis* and *Globodera pallida*, and these are widely distributed in the UK. They are morphologically very similar and were described as distinct species only in 1973. In the 1960s, potato cultivars were bred with resistance only to *G. rostochiensis*. This narrow resistance was deployed largely because it depended only on a single major gene (36) and was therefore easy to incorporate into agronomically acceptable cultivars. This was in contrast to the polygenic resistance to *G. pallida*. An integrated control program was devised which incorporated the use of nematicides and *G. rostochiensis* resistant cultivars, such as Maris Piper. These were grown in rotation with fully susceptible cultivars and nonhosts to slow the selection of *G. pallida*, which is capable of multiplying on Maris Piper. This recommended strategy was widely supported by the extension service but often ignored by growers who tended to overcrop Maris Piper, which was favored for its agronomic characteristics as well as its nematode resistance; the cultivar has been one of the most widely grown in the UK for almost 20 years. Maris Piper was grown in soils in which *G. pallida* was present but not detected using standard soil sampling and extraction techniques. As a consequence, *G. pallida* has increased in many soils and is now believed to be the dominant species of potato cyst nematode throughout the potato growing area. Before the use

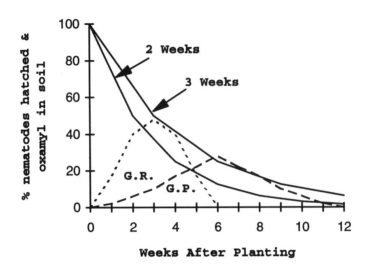

Figure 1. Hatching patterns for *Globodera pallida* (G.P.) and *G. rostochiensis* (G.R.) under a potato crop and decay curves for the nematicide oxamyl with 2- or 3-week half-lives.

of resistant cultivars, *G. pallida* was rare in the major ware potato-growing regions of Southeast England. Today, potato cyst nematodes cause direct and indirect yield losses totalling approximately £50 million per annum (equivalent to 9% of potato production) and growers spend £9 million per annum applying nematicides for their control.

Despite the fact that *G. pallida* appeared to be displacing *G. rostochiensis* (16), the assumption was made that this nematode would be effectively controlled by the methods applied to its sibling species, *G. rostochiensis*. In practice, *G. pallida* has proved more difficult to manage than *G.rostochiensis* for several reasons. The eggs of *G. pallida* hatch more slowly and over a longer period of time than those of *G. rostochiensis* (40). Also, the second stage juveniles of *G. pallida* utilize their lipid reserves more slowly and so can survive in soil for longer (29). Because of the different hatching patterns for the two species (Figure 1), oximecarbamate nematicides (which act as nematostats) applied at planting do not remain at toxic concentrations in soil long enough to control *G.*

pallida (40). There is considerable genetic variation in *G. pallida* within the UK and populations differ markedly in their ability to multiply on the partially resistant cultivars currently available, such as Santé and Morag, which limit nematode reproduction but do not prevent populations increasing (38). Overcropping with partially resistant cultivars is likely to result in rapid selection of more virulent nematode populations (39). Finally, in the absence of host plants, populations of *G. pallida* decline more slowly than those of *G. rostochiensis* and so crop rotations must be longer (41). Thus, despite the marked similarities between the two species of potato cyst nematode, there are major differences in physiology, behavior and genetic variation which affect the development of control strategies. More measures are required to control *G. pallida* than *G. rostochiensis* and cycles between susceptible crops are longer (Table 1).

Clearly, the development of integrated management strategies for nematode pests requires sensitive and accurate methods of identification, plants with broad resistance, and other sustainable control methods which, above all, must be acceptable to the farmer. A system which included the use of susceptible cultivars less favored than Maris Piper and of unacceptably long crop rotations was not widely taken up by growers and, as a consequence, potato cyst nematodes in the form of *G. pallida* have become more abundant and widespread within the UK. The following sections discuss three new technologies that have attracted considerable attention for the improvement of nematode management but, as yet, remain untried in the field.

NEMATODE DIAGNOSTICS

Resistant cultivars, biological control agents and other sustainable methods of nematode control are frequently useful only against particular species, races or pathotypes of nematode pests. Therefore, their effective deployment is dependent upon the accurate identification and quantification of the nematode pest present. There is a shortage of skilled nematode taxonomists even within western Europe and the United States, where work on the identification of plant parasitic nematodes has not attracted sufficient funds over the past 20 years to maintain expertise. Hence, at a time when there are considerable demands from industry and the consumer for more sensitive management methods, a key component of such strategies is in short supply. Therefore, there is an urgent need to train more nematode taxonomists and to develop simple diagnostic methods.

Table 1. The impacts of control methods on the population densities of *Globodera rostochiensis* and *Globodera pallida* in systems of integrated nematode management. The effect of each factor (F) is presented as the proportion of the population that remains after the factor is used.

$$G. \ rostochiensis \ \ F_o = 0.67 \ \ F_r = 0.25 \ \ F_{s/n} = 5$$

$$Pf_{84} = Pi_{100} \ (0.67)^1 \ (0.25)^1 \ (5)^1$$

$$G. \ pallida \ \ F_o = 0.8 \ \ F_r = 2 \ \ F_{s/n} = 20 \ \ F_t = 0.2 \ \ F_f = 0.2$$

$$Pf_{82} = Pi_{100} \ (0.8)^3 \ (2)^1 \ (20)^1 \ (0.2)^1 \ (0.2)^1$$

F_o = nonhost crop; F_r = resistant cultivar; $F_{s/n}$ = susceptible cultivar treated with nematicide; F_t = trap crop; F_f = fumigation; Pf = final population density (eggs g^{-1} soil); Pi = initial population density (eggs g^{-1} soil).

Although biochemical, immunological and molecular methods provide opportunities for simple, reliable and rapid diagnostic tests, few are in routine use outside research laboratories. An understanding of the biodiversity of soil nematode communities and the impact of releases of biological control agents, including entomopathogenic nematodes, on these communities is likely to remain dependent upon traditional taxonomy for the foreseeable future. Morphological characteristics, however, are not reliable for the separation of some species or races and pathotypes, for which other methods appear more promising.

Enzyme phenotypes have been recognized on electrophoresis gels and can be used to separate proteins into characteristic bands visualized by staining. For example, *Meloidogyne arenaria*, *M. incognita*, *M. javanica* and *M. hapla* can be routinely distinguished by their esterase and malate dehydrogenase phenotypes (11). Diagnostic proteins have also been identified for the separation of the two species of potato cyst nematodes (13). Such methods are not quantitative, are relatively insensitive and depend on careful standardisation in the laboratory. Hence, they have only limited application.

The development of immunological methods for identification of nematodes has concentrated on the use of monoclonal antibodies (MAbs) because of their greater potential for discrimination than that of polyclonal antisera (PAbs). However, PAbs tend to be more sensitive than MAbs and may be of value for the identification of certain genera especially in support of quarantine and certification schemes in which the detection level is critical. Some cyst and root-knot nematode species (9,31) have been identified using MAbs which have been incorporated in enzyme-linked immunosorbent assays (ELISA) to enable the quantification of these nematodes in roots (8) and from soil samples (12). These assays require laboratory facilities but simple dipstick assays using antibodies may be used to detect the presence of specific pests in the field. Methods using MAbs are potentially much more sensitive than biochemical protein assays and populations of potato cyst nematodes of about 1 to 5 eggs g^{-1} soil have been reliably detected (12). Careful selection of MAbs allows the two species of potato cyst nematodes to be distinguished and quantified by immunoassay (Figure 2). ELISA techniques are relatively cheap and may be partially automated. Hence, there is considerable potential to use such techniques for the routine assessment of populations of nematodes in soils, with the added likelihood that MAbs can be selected for their ability to recognize only live nematodes. MAbs are already available that distinguish *G. pallida* from *G. rostochiensis* and enable quantification of each species in mixed populations (12). If proved successful in a range of soils, these methods may supersede the use of more traditional methods and, because they are much faster, would enable more intensive sampling of fields. Such improvements in techniques may lead to the mapping of

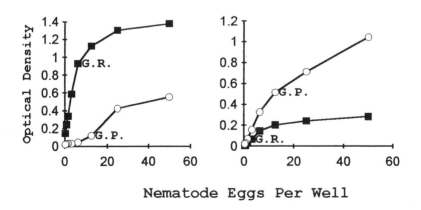

Figure 2. Differentiation and quantification of *Globodera pallida* (G.P.; –O–) and *G. rostochiensis* (G.R.;–■–) using two different MAbs in ELISA.

nematode distribution in fields and the application of nematicides to the patches where populations exceed damage thresholds; it has been predicted that targeted rather than broadcast treatments would reduce nematicide applications for potato cyst nematode control by 60%.

The separation of second stage juveniles of *G. rostochiensis* and *G. pallida* using morphological characters is difficult and often unreliable, whilst observation of differences between the two species in color changes during the maturation of adult females is dependent on critical timing. More reliable tests to separate the two species, based on the use of differential hosts, take at least 10 weeks to perform. Although immunological methods such as ELISA may be sufficiently sensitive to detect small (<1 egg g^{-1} soil) nematode infestations of either species in soil, they are unlikely to enable the separation of the pathotypes within species, which is essential for the effective management of resistant cultivars. For this, molecular techniques may be required.

Most recent research on new methods for nematode identification, especially in studies of phylogeny, has concentrated on the use of molecular technologies. Several analytical techniques to distinguish DNA sequences in different nematode species have been developed (7). The approach used depends on the needs of the study and the amount of test material available. The cleavage of DNA at specific recognition sites by an endonuclease may give rise to characteristic restriction fragment length polymorphisms (RFLPs) which can be separated on a gel and used to distinguish nematode species. Considerable quantities of DNA are required and so many nematodes or highly repetitive DNA sequences must be used. Specific DNA probes that hybridize in controlled conditions with samples of denatured (single-stranded) DNA have been used in dot blot assays. As a 5% sequence divergence is likely to prevent binding of the probe, such assays are sensitive, especially when highly repetitive sequences, such as occur in rDNA and mtDNA, are used. The use of the polymerase chain reaction (PCR) enables the DNA from a single nematode to be amplified in the laboratory for analysis, although the selection of suitable primers that flank the sequence to be increased is critical. Molecular diagnostics is a rapidly evolving technology and initial difficulties caused by the need to use radioisotopes to detect the small quantities of DNA available and the lack of quantitative procedures have now been largely, if not completely, overcome. However, the procedures for diagnostic tests are not as well developed as those using antibodies and are likely to remain laboratory based. Although DNA probes for the separation of nematode pathotypes have been reported (5), they have not been tested in studies of population genetics in the field. Determination of changes in the proportion of specific pathotypes in the nematode population, after cropping with different resistant cultivars, will require probes closely linked to the virulences genes. As several virulence genes, possibly on different chromosomes, are likely to be involved this is not a trivial task and the production of useful probes for pathotypes or races of any nematode will take much painstaking research. If successful, DNA probes could provide, for the first time, methods to manage effectively the development of virulence in field populations of nematodes and the deployment of appropriate resistant cultivars.

TRANSGENIC PLANTS FOR NEMATODE CONTROL

Most plant parasitic nematodes remain active after their emergence from eggs and repeatedly move to new feeding sites after they have killed the plant cells on which they have fed. Nematodes such as cyst and root-knot nematodes, which have sedentary stages in roots, induce the formation of complex feeding cells from which they derive nutrition. These nematodes cannot develop if they fail to induce these feeding cells. The cells are metabolically very active and are so modified to enable the rapid transfer of nutrients from the conducting tissue in the root to the nematode. These modified cells are similar to transfer cells which occur in healthy plants where nutrients must be transported rapidly. Hence, to support their development, these nematodes appear to regulate plant genes (3) by interactions that are poorly understood. Also, there are rather few sources of resistance traits that have been successfully utilized in commercial crops. As a consequence, many laboratories are examining the potential of molecular techniques to produce transgenic plants that express foreign genes that are toxic to nematodes or inhibit their development.

Several general approaches for the production of transgenic plants with novel resistances to nematodes are being examined (6). Firstly, the characterization of known resistance genes, such as the Mi resistance gene to root-knot nematodes (43) and the H_1 gene against *G. rostochiensis*, is well advanced (28). Once identified, these genes could be transferred to susceptible cultivars to confer resistance. Such an approach would avoid the need for long term backcrossing programs to remove unwanted traits associated with the resistance source. However, natural genes conditioning resistance to nematodes have been identified for only a few nematode species and, in all situations where these genes have been deployed, virulent populations capable of reproducing on "resistant" plants have developed. In several plant species, nematode resistance is polygenic, so the characterization of several genes may be necessary and, inevitably, will be time consuming.

The introduction of toxic genes such as those coding for Bt toxins or the expression of inhibitors to key enzymes such as proteinases have provided resistance to some insect pests and may be effective against nematodes. The expression of a proteinase inhibitor in potato roots did not prevent the establishment of *G. pallida* or *M. incognita* but the fecundity of adult females was reduced (1). Such an approach could give rapid results and may provide opportunities to develop plants with resistance to migratory nematodes. However, these genes may, depending on the promoters used, be overexpressed and as a consequence cause yield penalties in transgenic plants. There is, therefore, a need to identify specific promoters that limit the expression of introduced genes. Some promoters specific to roots and even individual nematode-feeding cells have been found (14). At least one novel gene providing resistance to root-knot nematodes has been transferred into a susceptible host and the transgenic plants have been field tested (25). So far, few nematicidal genes have been identified. These methods will be especially important for transferring nematode resistance to preferred local cultivars, if selected genes can be inserted without altering other plant characteristics.

As well as the expression of nematicidal genes, selected antibodies may be expressed within plants to inhibit the function of key proteins in the host parasite relationship (6). Salivary proteins have again been selected as susceptible to this approach for nematode control (18) but the feeding tubes considered essential for several sedentary nematodes and interference with the sites associated with host recognition may be other useful targets.

The use of antisense techniques to down-regulate genes in the feeding cells of sedentary nematodes is also being considered (6). Antisense RNA has been used widely to affect the regulation of plant genes. Such an approach is dependent on the identification of key genes critical to the relationship between the nematode and the host plant; it is assumed that the down-regulation of these genes will affect the development of the nematode and prevent or significantly reduce the number of females reaching maturity. Antisense techniques are impractical if there are several genes that must be down-regulated to have an effect. Also, the regulation should be confined to the site of action and not affect other cells and cause general changes in growth. Nematode-feeding cells, which are metabolically very active, are likely to contain the products of up-regulated genes whose down-regulation by antisense constructs could have marked effects on the nematode without influencing normal cells.

The use of feeding cell-specific promoters with these constructs would increase their efficacy. The use of the differential display technique (2) is a powerful tool for the identification of feeding cell-specific genes.

Molecular biology has much to offer plant nematologists through the development of plants with novel forms of resistance. Resistant plants are widely considered as the most practical and sustainable method for the control of many nematode pests. However, as experience with the potato cyst nematodes has demonstrated, there is a need to manage resistance genes with care and to combine their use with other control methods.

BIOLOGICAL CONTROL AGENTS

Several microorganisms have been considered as potential biological control agents for nematodes (Table 2). Although some fungi and bacteria have been commercialized, none has proved successful; the control achieved has been inconsistent and the necessary application rates have frequently been impractical for use on field crops. For control of migratory nematodes that remain active throughout their life, agents must develop adhesive spores or specialist traps to infect their hosts. Nonspecialist fungi in the rhizosphere are able to colonize only sedentary nematode stages. Rhizosphere bacteria may reduce the attractiveness of roots to nematodes by the alteration of root exudates, by the production of toxins that are directly nematicidal, or by the induction of resistance in roots to nematode invasion (17). Some bacteria, such as *Pseudomonas fluorescens*, have significantly reduced invasion by the beet cyst nematode, *Heterodera schachtii* (27), but protection has tended to be short-lived and no tested bacteria have prevented nematode populations from increasing. Hence, these bacteria may provide immediate but brief protection. In contrast, parasitic organisms have the potential to reduce both damage and nematode multiplication but tend to be slower acting.

Recently, endophytic fungi, including mycorrhizae, that colonize plant roots have been considered as potential biological control agents. These fungi may produce toxins or compete for space with endoparasitic nematodes and thereby significantly reduce nematode infestations (19,32). They have the advantage that they are able to reduce populations of both migratory and sedentary nematodes. These fungal isolates appear to be nonpathogenic to the plant and are often numerous in root tissue. However, in most cases the species involved are closely related to plant pathogens and may even be isolates of plant pathogenic species such as *Fusarium oxysporum*. They may therefore be difficult to register for mass release in soil.

Table 2. Some potential biological agents for the control of plant parasitic nematodes.

Type of agent		Species tested
Obligate parasites	-Bacteria	*Pasteuria penetrans*
	-Fungi	*Hirsutella rhosilliensis, Drechmeria coniospora*
Facultative parasites		*Paecilomyces lilacinus, Verticillium chlamydosporium,* Nematode-trapping fungi, e.g. *Arthrobotrys oligospora*
Endophytic fungi		*Fusarium oxysporum, Cylindrocarpon destructans,* Mycorrhizae
Rhizosphere bacteria		*Agrobacterium radiobacter, Bacillus sphaericus, B. subtilis, Pseudomonas fluorescens*
Soil fungi		*Trichoderma harzianum, Gliocladium virens*

Pasteuria spp. are obligate bacterial parasites producing adhesive spores that remain dormant in soil until they attach to the cuticle of passing nematodes. *Pasteuria penetrans* infects *Meloidogyne* spp. when spores that attach to the cuticle of second stage juveniles in soil germinate after the nematode has invaded a host root. The bacterium proliferates inside the developing nematode but apparently does not produce toxins because the nematode is not killed until it is adult. Infected female nematodes usually produce no eggs and the body is filled with about 2×10^6 spores. The spores are released into the soil when the nematode degenerates and the cycle is repeated. Populations of this bacterium are highly specific in their host range. Consequently, the spores are resistant and capable of surviving long periods in the absence of suitable hosts. In addition, populations differ in their virulence and careful selection of the

Table 3. Integrated control of *Meloidogyne* spp. on cucumber in Crete using solarization, a nematicide (oxamyl), and the bacterium *Pasteuria penetrans*.

Solarization	*P. penetrans*	Oxamyl +	Oxamyl −	Mean
+	+	356	553	455
+	−	294	875	585
−	+	1,206	794	1,000
−	−	1,496	2,844	2,157
	Mean	831	1,267	

Values are eggs g^{-1} soil after harvest. Data from Tzortzakakis and Gowen (37) used by permission from Butterworth-Heinemann journals, Elsevier Science Ltd.

bacterium is essential to ensure that appropriate strains are used against particular nematode species (10). The spores can be stored dry without significant loss of viability, which greatly eases the handling of the organism. The bacterium has not been cultured in vitro (4,42) and spores have to be produced in vivo (35) using methods that are labor intensive and unsatisfactory for large-scale production of inoculum.

Pasteuria penetrans in a naturally infested soil controlled *M. arenaria* on peanut in Florida (26) and the bacterium has considerable potential as a biological control agent. However, commercial use will largely depend on the development of a simple method of mass production. Spores produced in vivo were applied at rates of 2×10^4 to 5×10^4 spores g^{-1} soil to give significant control of *M. javanica* on tomato (33). However, the integration of *P. penetrans* with other control methods (Table 3) such as the nematicide, oxamyl, and pretreatment of the soil by solarization was necessary for effective control of *Meloidogyne spp.* on cucumber (37).

Nematophagous fungi as well as *P. penetrans* may increase to densities that eventually control specific pest nematodes including root-knot and cyst nematodes (20,34). For example, the cereal cyst nematode has been effectively controlled for the past 20 years in many soils in northern Europe by parasitic fungi (mainly *Nematophthora gynophila* and *Verticillium chlamydosporium*) that kill developing female nematodes on the roots of susceptible crops. Thus, these nematophagous fungi have provided sustainable control of a damaging nematode pest in an intensive cropping system. This natural regulation of the nematode is induced and is dependent on an initial period when the nematode host is prevalent and able to support the increase of the fungi. Hence, other control methods would be required to prevent major yield losses for the four to five seasons that it takes for the fungi to become effective. The addition of organic matter may enhance the activity of the nematophagous microflora in soils and increase nematode control (30) but it has proved difficult to manipulate these organisms with treatments that are practical for field crops.

Paecilomyces lilacinus is a facultative parasite of nematode females and eggs and is the most extensively field-tested biological control agent for nematodes. It is commercially produced as Bioact in the Philippines. In general, control has been variable and benefits from applications of this fungus have often been difficult to measure (21). Environmental conditions presumably have a significant effect on the efficacy of *P. lilacinus* and relatively large dose rates (1×10^6 spores g^{-1} soil) are needed for nematode control (15). There is a need for basic information on the epidemiology of the fungus and the key factors that affect its efficacy. Such information is being collected for *V. chlamydosporium*, a nematophagous fungus with a similar mode of action to *P. lilacinus*.

Verticillium chlamydosporium parasitizes cyst and root-knot nematode eggs and females in many tropical and temperate soils. As with *P. penetrans*, improved methods for the mass production of inoculum are required. Few chlamydospores, the most effective propagules for establishment of the fungus in soil, are produced in liquid media, which is the favored method for commercial production (22). Ability to grow in the rhizosphere differs markedly between fungal isolates (Table 4) and plant species but is essential for effective nematode control (23). The fungus rapidly colonizes the egg masses of root-knot nematodes and the females of cyst nematodes developing in the rhizosphere and parasitizes their eggs. Isolates of the fungus also differ in their pathogenicity and in their ability to produce chlamydospores. Hence, there is a need to screen isolates carefully for their potential as biological control agents. A semiselective medium has been developed which, combined with standard dilution plating techniques, may be used to monitor the distribution and changes in abundance of the fungus in soil and the plant rhizosphere.

Table 4. Differences between isolates of *Verticillium chlamydosporium* in their ability to grow in the rhizosphere of tomato plants and control *Meloidogyne arenaria*.

| Treatment | Number of colony-forming units (CFU) | | Nematode eggs and juveniles g^{-1} soil | Infected nematode eggs (%) |
	log CFU g^{-1} soil	$\times$ 10^3 CFU cm^{-2} root		
Control	0	0	137	0
Isolate 10	6	37	29	32
Isolate 35	5	0.1	124	0
Isolate 43	5	0.3	164	0
$\pm$ SED	0.3	8.0	9.4	

It is difficult, however, to assess the amount of fungal growth accurately because hyphae, conidia or chlamydospores will all give rise to similar colonies on the dilution plates. There is a need for an alternative method, such as the use of MAbs, to estimate the extent of hyphal growth in the rhizosphere but the amount of fungus on the roots (estimated from dilution plates) is usually related to the level of nematode control. However, if large galls are produced in response to nematode attack, these prevent the egg masses being exposed to fungal parasitism in the rhizosphere and a significant proportion of eggs remain embedded in roots and escape attack. Hence, effective nematode control with *V. chlamydosporium* is largely dependent on the combined use of the fungus with tolerant or poor hosts for the nematode or with other treatments which will decrease infestations in soil; these treatments will limit gall size and should maximize the number of egg masses exposed to the fungus and the control achieved. A single application (5 $\times$ 10^3 chlamydospores g^{-1} soil) of *V. chlamydosporium* at planting has provided effective control of small infestations of *M. hapla* on tomato plants (24) but more widespread field testing is necessary to assess the conditions in which it is effective and its potential as a commercial biological control agent.

CONCLUDING REMARKS

Although there are several exciting possibilities for improving the control of nematode pests, their management is likely to continue to rely on the integration of control methods including the judicious use of nematicides where appropriate. The new technologies described must be rigorously tested in the field before they are recommended to growers. Experience with potato cyst nematodes in the UK has clearly indicated the need for a detailed understanding of the biology and ecology of pest species in the development of sustainable management strategies and also the importance of grower support for such strategies.

LITERATURE CITED

1.   Atkinson, H.J., and Koritsas, V.M. 1993. Development of a novel strategy for nematode control based on proteinase inhibition. Page 192 in: Abstracts, 6th International Congress of Plant Pathology, Montreal, Canada. National Research Council of Canada, Ottawa, Canada.

2.   Bertioli, D.J., Schlichter, U.H.A., Adams, M.J., Burrows, P.R., Steinbiss, H.-H., and Antoniw, J.F. 1995. An analysis of differential display shows a strong bias towards high copy number mRNAs. Nucleic Acids Res. 23:4520–4523.

3.   Bird, D.McK. 1992. Mechanisms of the *Meloidogyne*-host interaction. Pages 51–59 in: Nematology from Molecule to Ecosystem. F.J. Gommers and P.W.Th. Maas, eds. Dekker and Huisman, Wildervank, The Netherlands.

4.   Bishop, A.H., and Ellar, D.J. 1991. Attempts to culture *Pasteuria penetrans in vitro*.   Biocontrol Sci. Technol. 1:101–114.

5.   Burrows, P.R. 1990. DNA hybridisation probes to identify pathotypes of *Globodera rostochiensis* and *G. pallida*. Nematologica 36:336–337. (Abstr.).

6.   Burrows, P.R., and Jones, M.G.K. 1993. Cellular and molecular approaches to the control of plant parasitic nematodes. Pages 609–630 in: Plant Parasitic Nematodes in Temperate Agriculture. K. Evans, D.L. Trudgill, and J.M. Webster, eds. CAB International, Wallingford, UK.

7.   Curran, J. 1992. Molecular taxonomy of nematodes. Pages 83–91 in: Nematology from Molecule to Ecosystem. F.J. Gommers and P.W.Th. Maas, eds. Dekker and Huisman, Wildervank, The Netherlands.

8.   Curran, J., and Robinson, M.P. 1993. Molecular aids to nematode diagnosis. Pages 545–564 in: Plant Parasitic Nematodes in Temperate Agriculture. K. Evans, D.L. Trudgill, and J.M. Webster, eds. CAB International, Wallingford, UK.

9.   Davies, K.G., and Lander, E.B. 1992. Immunological differentiation of root-knot nematodes (*Meloidogyne* spp.) using monoclonal and polyclonal antibodies. Nematologica 38:353–366.

10.   Davies, K.G., Leij, F.A.A.M. de, and Kerry, B.R. 1991. Microbial agents for the biological control of plant-parasitic nematodes in tropical agriculture. Trop. Pest Manage. 37:303–320.

11.   Esbenshade, P.R., and Triantaphyllou, A.C. 1990. Isozyme phenotypes for the identification of *Meloidogyne* species. J. Nematol. 22:10–15.

12.   Evans, K., Curtis, R.H., Robinson, M.P., and Yeung, M. 1995. The use of monoclonal antibodies for the identification and quantification of potato cyst nematodes. EPPO Bull. 25:357–365.

13.   Fleming, C.C., and Marks, R.J. 1983. The identification of the potato cyst nematodes *Globodera rostochiensis* and *G. pallida* by isoelectric focusing of proteins on polyacrylamide gels. Ann. Appl. Biol. 103:277–281.

14.   Goddijn, O.J.M., Lindsey, K., Van der Lee, F.M., Klap, J.C., and Sijmons, P.C. 1993. Differential gene expression in nematode-induced feeding structures of transgenic plants harbouring promoter-gusA fusion constructs. Plant J. 4:863–873.

15.   Gomes-Carneiro, R.M.D., and Cayrol, J.C. 1991. Relationship between inoculum density of the nematophagous fungus *Paecilomyces lilacinus* and control of *Meloidogyne arenaria* on tomato. Rev. Nématol. 14:629–634.

16.   Hancock, M. 1988. The management of potato cyst nematodes in UK potato crops. Aspects Appl. Biol. 17:29–36.

17. Hasky, K., and Sikora, R.A. 1994. Induced resistance—a mechanism induced systemically throughout the root system by rhizosphere bacteria towards the potato cyst nematode *Globodera pallida*. Page 67 in: Proceedings, 22nd International Symposium of the European Society of Nematologists, 7–12 August 1994, Gent, Belgium.

18. Hussey, R.S. 1992. Secretions of oesophageal glands in root-knot nematodes. Pages 41–50 in: Nematology from Molecule to Ecosystem. F.J. Gommers and P.W. Th. Maas, eds. Dekker and Huisman, Wildervank, The Netherlands.

19. Jatala, P. 1986. Biological control of plant-parasitic nematodes. Annu. Rev. Phytopathol. 24:453–489.

20. Kerry, B.R. 1987. Biological control. Pages 233–263 in: Principles and Practice of Nematode Control in Crops. R.H. Brown and B.R. Kerry, eds. Academic Press, Sydney, Australia.

21. Kerry, B.R. 1990. An assessment of progress toward microbial control of plant-parasitic nematodes. Ann. Appl. Nematol. 22:621–631.

22. Kerry, B.R., Irving, F., and Hornsey, J.C. 1986. Variation between strains of the nematophagous fungus, *Verticillium chlamydosporium* Goddard. I. Factors affecting growth *in vitro*. Nematologica 32:461–473.

23. Leij, F.A.A.M. de, and Kerry, B.R. 1991. The nematophagous fungus *Verticillium chlamydosporium* as a potential biological control agent for *Meloidogyne arenaria*. Rev. Nématol. 14:157–164.

24. Leij, F.A.A.M. de, Kerry, B.R., and Dennehy, J.A. 1993. *Verticillium chlamydosporium* as a biological control agent for *Meloidogyne incognita* and *M. hapla* in pot and micro-plot tests. Nematologica 39:115–126.

25. Opperman, C.H., Taylor, C.G., and Conkling, M.A. 1994. Root-knot nematode-directed expression of a plant root-specific gene. Science 263:221–223.

26. Oostendorp, M., Dickson, D.W., and Mitchell, D.J. 1991. Population development of *Pasteuria penetrans* on *Meloidogyne arenaria*. J. Nematol. 23:58–64.

27. Oostendorp, M., and Sikora, R.A. 1989. Seed treatment with antagonistic rhizobacteria for the suppression of *Heterodera schachtii* early root infection of sugar beet. Rev. Nématol. 12:77–83.

28. Pineda, O., Bonierbale, M.W., Plaisted, R.L., Brodie, B.B., and Tanksley, S.D. 1993. Identification of RFLP markers linked to the H_1 gene conferring resistance to the potato cyst nematode *Globodera rostochiensis*. Genome 36:152–156.

29. Robinson, M.P., Atkinson, H.J., and Perry, R.N. 1987. The influence of temperature on the hatching, activity and lipid utilization of second stage juveniles of the potato cyst nematodes, *Globodera rostochiensis* and *G. pallida*. Rev. Nématol. 10:349–354.

30. Rodriguez-Kabana, R., Morgan-Jones, G., and Chet, I. 1987. Biological control of nematodes: soil amendments and microbial antagonists. Plant Soil 100:237–247.

31. Schots, A., Gommers, F.J., Bakker, J., and Egberts, E. 1990. Serological differentiation of plant-parasitic nematode species with polyclonal and monoclonal antibodies. J. Nematol. 22:16–23.

32. Sikora, R.A. 1992. Management of the antagonistic potential in agricultural ecosystems for the biological control of plant parasitic nematodes. Annu. Rev. Phytopathol. 30:245–270.

33. Stirling, G.R. 1984. Biological control of *Meloidogyne javanica* with *Bacillus penetrans*. Phytopathology 74:55–60.

34. Stirling, G.R. 1991. Biological Control of Plant Parasitic Nematodes. CAB International, Wallingford, UK. 282 pp.

35. Stirling, G.R., and Wachtel, M.F. 1980. Mass production of *Bacillus penetrans* for the biological control of root-knot nematodes. Nematologica 26:308–312.

36. Stone, A.R. 1987. Genetic systems in plant pathogenic nematodes and their hosts. Pages 101–109 in: Genetics and Plant Pathogenesis. P.R. Day and G.J. Jellis, eds. Blackwell Scientific Publications, Oxford, UK.

37. Tzortzakakis, E.A., and Gowen, S.R. 1994. Evaluation of *Pasteuria penetrans* alone and in combination with oxamyl, plant resistance and solarization for the control of *Meloidogyne* spp. on vegetables grown in greenhouses in Crete. Crop Prot. 13:455–462.

38. Whitehead, A.G. 1991. The resistance of six potato cultivars to English populations of potato cyst-nematodes, *Globodera pallida* and *G. rostochiensis*. Ann. Appl. Biol. 118:357–369.

39. Whitehead, A.G. 1991. Selection for virulence in the potato cyst-nematode *Globodera pallida*. Ann. Appl. Biol. 118:395–402.

40. Whitehead, A.G. 1992. Emergence of juvenile potato cyst-nematodes *Globodera rostochiensis* and *G. pallida* and the control of *G. pallida*. Ann. Appl. Biol. 120:471–486.

41. Whitehead, A.G., Nichols, A.J.F., and Senior, J.C. 1994. The control of potato pale cyst-nematode (*Globodera pallida*) by chemical and cultural methods in different soils. J. Agric. Sci. Cambridge 123:207–218.

42. Williams, A.B., Stirling, G.R., Hayward, A.C., and Perry, J. 1989. Properties and attempted culture of *Pasteuria penetrans*, a bacterial parasite of root-knot nematode (*Meloidogyne javanica*). J. Appl. Bacteriol. 67:145–156.

43. Williamson, V.M., Ho, J-Y, and Ma, H.M. 1992. Molecular transfer of nematode resistance genes. J. Nematol. 24:234–241.

Chapter 8

BIOLOGY OF *AGROBACTERIUM* AND MANAGEMENT OF CROWN GALL DISEASE

L.W. Moore and M. Canfield

INTRODUCTION

An extensive research program on crown gall in the Pacific Northwest has been conducted at Oregon State University for the past 20 years. Special emphasis has been focused on management of the disease in woody nursery crops. Understanding of the population diversity and dynamics of *Agrobacterium* is considered a critical factor in this management process. Consequently, numerous *Agrobacterium* strains have been isolated from a large variety of plant hosts and habitats. Characterization of hundreds of these isolates revealed wide diversity among strains from the various plant hosts and planting sites. Failure to recognize *Agrobacterium* diversity leads to studies that include only a few strains. Care must be exercised to avoid making unwarranted extrapolations about natural conditions based on such small samplings lest they result in unsound assumptions and generalizations about the biology of agrobacteria. The importance of these considerations is reinforced by the recent plea for greater attention to the role and significance of microbial diversity in the health and maintenance of plant and animal species (42,83,109).

To assist our investigations into the diversity and structure of *Agrobacterium* populations, we have assembled a large culture collection (over 3,000 strains) and developed a comprehensive data base containing hundreds of well characterized strains from a variety of hosts and habitats. The practical importance of these data for disease control has already been mentioned, but added benefits include an enhanced understanding of ecological relationships among strains within the genus and between agrobacteria and other microorganisms. Included in these ecological considerations is the impact of genetic exchange on the population diversity and dynamics of *Agrobacterium*, and whether these population factors play a role in stabilization or destabilization of microbial community structure.   Of special interest in crown gall research is whether the planting site or plant host might impose an ecological selection pressure for particular populations of agrobacteria.   For example, in plant breeding programs where single gene resistance has been utilized, it is recognized that the host can influence the genetic diversity and population structure of the pathogen.   We are interested in whether the host can play a similar, though more subtle, role in the population structure of *Agrobacterium*.

As a prelude to further discussion of *Agrobacterium* ecology, some background information is provided to aid in the understanding of the taxonomy and classification of agrobacteria, events involved in the infection process, and methodology used for diagnosis and identification of field isolates. This information establishes a framework around which the remaining elements of the chapter are crafted.

BACKGROUND

Taxonomy And Classification

Historically, *Agrobacterium* species were described as *A. tumefaciens* or *A. radiobacter* depending upon their pathogenicity. A third taxonomic group was called *A. rhizogenes* because the infected hosts produced excessive roots rather than tumors, and *A. rubi* was established because of its association with *Rubus* galls (56). Thus, early taxonomy was based on the presence or absence of a host response rather than a characterization of the bacterium. Since then, elegant experiments have demonstrated that the infectious properties of the pathogens are based on plasmid-encoded genes which can be lost or transferred between strains (51). Because of the potential instability of plasmids, classification based on chromosomal genes is obviously preferable. The most widely used classification divides agrobacteria into three biovars based on their behavior in biochemical and physiological tests (14,56,73). There is now good evidence to support raising the three biovars to species status, with the names *A. tumefaciens* for biovar 1, *A. rhizogenes* for biovar 2 and *A. vitis* for biovar 3 based on priority of the first described organisms in each group (48,80). Both pathogenic and nonpathogenic forms are present in each of these "new" species. Although this scheme is based on sound biological differences, it is flawed because the species names *A.*

tumefaciens and *A. rhizogenes* have been used historically to name pathogens. To avoid confusion, biovar designations of *Agrobacterium* will be used throughout the chapter and individual strains referred to as pathogens or nonpathogens.

Tumor-inducing Plasmids

There have been extraordinary advances in our knowledge of the molecular events associated with infection of plants by *Agrobacterium* over the past two decades. These advances were stimulated by the discovery that infection occurs when genes from a tumor-inducing plasmid (pTi) carried by the pathogen are transferred to the host plant's genome. Genes within the transferred region (T-DNA) encode enzymes required for synthesis of plant auxin and cytokinin. Overexpression of the plant hormonal genes results in uncontrolled cell division and enlargement, leading to the tumor phenotype. Another region of the transferred DNA codes for the synthesis of opines, condensation products of organic acids and amino acids or disaccharides. The opines synthesized within tumors are available for agrobacteria as carbon and nitrogen sources. Genes for synthesis of enzymes used to degrade opines are also encoded on the pTi. There are substantial differences in the gene sequences between different Ti plasmids, e.g. those sequences delimiting host range, and synthesis and degradation of different opines. The virulence (vir) genes compose a second gene cluster on the pTi that are involved in the early interactions between plant and pathogen and include: (i) the pathogen's detection of a wound environment; (ii) specific surface interactions; and (iii) pTi DNA processing, transfer, and integration into the host genome (51,84,113).

DETECTION AND CHARACTERIZATION OF AGROBACTERIA

One problem in understanding the diversity within agrobacteria has been the inability to detect pathogens in their natural habitat. Ecological studies are aided by the ability to track the fate of organisms released into the environment but such studies have been limited by the lack of sensitive, specific detection methods. To investigate the potential for gene transfer between pathogenic agrobacteria and the nonpathogenic strain K84 in the environment, several regions of pTi DNA and the *Sma*IG DNA from pAgK84 were evaluated for use as ^{32}P-labeled probes (74). These DNA probes were screened for specificity and ability to discriminate between pathogens and nonpathogens. For detection of pathogens, *virFAB* from the virulence region (97) and the *iaa*M and *iaa*H (*tms1-tmr*) genes from T-DNA of pTi A6 (43) were the most useful. Of 108 pathogens tested in the initial screening for specificity, all hybridized to the *vir* probe and 84% hybridized to the T-DNA probe. None of the nonpathogens hybridized to either probe, including K84. DNA from an agrocin synthesis gene, *Sma*IG (39), hybridized to K84 but to no other pathogens or nonpathogens tested (74).

The specificity of the *virFAB* and *iaa*M*iaa*H probes was further tested on non-*Agrobacterium* soil-inhabiting bacteria. Twenty-five different strains of soil bacteria, including species of *Pseudomonas, Erwinia, Enterobacter, Alcaligenes, Streptomyces*, and *Bradyrhizobium*, were tested by colony hybridization. None of the 25 strains of soil bacteria hybridized with *virFAB* (74). Additional testing with digoxygenin-labeled probes (67) has since been conducted on 359 pathogenic strains representing all three biovars from 47 different plant hosts; over 90% of the strains hybridized to the probes while 116 nonpathogens tested did not (64). These probes are superior to others reported in the literature. The radioactive ^{32}P-labeled probes pTHE17 (containing several T-DNA regions, and derived from *Agrobacterium* strain C58) and LBA4404 (containing virulence region genes, and derived from *Agrobacterium* strain Ach5) hybridized to the DNA from 28 pathogens tested, but also hybridized to DNA from two of 22 nonpathogens, K84 and HLB-2 (18). Auxin and cytokinin genes *tms1tms2* and *tmr* from pTi A6 did not discriminate well between pathogens and nonpathogens isolated from grape (34,87).

Several methods have been evaluated for isolating *Agrobacterium* from tumors, the rhizosphere, and soil. The most effective have been the use of semiselective media such as Kerr 1A and 2E (15) for biovars 1 and 2, respectively, and Roy and Sasser's medium (91) for biovar 3. Nonselective media such as modified mannitol glutamate (MG) (52) have been used to isolate unusual strains and new biovars even though competition from other bacteria, e.g. pseudomonads, can be a problem. Opines in modified MG proved helpful for isolating pathogens from apple tumors (21,23).

A monoclonal antibody specific to biovar 3 strains was effective for detection of these bacteria from grape. However, the antibody does not distinguish between tumorigenic and nontumorigenic strains. It is still a very good diagnostic tool due to the fact that all biovar 3 strains, whether tumorigenic or not, produce a polygalacturonase that results in root necrosis (16,90). Strain-specific polyclonal antisera to lipopolysaccharide (LPS) moieties of *Agrobacterium* (11) were helpful in tracking a specific strain in the environment, but there is as yet no general antiserum for detection of biovars 1 and 2. One polyclonal antibody that reacted with all agrobacteria strains tested also reacted with *Rhizobium* species (8).

The majority of agrobacteria isolated from soil, the rhizosphere, and occasionally from tumors, are nonpathogenic (23,53); hence, DNA probes have proven to be extremely efficient in screening for less numerous pathogenic strains. To verify that probe-positive strains are pathogenic, isolates that hybridize to digoxygenin-labeled *virFAB* and *iaaMiaaH* are typically inoculated to tomato plants or the host of origin. More than 4,000 strains isolated from 12 different hosts during 1993 to 1995 were assayed by these methods. Over 95% of the strains that hybridized to the probes were pathogenic when tested on plants while none of the probe-negative strains were pathogenic (M. Canfield and L.W. Moore, unpublished). However, even with the positive relationship between hybridization and pathogenicity, some confirmatory plant inoculations are advisable. For example, a strain with mutations within the pTi could react positively with the probes but not be pathogenic. Use of colony hybridization as a screening assay for pathogenicity has eliminated much of the labor-intensive plant inoculations that were needed in the past because probe-negative strains are no longer tested. In some cases, over 90% of isolates from tumors and 100% from soil are nonpathogens.

Another important procedural advance in evaluating the pathogenicity of agrobacteria has been the development and use of micropropagated woody plants in tissue culture as assay hosts. There are several advantages of using micropropagated plants. Because the plants are axenic, there is more certainty that symptoms observed are a direct result of the inoculated strain and hundreds of plantlets can be maintained year-round in a very small area. Agrobacteria that are host specific or have a narrow host range can be easily tested on the host of origin if it is available in tissue culture. Tumors are detectable as early as 2 weeks after inoculation of woody plant species in tissue culture (Figure 1). This contrasts sharply with inoculations on greenhouse-grown woody plants where tumors may take up to 6 months to develop.

Armed with these improved detection methods, standardized procedures for isolation and characterization of agrobacteria have been developed. For each reported case of crown gall, at least five different infected plants are collected from the same site. Isolations are made onto selective media and nonselective MG medium with 0.4 g liter^{-1} yeast extract, trace minerals (49) and 2% cycloheximide (referred to as MGYS plus cycloheximide medium). Colonies with the appearance and morphology of agrobacteria, i.e. glossy, opaque, domed, and mucoid with entire margins, are selected and restreaked at least once onto MG plus 1 g liter^{-1} yeast extract (MGY) before spotting colonies onto duplicate nylon membranes. DNA from the lysed colonies is bound to the membrane and then incubated with digoxygenin-labeled *virFAB* and *iaaMiaaH* DNA. At least 100 colonies per membrane can be screened in this way. Those colonies that hybridize to the probes are considered to be pathogenic. To confirm pathogenicity, a subgroup of all those that react positively with The probes are further purified and inoculated to tomato plants or tissue culture plantlets.

Once pathogenicity has been determined, variability within a population can be examined by testing for opine utilization and sensitivity to various antagonistic organisms. However, these characteristics, which are usually due to pTi genes, are not useful for taxonomic identification because they

Figure 1. Wild blackberry plantlet in tissue culture showing tumor development 1 month after inoculation with a wild blackberry strain, *Agrobacterium* B230/85.

can be exchanged between strains or lost entirely. Identification of strains, therefore, must be based on the more stable chromosomal traits.

Identification Of Strains

Standard biochemical and physiological tests, even though time consuming, are important steps in identifying unknown isolates (73). Other more rapid methods for identification of bacteria are carbon source utilization and fatty acid analysis, but these are much more expensive to conduct (10). All of these tests are important because they depend upon the expression of chromosomal genes. Several new DNA probes have been developed for detection and classification of bacteria, based on variations in conserved sequences of rRNA. Palleroni (83), however, emphasizes the following limitation to this strategy; the reliability of rRNA probes and other techniques depends upon the strains selected as sources of nucleic acids. If these strains are not representative of the population, then the results may be misleading. Palleroni emphasizes that biochemical versatility is what distinguishes bacteria. This can only be determined by physiological tests with live bacteria and not by the use of one particular probe. Studies of the whole cell are important, including the analysis of biochemical properties. The only way biochemical diversity can be determined is to examine a large number of bacteria from various hosts and habitats. The isolation and characterization methods that are used in our laboratory are consistent with Palleroni's precautions.

Adoption of the methods described above has considerably enhanced our ability to tackle investigations into the ecology of agrobacteria. These in turn can lead to a better understanding of the epidemiology of crown gall.

ECOLOGICAL CONSIDERATIONS

The explosion of molecular information about the infection process of *Agrobacterium* strains under controlled conditions contrasts sharply with the paucity of knowledge about the ecology of agrobacteria. Data from laboratory studies performed with pure cultures may not have validity within natural microbial communities (35). While several authors have addressed the need to study *Agrobacterium* ecology, relatively little effort has been made to investigate the population dynamics and diversity among species and strains of this genus in tumors and the soil.

There are several factors that may be important for survival and dissemination of agrobacteria in their natural habitats. Questions need to be raised concerning the survival advantages of pathogens within tumors compared to the rhizosphere and soil. The tumor environment might be advantageous for agrobacteria with a pTi due to the presence of opines.

Role Of Opines

Octopine and nopaline were the first opines to be discovered and described (105). Other opines discovered since then include mannopine, agrocinopine, succinamopine and vitopine (29,38,103). The presence of opines in plant tumors has led to speculation that these compounds are advantageous for pathogenic agrobacteria because of their usefulness as carbon and nitrogen sources (106). Recent studies have shown, however, that many other microbes such as *Pseudomonas* and coryneforms can also use opines (75,110). Pseudomonads that utilize opines have been isolated along with agrobacteria from apple tumors (21,23) suggesting that they can compete for opines with agrobacteria in tumors. Competition experiments were conducted in chemostats between octopine-utilizing agrobacteria and fluorescent pseudomonads under several different growth conditions. The *Pseudomonas* strains were able to outcompete the agrobacteria, suggesting that they would survive better than agrobacteria within tumors (5,6). In addition to nutrient competition, antagonism between strains could favor the survival of some strains. In petri plate assays, many *Pseudomonas* strains produced antibiotic-like compounds that inhibited the growth of agrobacteria isolated from the same tumors. Thus, antibiotic production would seem to provide the pseudomonads with an additional competitive advantage (23).

In the case where several strains capable of using different opines are present in the same tumor, the particular opine or opines synthesized could favor one species over another. Pseudomonads that used octopine and nopaline were isolated from a tumor along with agrobacteria that could use mannopine and nopaline. If the opines present in the tumor were octopine or nopaline, then pseudomonads might be favored. However, if the tumor synthesized mannopine, agrobacteria might be favored since no pseudomonads capable of utilizing mannopine have so far been discovered (23,76). Some opines are important in conjugal transfer of plasmids between strains of agrobacteria and this could lead to an increased number of pathogens within tumors. Investigations in the laboratory and in planta have demonstrated unequivocally that plasmid transfer occurs (44,54). The frequency of plasmid transfer can be very high in the presence of opines that act as conjugal inducers (36,37,85). Opines also induce production of a conjugal factor that greatly increases the frequency of plasmid transfer in vitro (115). Genetic exchange of plasmids among agrobacteria in nature could obviously lead to new genotypes that add to the diversity found among tumor populations.

The tumor environment differs markedly from that in the soil, where agrobacteria exist in a starved condition. The pTi may no longer confer a competitive advantage so that the nonpathogenic agrobacteria take on new importance in understanding the ecology and epidemiology of agrobacteria.

Nonpathogenic *Agrobacterium*

Among natural populations of *Agrobacterium*, with few exceptions, nonpathogenic agrobacteria outnumber pathogens in soil by ratios of 13:1 to 500:1 (53,93). Theoretically, the population of pathogenic agrobacteria in soil might increase during an epidemic year, thus increasing the risk of crown gall in subsequent years. Epidemics occur sporadically, however, and this sporadic occurrence coupled with a general inability to identify sources of primary inoculum, particularly in soil, has made it difficult to establish a relationship between populations of pathogenic agrobacteria in soil and disease incidence.

It is possible that the pTi reduces the fitness of pathogenic *Agrobacterium* in the soil environment. On the other hand, it may be only the genes directly involved in pathogenicity that reduce the fitness of strains outside the tumor habitat. Consequently, loss of the plasmid might increase fitness, or significant mutations in the plasmid could lead to a mutated pTi that confers a fitness advantage on its bacterium. If the argument that pathogenic agrobacteria are less ecologically fit is valid, then the pTi could carry genes that make the bacterium susceptible to physical or biological stresses in soil. If true, then nonpathogenic strains would predominate.

Some nonpathogenic agrobacteria isolated from soil are probably pathogens that have lost the pTi or have a mutation in one of several key genes required for infection. No attempt to investigate the population diversity of nonpathogenic agrobacteria has been made, especially with regard to the fate of genes involved in pathogenicity. Regardless, the rather incomplete and incidental observations that have been made of *Agrobacterium* populations in nature suggest that we should expect to see phenotypic diversity. More complete studies of *Rhizobium* and *Bradyrhizobium* spp., close relatives of agrobacteria, indicate considerable diversity within natural populations. Genomic rearrangements were common in *Rhizobium phaseoli* (40) and great heterogeneity was found among indigenous *Bradyrhizobium* spp. isolated from four different hosts in Hawaii (107). A high ratio of nonsymbiotic to symbiotic *Rhizobium leguminosarum* was found among strains isolated from bean roots (94), similar to observations of different ratios of nonpathogenic to pathogenic *Agrobacterium* in their natural habitats. These similarities between members of the Rhizobiaceae lead us to postulate that population heterogeneity is the norm for *Agrobacterium* spp. in their natural habitat.

Most research associated with crown gall disease has focused on the pathogens with little attention paid to the nonpathogenic agrobacteria. An exception that has been studied extensively is *Agrobacterium* strain K84, because of its importance as a biological control agent (79). One plasmid from K84 has been well characterized, i.e. pAtK84b (95). This plasmid carries the nopaline-catabolizing (*noc*) gene. Significantly, this plasmid was 80% homologous with the *noc* and origin of replication regions of pTiC58, but had no detectable homology with the *vir* and T-DNA regions (30). For these reasons, pAtK84b could be considered a disarmed Ti plasmid. The question of how prevalent such plasmids are in the population needs to be addressed.

A more careful attempt to examine plasmids within tumor populations was recently completed. Phenotypic comparisons were made of more than 100 agrobacteria strains isolated from tumors on cherry trees previously treated with K84 (65,66). Each isolate was characterized relative to biovar, plasmid content, and the presence or absence of key T-DNA and *vir* region sequences. Although all the nonpathogens carried large plasmids similar in size to pTi, they did not hybridize to *virFAB* and *iaaMiaaH* probes. Similar results were observed among 300 isolates from galls of raspberry plants that had been treated with K84 before planting (65).

In a previous study, 12 nonpathogenic strains were examined for homology of their plasmids with pTi (68). There was as much as 51% homology between plasmids from the nonpathogens and the pTi of the nopaline-utilizing strain C58 (5 to 51%), but less homology occurred with the octopine-utilizing strain pTiA6 (1 to 24%).

The homology between pTi and plasmids from nonpathogens indicates that horizontal gene transfer of plasmids may be occurring in nature. In order to understand the similarities and differences among *Agrobacterium* plasmids, large numbers of strains need to be examined. pTis from *Agrobacterium* strains C58 and A6 were compared and were shown to share 30% overall homology, but 80 to 85% homology embedded within nonhomologous regions, suggesting that they are mosaic structures. Recent investigations have revealed that considerable heterogeneity exists among pTis found in different pathogenic *Agrobacterium* strains isolated from grape (82). There are several ways that diversity may arise.

Mechanisms For Heterogeneity

Genetic exchange, spontaneous mutations and transposition can contribute to genotypic change in *Agrobacterium* strains (32,54,82). The frequency, significance and contribution of these genetic changes to species richness and population dominance in the environment are unknown. If nonpathogens are more stable, as indicated by their preponderance in soil, there may be a genetic drift towards populations of nonpathogens. Genetic exchange of plasmids among agrobacteria in nature could obviously lead to new genotypes and the frequency of plasmid transfer can be high in tumors due to the presence of opines that act as conjugal inducers (36,37,85). In other environments, such as the rhizosphere or soil, there is no conclusive evidence for conjugal transfer among *Agrobacterium*. Variability arising from agrobacteria pTi exchange may therefore arise only in tumors and be disseminated from there into the rhizosphere and soil.

Spontaneous mutations can also lead to population heterogeneity. When agrocin 84, the antibiotic produced by K84, was used as a selection pressure, a spontaneous mutation to avirulence was detected at a frequency of 10^{-3} to 10^{-4} among several pathogenic strains of *Agrobacterium* (32), a frequency three to four orders of magnitude higher than the commonly observed spontaneous mutation rate of 10^{-6} to 10^{-7}. Pathogenesis is controlled by several key genes, e.g. those involved in wound signal reception, site-specific attachment, *vir* induction, and T-DNA functions. Mutations in any one of these genes could lead to loss of pathogenicity (78), attenuated virulence (7), or change in host range (108).

Certain pathogenic strains of *Agrobacterium* mutate to a nonpathogenic form in the presence of acetosyringone, a plant wound compound. Strain C58, grown in liquid minimal medium adjusted to pH 5.8 in the presence of 60 μM acetosyringone, produced mutants that were avirulent. The virulence of some of the mutants could be restored by complementation with *virA* and of others by *virG*, regions that encode enzymes involved in signal recognition and transduction (41).

The loss of virulence in the presence of acetosyringone indicated that plant compounds might present a selection pressure for mutants that occur naturally within tumors. It could also explain the preponderance of nonpathogens isolated from tumors of some plant hosts such as apple and

Table 1. Source and phenotype of *Agrobacterium* strains used to assay for presence of avirulent mutants in tumors of inoculated plants.

Strain	Host of origin	Biotype	Opine utilization
C58	Cherry	I	Nopaline
B49C/83	Apple	II	Nopaline and mannopine
D10B/87	Apple	II	Nopaline
111/85	Cherry	II	Nopaline
I22/85	Cherry	II	Nopaline
B209B/85	Blackberry	ND[a]	ND
B230/85	Blackberry	ND	ND

[a] Not determined. Reprinted by permission of Kluwer Academic Publishers, Dordrecht.

blueberry. To test the possibility of plant selection for nonpathogenic mutants, several pathogenic *Agrobacterium* strains were inoculated to apple, cherry, peach, pear and blackberry in tissue culture. These strains had been isolated from several different hosts including apple and cherry (Table 1). One strain, D10B/87, isolated from an apple tumor, produced a large number of mutants when inoculated to apple in tissue culture (3). Ninety per cent of the bacteria recovered from apple plantlet tumors were not pathogenic. None of the strains isolated from other hosts mutated on any of the tissue culture plantlets (Table 2). Examination of 14 mutant strains of D10b/87 revealed that they retained most of the pTi genes and that pathogenicity could be restored by complementation with a wild-type *virG*.

Differences were apparent in mutants from different tumors. In one case, mutants had deletions which encompassed all of the *vir* region and part of the T region. Other mutants had point mutations at position -63 relative to the translational start site and in the C-terminal coding region

Table 2. Properties of bacteria recovered from surface or inner tissue of tumors induced on various apple rootstocks in vitro.

	Proportion of clones (%)			
	Nopaline-utilizing		Avirulent	
Rootstock	Surface	Inner tissue	Surface	Inner tissue
Mark	ND[a]	100	ND	90
Ottawa 3	100	100	99	98
P106	100	100	68	95
BUD 116	99	100	82	90

[a] Not determined. Reprinted by permission of Kluwer Academic Publishers, Dordrecht.

of virG (4). This appears to be the first example of a plant selecting for mutant strains and would be a useful strategy by the plant for limiting the spread of pathogens. Competition studies were conducted using a wild-type and a mutated apple pathogen inoculated to apple plantlets. The wild-type strain was almost undetectable after 5 weeks while the mutant strain numbers remained unchanged (4). This supports the hypothesis that the nonpathogenic mutants have a selective advantage over wild-type cells in the plant environment.

Nonpathogenic mutant strains may continue to represent a threat to plants, however, because of the phenomenon of strain complementation. Under laboratory conditions, mixtures of pathogenic and nonpathogenic *Agrobacterium* strains were more infectious when coinoculated to plants (61,62).

Transposition is a third mechanism for introducing variability into *Agrobacterium* strains. The *iaa* genes in *Agrobacterium* strain Ach 5 are homologous to those found in *Pseudomonas syringae* subsp. *savastanoi* and each is bordered by an IS-like element, indicating a transpositional mechanism (114). Insertion sequences close to the *iaa* genes have been found in several *Agrobacterium* strains (82).

The mechanisms described above help explain the diversity among *Agrobacterium* that is increasingly being reported. Among a group of 80

Spanish isolates from 13 hosts and several geographic regions, there were differences in biovars, opine utilization patterns, sensitivity to agrocin 84, and the numbers and sizes of plasmids (1). Tumor-inducing plasmids from three different biotypes of *Agrobacterium* from grape differed in their host range and opine utilization. Among the strains examined were nonpathogens that utilized octopine (57). The DNA in the vitopine pTi from grape *Agrobacterium* strain S4 has little homology with known T-DNA genes. The virulence genes *virD* of pTiS4 are homologous to *virD* genes in pTiC58 but not to *virD* genes of pTiAch5 (45). Genes that code for tartrate utilization are found on plasmids in many grape isolates. These plasmids differ in size, transfer frequency and ability to self conjugate (104).

All of the examples cited have related to changes in *Agrobacterium* plasmids. Despite the role of the pTi in determining pathogenicity in *Agrobacterium,* many fitness determinants would be expected on the chromosome since it contains the majority of the genome of bacterial cells. Even in pathogenic agrobacteria, some chromosomally inherited characters are required for pathogenicity, e.g. genes that govern the attachment of virulent bacterial cells to the surface of the host and of *vir* gene induction (26,47,89). Thus, to investigate ecological fitness, there is a need to determine whether there are genetic elements in soilborne nonpathogens that are essential for robustness in soil and whether these chromosomal characters are absent in pathogenic strains.

There are chromosomally encoded regions that can lead to a variety of nonpathogenic agrobacteria. Although most bacteria described have been grouped within three major biovars based on chromosomally inherited characters, major sub-biovar differences between pathogenic and nonpathogenic strains of *Agrobacterium* have recently been identified. Using DNA and protein analysis, biovar 1 strains could be divided into five different chromosomal subgroups. Most importantly to this discussion, distinct pTi types occurred in specific subgroups and the pTi-containing pathogenic strains were found in chromosomal backgrounds different from those in nonpathogenic strains. This suggested a preferential association of the pTi with particular genomes (13,69,77). Further cladistic analysis of within-biovar variation among strains isolated from multiple sites needs to be done to determine whether this is a general phenomenon.

The general picture emerging from these studies of agrobacteria isolated from the environment is one of considerable genetic complexity. This complexity must be reckoned with to manage crown gall disease. A better understanding of the relationships between agrobacteria will undoubtedly lead to new and improved methods of disease management.

MANAGEMENT OF CROWN GALL DISEASE

Various strategies are available for control of any plant disease. Among these are application of biological antagonists, physical and chemical methods for prevention of infection, avoidance of the pathogens, and selection of disease resistant plant material. Each of these is discussed in the next sections. For additional considerations see Moore and Cooksey (72).

Biological Control

By far the most successful method for preventing crown gall has been the use of the nonpathogenic *Agrobacterium* strain K84. The effectiveness and ease of application of K84 since its discovery in Australia in 1972 (79) have led to its widespread usage. A suspension of the bacteria can be made at a concentration of 10^8 colony-forming units (CFU) ml^{-1} and used as a dip or spray for root-pruned plants prior to planting at field sites. The primary basis for plant protection by K84 is the production of agrocin 84, encoded on a conjugative plasmid, pAgK84 (95). Sensitivity of pathogens to agrocin 84 can be demonstrated by standard petri plate assays. There is some evidence to suggest that competitive attachment of K84 to wound sites may also play a role in protecting plants. Mutant K84 strains lacking the ability to produce agrocin still reduce the incidence of crown gall (31,63).

Although K84 is usually applied as a preplanting dip at nurseries, adding K84 to soil was effective in decreasing the incidence of crown gall disease in Spain on plum and peach rootstocks (63). An Oregon study showed that K84 can persist on ryegrass roots for at least 2 years. Isolations from infested soil showed that K84 was established at a concentration of 10^5 CFU ml^{-1} in the rhizosphere of cherry and could still be detected on the roots 2 years after inoculation. Rhizosphere populations were established on 11 other herbaceous plants besides ryegrass. Ryegrass roots were used as traps to locate strain K84 40 cm from the site of infestation, indicating that the bacterium was disseminated through the soil (99).

Growers have been concerned that K84 might not be compatible with several fungicides applied to protect dormant trees in cold storage. Therefore, the fungicides Banrot, Benlate, Captan, Truban and Zyban were all tested with K84 by treating Colt and Mazzard cherry rootstocks and domestic apple seedlings with the fungicides before treatment with K84. Crown gall incidence was high among trees treated with only the fungicide and challenged with pathogenic *Agrobacterium*. In contrast, reduction in the incidence of crown gall was statistically the same whether trees were treated only with K84 or treated with a combination of the fungicide and K84 before challenging with the pathogen (L.W. Moore and M. Canfield, unpublished).

In spite of the success of K84, there are potential and real problems associated with its application. For example, the process of dipping the rootstocks in a suspension of K84 could wash other bacterial or fungal pathogens from the plant roots into the K84 suspension and contaminate or infect subsequent rootstocks. Not all pathogenic agrobacteria are sensitive to agrocin 84 and therefore some infections occur even when K84 has been applied (2,19). This appears to be the major reason for the ineffectiveness of K84 on grape. Most of the pathogens isolated from grape tumors are biovar 3, all of which are insensitive to agrocin 84 (55). Although biovar 3 pathogens infect grape vines almost exclusively, it doesn't follow that agrocin sensitivity of a pathogen is host related. Agrocin insensitivity of pathogenic *Agrobacterium*, however, is an important consideration when determining the effectiveness of K84 at a particular nursery site. Where possible, obtain numerous infected plants from a nursery, isolate a large number of pathogens from multiple tumors, and screen them for sensitivity to K84 to determine the probability of success of K84 at that site.

Surprisingly, in some cases, K84 is effective even when challenged with insensitive pathogens (9,63,70,92). K84 decreased the incidence of crown gall on several host species, even though the pathogens used were insensitive to agrocin 84. This may result, in part, from competitive binding at wound sites. A third hazard associated with the use of K84 is the possibility of conjugal transfer of the agrocin plasmid between K84 and pathogenic strains leading to agrocin-producing strains that are pathogenic and insensitive to K84. In a study in which K84 was coinoculated with the rifampicin-resistant pathogen B49C/83, rif-resistant strains were isolated that produced agrocin 84, indicating that they were transconjugants (74).

Because of the risk of losing protection by K84 through conjugation with pathogens, a mutant strain has been constructed by deleting the transfer (*Tra*) region of pAgK84 (50). This *tra*-deletion mutant (K1026) has been used successfully to control crown gall in Australia and Spain (112). Both K84 and K1026 colonized the roots of Montclar and Nemaguard peach seedlings and populations of the bacteria were similar over a 3-week period. In a biological control experiment using the two strains, no tumors developed on GF677 and Adufuel peach × almond rootstocks inoculated with K84-resistant strains previously treated with K1026. A few tumors did develop on the K84-treated trees and nine strains isolated from these tumors were pathogenic agrocin-producing strains, indicating that they were transconjugants (112). These results show the superiority of K1026 over K84. However, this mutant strain has not yet been approved for use in the United States.

In some cases where K84 does not work, other biological control agents have been effective. The nonpathogenic strains HLB-2 and E26 of biovar 1 of *Agrobacterium* have been used in China to reduce tumor formation in grape (28,59,60). HLB-2 has also been used in one study in the United States and successfully limited tumor initiation in the grape cultivar Chancellor (86). In South Africa, five nonpathogenic biovar 3 strains isolated from grape reduced tumor incidence and size on the grape cultivar Jaquez (98).

Alternative Methods Of Control

Heat therapy used to promote healing of pruning wounds reduced the incidence of crown gall in Maheleb and Mazzard cherry and Myrobalan plum seedlings. When the seedlings were inoculated with pathogenic agrobacteria after incubation in an insulated box for 3 weeks at 24°C, the incidence of galled seedlings was 0 to 6% compared to 100% in unheated controls (71). Hot water treatments of dormant grape cuttings at 50°C for 30 minutes dramatically reduced populations of *Agrobacterium* biovar 3 in scion cultivars Chardonnay and Zante Currant and rootstock cultivars Ramsey and K51-40. Grafted vines planted into the field had a tumor incidence of 2% compared to 60% on unheated controls (81). In another report, a treatment at 27°C was able to reduce tumorigenesis induced by biotype 3 strains but not by biotype 1 and 2 strains (27). Solarization in both sandy and silty soils reduced populations of pathogenic agrobacteria without affecting populations of beneficial microorganisms such as fluorescent pseudomonads, bacillus and various Actinomycetes (88).

Chemical treatments reduced incidence of crown gall in apple, a host in which K84 has been ineffective. Benzoyl peroxide, Copac E, streptomycin sulfate with streptomycin-resistant K84, and Terramycin were compared as preplanting dips at nurseries in Oregon and Washington using three different rootstocks. Terramycin and Copac E consistently gave the best results in reducing tumor incidence (24).

Soil fumigation with Vorlex at a rate of 38 liter ha^{-1} 2 months before planting indexed grapevines reduced populations of agrobacteria. Only 16% of Chancellor grapevines planted into the fumigated soil developed crown gall compared to 100% in nonfumigated soil (87). In a Canadian study, soil fumigation with metham-sodium had no effect on crown gall of apple at a rate of 278 kg ha^{-1} but was effective in reducing crown gall when applied at a rate of 640 kg ha^{-1} (111).

Starting with pathogen-free plant tissue is important in preventing crown gall. In plants where agrobacteria can become systemic, such as grape (58), shoot tip culture is one way to eliminate bacteria. The rapid growth of grapevines in the spring may outdistance the bacteria that are resident in vascular tissue. Using the shoot tips and micropropagating them in tissue culture is one strategy for producing pathogen-free material (17).

Selection of resistant rootstocks is another method for reducing incidence of crown gall. Several grape rootstocks have been tested in Hungary and *Vitis riparia* cv. Gloire de Montpellier was the most resistant of 14 genotypes tested when inoculated with biovar 1 and biovar 3 strains (102). Similar evaluations in New York on 19 *Vitis* genotypes and their hybrids revealed that one hybrid, *V. riparia* × *V. rupestris* C3309 was also highly resistant to infection by three strains of biovar 3. Caution must be observed when generalizing from these results, for two reasons. First, studies on resistance by independent investigators may involve the use of different groups of pathogenic agrobacteria. Because of the variability found between strains, resistance ratings may differ from study to study. A case in point is the *Vitis* genotype *V. berlandieri* × *V. riparia* T5C. In two out of three studies, it was rated as very susceptible to crown gall, and in a third study it appeared to be less susceptible based on the percentage of wound sites that became infected (46,100,102). The best recommendation for evaluation of resistance would be to test rootstocks with strains isolated from grapevines in the region in which the plants will be grown. A second precaution is to use pathogen-free scion wood when plants are grafted. Even though resistant rootstocks may be symptomless, they can still harbor populations of pathogens and serve as a conduit to susceptible scion wood (100,102).

Variability among the agrobacteria, plant hosts, and planting sites must be reckoned with in any attempt to control crown gall disease. This point cannot be overemphasized.

VARIABILITY

Numerous examples of variability have been encountered in our study of crown gall over the past 20 years. They include differences found between various plant hosts and planting sites, differing ratios of nonpathogens to pathogens, and the diversity found among individuals within a single isolation. One type of variability that occurs between hosts is that seen in symptom development. Tumor morphology and location of tumors on plants can be quite different (Figure 2). The following "case studies" have been chosen to illustrate other kinds of variability encountered in the study of crown gall. They are described in some detail so that the importance of variability can be appreciated.

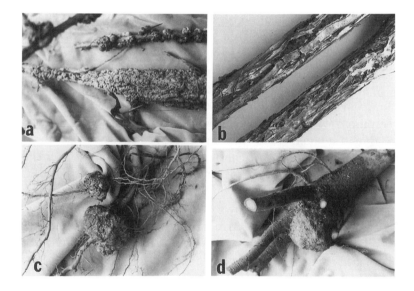

Figure 2. Examples of variability in tumor morphology and location on plant: aerial galls on (a) blackberry and (b) blueberry; root galls on (c) apple and (d) walnut rootstocks.

Apple

This host has proven to be one of the most difficult ever encountered in the study of crown gall. Although the incidence of crown gall on apple is usually low, in some years growers experience large numbers of infected trees. Most attempts to isolate pathogenic agrobacteria from apple tumors have been unsuccessful. Few to no pathogens have been detected, and most that were recovered were only weakly virulent in field inoculations of apple (23). K84 has not been effective in reducing the disease incidence except for one field trial in Poland (96). In the absence of pathogens for use as challenge organisms, it is difficult to evaluate new control measures. Several hypotheses were generated to explain the difficulty in finding pathogens. Among these were: (i) the isolation methods were not adequate to find the pathogenic agrobacteria, (ii) competition within tumors reduced the number of agrobacteria, (iii) bacteria were mutating to avirulence, or (iv) the tumors were not caused by agrobacteria. The hypothesis that our isolation methods were not appropriate led to the use of nopaline, mannopine, or octopine in isolation media. Two new strains from these opine media, B49C/873 and D10B/87, gave good results when inoculated to three cultivars of apple in field experiments. One measure of confidence that the strains were actually infecting apple in field tests was the change in tumor incidence as a function of bacterial inoculum concentration. When three different concentrations of strains B49C/83 and D10B/87 were used to inoculate apple trees in field tests the incidence of galled trees increased proportionately (Figure 3). The highest concentration (10^9 CFU ml^{-1}) was the one chosen for field experiments in disease control. These more highly virulent strains have proven useful as challenge organisms to test new methods of controlling crown gall on apple.

Because K84 was not effective in reducing crown gall incidence in apple, other biological control agents were tried. Several opine-utilizing pseudomonads and an Actinomycete antagonistic to agrobacteria in vitro reduced tumor incidence on tomato seedlings in greenhouse tests. However, when they were used as preplanting dips in field studies there was no reduction of crown gall. Because biological control agents failed, chemical methods were tried at field sites in Oregon and Washington. Copac E and Terramycin were the most effective of four chemical treatments in preventing crown gall (Figure 4) (24).

Persistence and changes in methodology have been effective in identifying pathogens and finding methods of control. Interestingly, in studying this particular host pathogen-interaction, evidence was found to

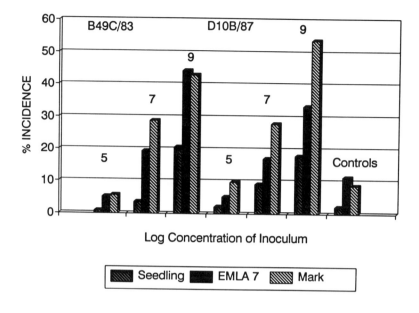

Figure 3. Incidence of crown gall disease 6 months after application of three concentrations of pathogenic *Agrobacterium* strains B49C/83 and D10B/87. Each concentration was applied as a preplanting dip to 100 trees each from three different kinds of apple rootstock: seedling, EMLA 7 and Mark.

support our hypothesis that low numbers of pathogens in tumors may result from mutation of the infecting bacteria. Strain D10B/87 inoculated to apple plantlets in tissue culture led to recovery of over 90% nonpathogenic isolates after 2 months (see earlier section on mechanisms for heterogeneity, page 163).

Euonymus

One nursery reported losses of up to 25% in container-grown euonymus plants in 1992. The tumors were seen predominantly at pruning wounds on young branches. Bacteria were isolated from tumors on four cultivars of euonymus. Twenty of 40 isolates tested were pathogenic on tomato and euonymus plants and 12 of the pathogens were

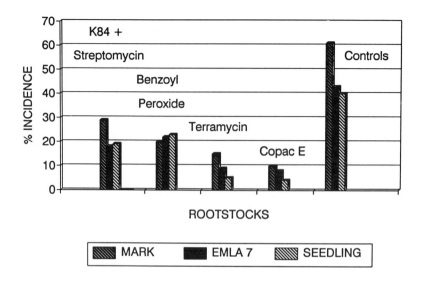

Figure 4. Effectiveness of four treatments to reduce crown gall on apple rootstocks Mark, EMLA 7 and seedling. One hundred trees of each rootstock were used per treatment. Treatments were made immediately before planting at a Washington field site. The treatments were: (a) rootstocks dipped first in a suspension of *Agrobacterium* K84 resistant to streptomycin, followed by a dip in 2.25 g liter^{-1} streptomycin, (b) rootstocks dipped in 4.8 g liter^{-1} benzoyl peroxide, (c) rootstocks dipped in 17.5 g liter^{-1} Terramycin and (d) rootstocks dipped in 5% Copac E. Incidence of crown gall was recorded 6 months after treatment.

sensitive to agrocin 84 in vitro. All of the pathogens were identified as biotype 2. Potted plants were wounded in a preliminary greenhouse study by pruning new shoots, then treated with a suspension of K84 at a concentration of 10^8 CFU ml^{-1}. The inoculated wounds were challenged 1 hour later with a mixture of four strains of euonymus pathogens, two sensitive and two insensitive to agrocin 84, at a concentration of 10^7 CFU ml^{-1}. Four cultivars of euonymus were used for the biological control experiments. Plants were observed monthly for 6 months. In three of the four cultivars, the incidence of crown gall was lower on K84-treated plants challenged with pathogens than on those inoculated only with the pathogens. Average incidence of crown gall on the three cultivars treated with K84 was 32% compared to 60% for the pathogen-only inoculations. The fourth cultivar had a higher incidence of crown gall on K84-treated plants (51%) than on plants inoculated only with the pathogens (37%), perhaps indicating that this cultivar is more susceptible to these pathogens, or that the activity of K84 was modified (L.W. Moore and M. Canfield, unpublished). Thus, K84 was able to reduce crown gall incidence even though the challenge organisms were a mixture of sensitive and insensitive strains. This provides an additional example of the effectiveness of K84 even when challenged by insensitive pathogens.

Blueberry

This host is similar to apple in that most strains isolated are nonpathogenic. In one case, 96 isolates from five different tumors had the colony morphology of *Agrobacterium*. One tumor sample yielded a high number of nopaline-utilizing strains (45 of 45 tested), indicating that they may possess a pTi. However, only one of these strains was pathogenic. Another pathogenic strain that utilized octopine was isolated from a second tumor, but none of the other 15 isolates from this tumor was pathogenic, including one other octopine-utilizing strain (L.W. Moore and M. Canfield, unpublished). It is possible that nonpathogenic mutants develop in tumors of blueberry, or are selected for, after infection occurs (as reported earlier with some apple strains of pathogenic agrobacteria). Isolations of pathogenic agrobacteria from blueberry tumors have been reported, but the strains identified were host specific (33). In 1995, agrobacteria were isolated from blueberry plants with atypical crown gall symptoms (Figure 2b). Among the 51 isolates tested, seven hybridized to Ti plasmid DNA probes. All seven isolates were pathogenic when tested on both tomato and blueberry plants. The strains were identified as biovar 2 agrobacteria (25).

Walnut

Over 300 isolates were obtained from eight tumor samples taken from four different locations within a California nursery in 1994. Most isolates (60%) from three of the four locations were pathogenic, but only one pathogen was isolated from the fourth location (Table 3).

Walnut strains were identified using classical biochemical tests and 17 of the 18 pathogens characterized were biovar 2; the remaining pathogen was biovar 1. They varied widely in their opine utilization patterns, some using mannopine or nopaline, while others used both mannopine and nop-

Table 3. Variability in population numbers of pathogenic *Agrobacterium* isolates from different walnut tumors sampled from multiple locations in a single nursery.

Field location[a]	Gall[b]	No. of isolates[c]	Probe response[d]	Pathogenicity[e]
A	1	30	30	6/6
	2	15	8	4/4
B	1	33	6	3/3
	2	70	52	12/12
C	1	53	2	1/2
	2	63	60	9/9
D	1	2	0	0/0
	2	36	1	1/1

[a] Galled trees were obtained from four different locations (A, B, C and D) in the nursery.
[b] Single galls were sampled, each from a different tree.
[c] Number of *Agrobacterium*-like colonies selected from the isolation media for further processing.
[d] Number of isolates that reacted positively with Ti plasmid T-DNA probes *virFAB* and *tms1tmr*.
[e] Number of probe-positive isolates that were pathogenic when inoculated to tomato plants. Numerator is number of pathogenic isolates, and denominator is the total number of isolates assayed.

aline, or both nopaline and succinamopine. They differed in their sensitivity to three antibiotic-producing agrobacteria, strains K84, E26, and HLB-2. Three of the strains were sensitive to K84, 13 to HLB-2, and two to E26. Nonpathogens were even more variable; of eight strains tested, two were biovar 1, two were biovar 2, and four could not be identified to any of the known biovars.

Because of the large number of pathogens found among the walnut strains, and in an effort to identify the source of pathogens, isolations were made from germinating seedlings that showed no symptoms. Five different lots of five seedlings each were divided into tissue groups of shoots, seeds or roots. Five strains, all isolated from one sample, hybridized to DNA probes and four of these were pathogenic. Two of these pathogens were biovar 1 and two were biovar 2. Among another 32 nonpathogens characterized, 19 were biovar 1, one was biovar 2, and 11 did not fit any known classification. The walnut strains also varied in their opine utilization patterns and sensitivity to the three antagonists. Two of the pathogens used octopine and two used both nopaline and succinamopine. Among the nonpathogens, three of the 36 tested used nopaline. Most of the pathogens and nonpathogens tested were biotype 2. Two of the strains were sensitive to K84, two to HLB-2, and 17 to E26 (M. Canfield and L.W. Moore, unpublished). This is a good example of the diversity that can be found within one host at a particular nursery, and shows the importance of isolating from several tumors and several areas within a nursery. The detection of pathogens on the seedling tissues indicates that agrobacteria can be present on walnut from an early stage of plant development and would be a threat when plants are pruned or wounded. Because so many of the pathogens isolated from both the tumors and seedlings were insensitive to K84, this strain would probably not be effective in preventing crown gall on walnut at this particular nursery. *Agrobacterium* strains HLB-2 and E26 would be better choices for control, but are not yet licensed for use in the United States.

Grape

Crown gall on grape offers a special problem because of the ability of pathogens to survive within the vascular system of the vines. Wounds such as those caused by cold injury are prime sites for infection and subsequent tumor formation. Recently, crown gall of grape has become a more serious problem in Oregon and California because of the recurrence of phylloxera (*Daktulosphaira vitifoliae*), an insect infestation that can kill susceptible plants. Over 95% of the grapevines planted in Oregon are grown on their own roots, and all are susceptible to damage from the insect (101). Vineyard owners are replacing old vines with plants grafted onto phylloxera-resistant rootstocks. However, nurseries producing grafted plants report a high incidence of crown gall at graft unions (Figure 5).

A survey of Oregon vineyards has been initiated to characterize agrobacteria that are found in both tumors and symptomless plants. The best methods for preventing crown gall in grape appear to be selection of rootstocks least susceptible to crown gall, the use of scion wood free of

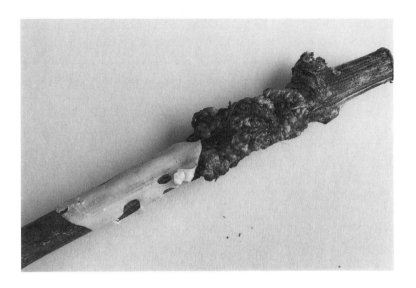

Figure 5. Tumor development at the graft union of grapevines propagated at a nursery.

the bacteria, and possibly heat treatments to kill systemically borne bacteria in scion and rootstock wood before grafting. If grapevines have to be planted on sites where crown gall has been frequent, soil treatments such as solarization or fumigation would be helpful.

Cherry

K84 has successfully prevented crown gall in cherry and other *Prunus* species. There are however, occasional failures after treatment. In 1989, the cause of crown gall on cherry trees treated with K84 prior to planting was evaluated. Agrobacteria were isolated and characterized in the usual way. The majority of the pathogens were insensitive to agrocin 84, providing a possible explanation of the failure to control the disease with K84. There were nine phenotypes among the bacteria isolated from the tumors including one with the same features as K84. Other phenotypes included nonpathogens, and pathogens that produced an antibiotic similar to agrocin 84. These latter strains were of most concern as they could be transconjugants formed within the tumors (65).

Wild Blackberry

Interest in obtaining *Agrobacterium* isolates from this host was aroused by curiosity about the kinds of agrobacteria present in crown gall on wild hosts in agriculturally undisturbed areas.   An earlier investigation of a natural savanna in Minnesota that has never been cultivated revealed that agrobacteria were readily isolated from the rhizosphere of plants growing there.   No tumors were observed on any of the savanna plants and none of the agrobacteria isolated were pathogenic (12).   In the case of wild blackberry in Oregon, aerial tumors are commonly seen and several were obtained for isolations onto selective media and media containing opines. Even though most of the colonies exhibited morphology that was not typical of agrobacteria, they were pathogenic when tested on tomato plants and hybridized to the *virFAB* and *iaa*M*iaa*H probes.   The bacteria grew so slowly and formed such small colonies that it was difficult to purify the isolates.   A new medium was devised which supported more vigorous growth and the strains were further characterized (Figure 6).

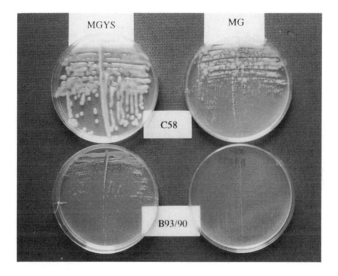

Figure 6. Comparative growth of blackberry isolates on two media. Growth of *Agrobacterium* strain C58 is shown in the upper petri plates on both mannitol glutamate (MG) (50) and MGYS, a modified MG medium containing yeast extract and various mineral salts.   In contrast, strain B93/90, a blackberry pathogen (shown in the lower two petri plates) grew only on MGYS.

Most of the pathogens isolated from blackberry did not fit into any of the established biovars of agrobacteria but resembled each other in physiological and biochemical characteristics. These strains exemplify another type of population diversity found among the agrobacteria (22). The blackberry isolates are of interest because of their evolutionary significance and as a possible source of primary inoculum for plants in nurseries.

Although little has been published about the occurrence of crown gall in natural ecosystems undisturbed by agricultural cultivation and tillage, galled plants have been observed regularly on wild roses, alder trees and conifers in Oregon (L.W. Moore, personal observation). Recently, pathogenic strains, mostly biovar 1, were isolated from naturally occurring tumors found on the common rangeland weeds, Russian knapweed and leafy spurge (20). The origin of these pathogens is unknown, but *Agrobacterium* strains associated with plants in areas never touched by agricultural cultivation are interesting because of their potential contribution to a better understanding of the evolutionary processes affecting the population structure of *Agrobacterium*. Isolations from wild grape have been recommended for just this reason. Even though tumors might not be found on wild grape, agrobacteria could be present in the vascular system, as in domestic varieties (82). A more systematic search would likely reveal many plant species in the wild that are infected by *Agrobacterium*.

SUMMARY

Crown gall disease occurs on many woody crops in the Pacific Northwest of the United States. Prediction of disease potential is difficult because the epidemiology of crown gall is not well understood. This is attributable, in part, to the variability observed in disease development among host plants, planting sites and the infecting *Agrobacterium* strains. Diversity is the norm among the hundreds of *Agrobacterium* strains isolated and characterized so far, but the root cause of this heterogeneity is not well understood. Soil typically contains much greater quantities of nonpathogens than pathogens, even around galled plants, and the relationship between these strains needs to be investigated. Such investigations have been hindered in the past by the lack of efficient, specific diagnostic tools to process large numbers of strains from divergent habitats and populations. The development and use of DNA probes and plant tissue culture have greatly improved the opportunity to detect and characterize these large numbers of agrobacteria from soil and tumors. Using these methods to conduct more exacting studies on the ecology and evolution of agrobacteria will yield data that can ultimately be applied to disease management. At present, *Agrobacterium* strain K84 is the most effective method for prevention of crown gall. In cases where K84 fails, other useful methods include Terramycin preplanting dips, heat treatments of planting material, selection of resistant plants, use of pathogen-free planting material, and soil treatments such as solarization or fumigation.

Future research will be focused on examination of *Agrobacterium* populations in natural habitats, and the relationship of these populations to disease development, together with comparisons of the genotypic and phenotypic diversity of pathogens and nonpathogens in the agroecosystem. Information developed on the relationship of crown gall incidence to populations, and on the variations in these populations will help in making choices for disease control measures in the future.

LITERATURE CITED

1. Albiach, M.R., and Lopez, M.M. 1992. Plasmid heterogeneity in Spanish isolates of *Agrobacterium tumefaciens* from thirteen different hosts. Appl. Environ. Microbiol. 58:2683–2687.

2. Alconero, R. 1980. Crown gall of peaches from Maryland, South Carolina, and Tennessee and problems with biological control. Plant Dis. 64:835–838.

3. Bélanger, C., Canfield, M.L., Moore, L.W., and Dion, P. 1993. Detection of avirulent mutants of *Agrobacterium tumefaciens* in crown-gall tumors produced *in vitro*. Pages 97–101 in: Advances in Molecular Genetics of Plant-Microbe Interactions, Vol. 2. E.W. Nester and D.P.S. Verma, eds. Kluwer Academic Publishers, Dordrecht, The Netherlands.

4. Bélanger, C., Canfield, M.L., Moore, L.W., and Dion, P. 1995. Genetic analysis of nonpathogenic *Agrobacterium tumefaciens* mutants arising in crown gall tumors. J. Bacteriol. 177:3752–3757.

5. Bell, C.R., Cummings, N.E., Canfield, M.L., and Moore, L.W. 1990. Competition of octopine-catabolizing *Pseudomonas* spp. and octopine-type *Agrobacterium tumefaciens* for octopine in chemostats. Appl. Environ. Microbiol. 56:2840–2846.

6. Bell, C.R., Moore, L.W., and Canfield, M.L. 1990. Growth of octopine-catabolizing *Pseudomonas* spp. under octopine limitation in chemostats and their potential to compete with *Agrobacterium tumefaciens*. Appl. Environ. Microbiol. 56:2834–2839.

7. Binns, A.N., and Thomashaw, M.F. 1988. Cell biology of *Agrobacterium* infection and transformation of plants. Annu. Rev. Microbiol. 42:575–606.

8. Bouzar, H. 1988. Serological identification of common and strain-specific antigens in *Agrobacterium*. Ph.D. thesis, Oregon State University, Corvallis, OR.

9. Bouzar, H., Daouzli, N., Krimi, Z., Alim, A., and Khemici, E. 1991. Crown gall incidence in plant nurseries of Algeria, characteristics of *Agrobacterium tumefaciens* strains, and biological control of strains sensitive and resistant to agrocin 84. Agronomie 11:901–908.

10. Bouzar, H., Jones, J.B., and Hodge, N.C. 1993. Differential characterization of *Agrobacterium* species using carbon-source utilization patterns and fatty acid profiles. Phytopathology 83:733–739.

11. Bouzar, H., and Moore, L.W. 1987. Complementary methodologies to identify specific *Agrobacterium* strains. Appl. Environ. Microbiol. 53:2660–2665.

12. Bouzar, H., and Moore, L.W. 1987. Isolation of different *Agrobacterium* biovars from a natural oak savanna and tallgrass prairie. Appl. Environ. Microbiol. 53:717–721.

13. Bouzar, H., Ouadah, D., Krimi, Z., Jones, J.B., Trovato, M., Petit, A., and Dessaux, Y. 1993. Correlative association between resident plasmids and the host chromosome in a diverse *Agrobacterium* soil population. Appl. Environ. Microbiol. 59:1310–1317.

14. Bradbury, J.F. 1986. Guide to Plant Pathogenic Bacteria. CAB International, Wallingford, UK. 332 pp.

15. Brisbane, P.G., and Kerr, A. 1983. Selective media for three biovars of *Agrobacterium*. J. Appl. Bacteriol. 54:425–431.

16. Burr, T.J., Bishop, A.L., Katz, B.H., Blanchard, L.M., and Bazzi, C. 1987. A root-specific decay of grapevine caused by *Agrobacterium tumefaciens* and *A. radiobacter* biovar 3. Phytopathology 77:1424–1427.

17.   Burr, T.J., Katz, B.H., Bishop, A.L., Meyers, C.A., and Mittak, V.L. 1988. Effect of shoot age and tip culture propagation of grapes on systemic infestations by *Agrobacterium tumefaciens* biovar 3. Am. J. Enol. Vitic. 39:67–70.

18.   Burr, T.J., Norelli, J.L., Katz, B.H., and Bishop, A.L. 1990. Use of Ti plasmid DNA probes for determining tumorigenicity of *Agrobacterium* strains. Appl. Environ. Microbiol. 56:1782–1785.

19.   Burr, T.J., Reid, C.L., Katz, B.H., Tagliati, M.E., Bazzi, C., and Breth, D.I. 1993. Failure of *Agrobacterium radiobacter* strain K-84 to control crown gall on raspberry. HortScience 28:1017–1019.

20.   Caesar, A.J. 1994. Pathogenicity of *Agrobacterium* species from the noxious rangeland weeds *Euphorbia esula* and *Centaurea repens*. Plant Dis. 78:796–800.

21.   Canfield, M.L., and Moore, L.W. 1990. Isolation of *Agrobacterium tumefaciens* from apple rootstocks with crown gall disease. Pages 829–833 in: Proceedings, 7th International Conference on Plant Pathogenic Bacteria, Budapest, Hungary. Z. Klement, ed. Akadémiai Kiadó, Budapest, Hungary.

22.   Canfield, M.L., and Moore, L.W. 1989. Unusual characteristics of *Agrobacterium tumefaciens* isolated from wild blackberry. Phytopathology 79:1181. (Abstr.).

23.   Canfield, M.L., and Moore, L.W. 1991. Isolation and characterization of opine-utilizing strains of *Agrobacterium tumefaciens* and fluorescent strains of *Pseudomonas* spp. from rootstocks of *Malus*. Phytopathology 81:440–443.

24.   Canfield, M.L., Pereira, C., and Moore, L.W. 1992. Control of crown gall in apple (*Malus*) rootstocks using Copac E and Terramycin. Phytopathology 82:1153. (Abstr.).

25.   Canfield, M.L., Putnam, M.L., White, T.J., and Moore, L.W. 1995. Isolation of *Agrobacterium tumefaciens* from blueberry (*Vaccinium corymbosum*). Phytopathology 85:1194. (Abstr.).

26.   Cangelosi, G.A., Ankenbauer, R.G., and Nester, E.W. 1990. Sugars induce the *Agrobacterium* virulence genes through a periplasmic binding protein and a transmembrane signal protein. Proc. Natl. Acad. Sci. USA 87:6708–6712.

27.   Charest, P.J., and Dion, P. 1985. The influence of temperature on tumorigenesis induced by various strains of *Agrobacterium tumefaciens*. Can. J. Bot. 63:1160–1167.

28. Chen, X., and Ziang, W. 1986. A strain of *Agrobacterium radiobacter* inhibits growth and gall formation by biotype III strains of *Agrobacterium tumefaciens* from grapevine. Acta Microbiol. Sin. 26:193–199.

29. Chilton, W.S., and Chilton, M.-D. 1984. Mannityl opine analogs allow isolation of catabolic pathway regulatory mutants. J. Bacteriol. 158:650–658.

30. Clare, B.G., Kerr, A., and Jones, D.A. 1990. Characteristics of the nopaline catabolic plasmid in *Agrobacterium* strains K84 and K1026 used for biological control of crown gall disease. Plasmid 23:126–137.

31. Cooksey, D.A., and Moore, L.W. 1982. Biological control of crown gall with an agrocin mutant of *Agrobacterium radiobacter*. Phytopathology 72:919–921.

32. Cooksey, D.A., and Moore, L.W. 1982. High frequency spontaneous mutations to Agrocin 84 resistance in *Agrobacterium tumefaciens* and *A. rhizogenes*. Physiol. Plant Pathol. 20:129–135.

33. Demaree, J.B., and Smith, N.R. 1952. Blueberry galls caused by a strain of *Agrobacterium tumefaciens*. Phytopathology 42:88–90.

34. Dong, L.-C., Sun, C.-W., Thies, K.L., Luthe, D.S., and Graves Jr., C.H., 1992. Use of polymerase chain reaction to detect pathogenic strains of *Agrobacterium*. Phytopathology 82:434–439.

35. Duncan, A.V., Stokes, H.W., and Daggard, G. 1992. Genetic exchange in natural microbial communities. Pages 383–429 in: Advances in Microbial Ecology, Vol. 12. K.C. Marshall, ed. Plenum Press, New York, NY.

36. Ellis, J.G., Kerr, A., Petit, A., and Tempe, J. 1982. Conjugal transfer of nopaline and agropine Ti-plasmids—the role of agrocinopines. Mol. Gen. Genet. 186:269–274.

37. Ellis, J.G., and Murphy, P.J. 1981. Four new opines from crown gall tumors—their detection and properties. Mol. Gen. Genet. 181:36–43.

38. Ellis, J.G., Murphy, P.J., and Kerr, A. 1982. Isolation and properties of transfer regulatory mutants of the nopaline Ti-plasmid pTiC58. Mol. Gen. Genet. 186:275–281.

39. Farrand, S.K., Slota, J.E., Shim, J.-S., and Kerr, A. 1985. Tn5 insertions in the agrocin 84 plasmid: the conjugal nature of pAgK84 and the locations of determinants for transfer and agrocin 84 production. Plasmid 13:106–117.

40. Flores, M., González, V., Pardo, M.A., Leija, A., Martínez, E., Romero, D., Piñero, D., Dávila, G., and Palacios, R. 1988. Genomic instability in *Rhizobium phaseoli*. J. Bacteriol. 170:1191–1196.

41. Fortin, C., Nester, E.W., and Dion, P. 1992. Growth inhibition and loss of virulence in cultures of *Agrobacterium tumefaciens* treated with acetosyringone. J. Bacteriol. 174:5676–5685.

42. Fox, P.C., Vasil, V., Vasil, I.K., and Gurley, W.B. 1992. Multiple *ocs*-like elements required for efficient transcription of the mannopine synthase gene of T-DNA in maize protoplasts. Plant Mol. Biol. 20:219–233.

43. Garfinkel, D.J., Simpson, R.B., Ream, L.W., White, F.F., Gordon, M.P., and Nester, E.W. 1981. Genetic analysis of crown gall: fine structure map of the T-DNA by site-directed mutagenesis. Cell 27:143–153.

44. Genetello, C., Van Larebeke, N., Holsters, M., De Picker, A., Van Montagu, M., and Schell, J. 1977. Ti plasmids of *Agrobacterium* as conjugative plasmids. Nature (Lond.) 265:561–56.

45. Gérard, J.-C., Canaday, J., Szegedi, E., Salle, H. de la, and Otten, L. 1992. Physical map of the vitopine Ti plasmid pTiS4. Plasmid 28:146–156.

46. Goodman, R.N., Grimm, R., and Frank, M. 1993. The influence of grape rootstocks on the crown gall infection process and on tumor development. Am. J. Enol. Vitic. 44:22–26.

47. Hawes, M.C., and Pueppke, S.G. 1989. Reduced rhizosphere colonization ability of *Agrobacterium tumefaciens* chromosomal virulence (*chv*) mutants. Plant Soil 113:129–132.

48. Holmes, B., and Roberts, P. 1981. The classification, identification and nomenclature of agrobacteria: incorporating revised descriptions for each of *Agrobacterium tumefaciens* (Smith & Townsend) Conn 1942, and *Agrobacterium rhizogenes* (Riker *et al.*) Conn 1942, and *Agrobacterium rubi* (Hildebrand) Starr & Weiss 1943. J. Appl. Bacteriol. 50:443–467.

49. Hooykaas, P.J.J., Klapwijk, P.M., Nuti, M.P., Schilperoort, R.A., and Rörsch, A. 1977. Transfer of the *Agrobacterium tumefaciens* Ti plasmid to avirulent agrobacteria and to *Rhizobium* ex planta. J. Gen. Microbiol. 98:477–484.

50. Jones, D.A., Ryder, M.H., Clare, B.G., Farrand, S.K., and Kerr, A. 1988. Construction of a Tra⁻ deletion mutant of pAgK84 to safeguard the biological control of crown gall. Mol. Gen. Genet. 212:207–214.

51. Kado, C.I. 1991. Molecular mechanisms of crown gall tumorigenesis. Crit. Rev. Plant Sci. 10:1–32.

52. Keane, P.J., Kerr, A., and New, P.B. 1970. Crown gall of stone fruit. II. Identification and nomenclature of *Agrobacterium* isolates. Aust. J. Biol. Sci. 23:585–595.

53. Kerr, A. 1969. Crown gall of stone fruit. I. Isolation of *Agrobacterium tumefaciens* and related species. Aust. J. Biol. Sci. 22:111–116.

54. Kerr, A. 1971. Acquisition of virulence by non-pathogenic isolates of *Agrobacterium radiobacter*. Physiol. Plant Pathol. 1:241–246.

55. Kerr, A., and Panagopoulos, C.G. 1977. Biotypes of *Agrobacterium radiobacter* var. *tumefaciens* and their biological control. Phytopathol. Z. 90:172–179.

56. Kersters, K., Deley, J., Sneath, P.H.A., and Sackin, M. 1973. Numerical taxonomic analysis of *Agrobacterium*. J. Gen. Microbiol. 78:227–239.

57. Knauf, V.C., Panagopoulos, C.G., and Nester, E.W. 1983. Comparison of Ti plasmids from three different biotypes of *Agrobacterium tumefaciens* isolated from grapevines. J. Bacteriol. 153:1535–1542.

58. Lehoczky, J. 1968. Spread of *Agrobacterium tumefaciens* in the vessels of the grapevine, after natural infection. Phytopathol. Z. 63:239–246.

59. Li, H.P., Goth, R.W., and Barksdale, T.H. 1988. Evaluation of resistance to bacterial wilt in eggplant. Plant Dis. 72:437–439.

60. Liang, Y., Di, Y., Zhao, J., and Ma, D. 1990. Strain E26 of *Agrobacterium radiobacter* efficiently inhibits crown gall formation on grapevines. Chinese Sci. Bull. 35:1055–1056.

61. Lippincott, B.B., Margot, J.B., and Lippincott, J.A. 1977. Plasmid content and tumor initiation complementation by *Agrobacterium tumefaciens* IIBNV6. J. Bacteriol. 132:824–831.

62. Lippincott, J.A., and Lippincott, B.B. 1970. Enhanced tumor initiation by mixtures of tumorigenic and nontumorigenic strains of *Agrobacterium*. Infect. Immun. 2:623–630.

63. Lopez, M.M., Gorris, M.T., Salcedo, C.I., Montojo, A.M., and Miro, M. 1989. Evidence of biological control of *Agrobacterium tumefaciens* strains sensitive and resistant to agrocin 84 by different *Agrobacterium radiobacter* strains on stone fruit trees. Appl. Environ. Microbiol. 55:741–746.

64. Lu, S., Canfield, M., Haas, J.H., Manuilis, S., Ream, W., and Moore, L.W. 1993. Use of sensitive nonradioactive methods to detect *Agrobacterium tumefaciens* in crown gall tumors of naturally infected woody plants. Page 45 in: Abstracts, 6th International Congress of Plant Pathology, Montreal, Canada. National Research Council of Canada, Ottawa, Canada.

65. Lu, S.-F. 1994. Isolation of putative pAgK84 transconjugants from commercial cherry and raspberry plants treated with *Agrobacterium radiobacter* strain K84. M.Sc. thesis, Oregon State University, Corvallis, OR.

66. Lu, S.F., and Moore, L.W. 1992. Populations of *Agrobacterium* in tumors of cherry trees treated with *A. radiobacter* strain K84. Phytopathology 82:1124. (Abstr.).

67. Martin, G.C., Miller, A.N., Castle, L.A., Morris, J.W., Morris, R.O., and Dandekar, A.M. 1990. Feasibility studies using β-glucuronidase as a gene fusion marker in apple, peach, and radish. Am. Soc. Hortic. Sci. 115:686–691.

68. Merlo, D.J., and Nester, E.W. 1977. Plasmids in avirulent strains of *Agrobacterium*. J. Bacteriol. 129:76–80.

69. Michel, M.-F., Brasileiro, A.C.M., Depierreux, C., Otten, L., Delmotte, F., and Jouanin, L. 1990. Identification of different *Agrobacterium* strains isolated from the same forest nursery. Appl. Environ. Microbiol. 56:3537–3545.

70. Moore, L. 1977. Prevention of crown gall on *Prunus* roots by bacterial antagonists. Phytopathology 67:139–144.

71. Moore, L.W., and Allen, J. 1986. Controlled heating of root-pruned dormant *Prunus* spp. seedlings before transplanting to prevent crown gall. Plant Dis. 70:532–536.

72. Moore, L.W., and Cooksey, D.C. 1981. Biology of *Agrobacterium tumefaciens*: plant interactions. Pages 15–46 in: The Biology of Rhizobiaceae. Supplement 13, International Review of Cytology. K. Giles, ed. Academic Press, New York, NY.

73. Moore, L.W., Kado, C.I., and Bouzar, H. 1988. *Agrobacterium*. Pages 16–36 in: Laboratory Guide for Identification of Plant Pathogenic Bacteria. 2nd edition. N.W. Schaad, ed. The American Phytopathological Society, St. Paul, MN.

74. Moore, L.W., Loper, J., Stockwell, V., and Kawalek, M. 1991. Novel methods for tracking *Agrobacterium radiobacter* K84 and plasmid-borne genes in agricultural ecosystems. U.S. Environmental Protection Agency Report 600/391/042. 93 pp.

75. Nautiyal, C.S., and Dion, P. 1990. Characterization of the opine-utilizing microflora associated with samples of soil and plants. Appl. Environ. Microbiol. 56:2576–2579.

76. Nautiyal, C.S., Dion, P., and Chilton, W.S. 1991. Mannopine and mannopinic acid as substrates for *Arthrobacter* sp. strain MBA209 and *Pseudomonas putida* NA513. J. Bacteriol. 173:2833–2841.

77. Nesme, X., Ponsonnet, C., Picard, C., and Normand, P. 1992. Chromosomal and pTi genotypes of agrobacterium strains isolated from *Populus* tumors in two nurseries. FEMS Microbiol. Ecol. 101:189–196.

78. Nester, E.W., Garfinkel, D.J., Gelvin, S.B., Montoya, A.L., and Gordon, M.P. 1981. A mutational and transcriptional analysis of a tumor inducing plasmid of *Agrobacterium tumefaciens*. Pages 467–476 in: Molecular Biology, Pathogenicity, and Ecology of Bacterial Plasmids. S.B. Levy, R.C. Clowes, and E.L. Koenig, eds. Plenum Press, New York, NY.

79. New, P.B., and Kerr, A. 1972. Biological control of crown gall: field measurements and glasshouse experiments. J. Appl. Bacteriol. 35:279–287.

80. Ophel, K., and Kerr, A. 1990. *Agrobacterium vitis* sp. nov. for strains of *Agrobacterium* biovar 3 from grapevines. Intern. J. System. Bacteriol. 40:236–241.

81. Ophel, K., Nicholas, P.R., Magarey, P.A., and Bass, A.W. 1990. Hot water treatment of dormant grape cuttings reduces crown gall incidence in a field nursery. Am. J. Enol. Vitic. 41:325–329.

82. Otten, L., Canaday, J., Gérard, J.-C., Fournier, P., Crouzet, P., and Paulus, F. 1992. Evolution of agrobacteria and their Ti plasmids—a review. Mol. Plant-Microbe Interact. 5:279–287.

83. Palleroni, N.J. 1994. Some reflections on bacterial diversity. ASM News. 60:537–540.

84. Peters, N.K., and Verma, D.P.S. 1990. Phenolic compounds as regulators of gene expression in plant-microbe interactions. Mol. Plant-Microbe Interact. 3:4–8.

85. Petit, A., Tempe, J., Kerr, A., Holsters, M., Van Montagu, M., and Schell, J. 1978. Substrate induction of conjugative activity of *Agrobacterium tumefaciens* Ti plasmids. Nature (Lond.) 271:570–571.

86. Pu, X.-A., and Goodman, R.N. 1993. Tumor formation by *Agrobacterium tumefaciens* is suppressed by *Agrobacterium radiobacter* HLB-2 on grape plants. Am. J. Enol. Vitic. 44:249–254.

87. Pu, X.-A., and Goodman, R.N. 1993. Effects of fumigation and biological control on infection of indexed crown gall free grape plants. Am. J. Enol. Vitic. 44:241–248.

88. Raio, A., Zoina, A., Canfield, M.L., and Moore, L.W. 1994. Effect of soil solarization on *Agrobacterium tumefaciens* populations. Phytopathology 84:1135. (Abstr.).

89. Robertson, J.L., Holliday, T., and Matthysse, A.G. 1988. Mapping of *Agrobacterium tumefaciens* chromosomal genes affecting cellulose synthesis and bacterial attachment to host cells. J. Bacteriol. 170:1408–1411.

90. Rodriguez-Palenzuela, P., Burr, T.J., and Collmer, A. 1991. Polygalacturonase is a virulence factor in *Agrobacterium tumefaciens* biovar 3. J. Bacteriol. 173:6547–6552.

91. Roy, M.A., and Sasser, M. 1983. A medium selective for *Agrobacterium tumefaciens* biotype 3. Phytopathology 73:810. (Abstr.).

92. Schroth, M.N., and Moller, W.J. 1976. Crown gall controlled in the field with a nonpathogenic bacterium. Plant Dis. Rep. 60:275–278.

93. Schroth, M.N., Weinhold, A.R., McCain, A.H., Hildebrand, D.C., and Ross, N. 1971. Biology and control of *Agrobacterium tumefaciens*. Hilgardia 40:537–552.

94. Segovia, L., Piñero, D., Palacios, R., and Martínez-Romero, E. 1991. Genetic structure of a soil population of nonsymbiotic *Rhizobium leguminosarum*. Appl. Environ. Microbiol. 57:426–433.

95. Slota, J.E., and Farrand, S.K. 1982. Genetic isolation and physical characterization of pAgK84, the plasmid responsible for agrocin 84 production. Plasmid 8:175–186.

96. Sobiczewski, P., Karczewski, J., and Berczynski, S. 1991. Biological control of crown gall *Agrobacterium tumefaciens* in Poland. Fruit Sci. Rep. 28:125–132.

97. Stachel, S.E., and Nester, E.W. 1986. The genetic and transcriptional organization of the *vir* region of the A6 Ti plasmid of *Agrobacterium tumefaciens*. EMBO J. 5:1445–1454.

98. Staphorst, J.L., Van Zyl, F.G.H., Strijdom, B.W., and Groenewold, Z.E. 1985. Agrocin-producing pathogenic and nonpathogenic biotype-3 strains of *Agrobacterium tumefaciens* active against biotype-3 pathogens. Curr. Microbiol. 12:45–22.

99. Stockwell, V.O., Moore, L.W., and Loper, J.E. 1993. Fate of *Agrobacterium radiobacter* K84 in the environment. Appl. Environ. Microbiol. 59:2112–2120.

100. Stover, E. 1994. Promising new methods for controlling crown gall in grapes. Grape Res. News. 5:1–3.

101. Strik, B. 1992. Assessing rootstocks for winegrape production in Oregon. Pages 43–48 in: The Oregon Winegrape Growers' Guide. T. Casteel, ed. The Oregon Wine Grape Growers' Association, Portland, OR.

102. Süle, S., Mozsar, J., and Burr, T.J. 1994. Crown gall resistance of *Vitis* spp. and grapevine rootstocks. Phytopathology 84:607–611.

103. Szegedi, E., Czakó, M., Otten, L., and Koncz, C.S. 1988. Opines in crown gall tumours induced by biotype 3 isolates of *Agrobacterium tumefaciens*. Physiol. Mol. Plant Pathol. 32:237–247.

104. Szegedi, E., Otten, L., and Czakó, M. 1992. Diverse types of tartrate plasmids in *Agrobacterium tumefaciens* biotype III strains. Mol. Plant-Microbe Interact. 5:435–438.

105. Tempé, J., and Goldmann, A. 1982. Occurrence and biosynthesis of opines. Pages 427–449 in: Molecular Biology of Plant Tumors. G. Kahl and J.S. Schell, eds. Academic Press, New York, NY.

106. Tempé, J., and Petit, A. 1982. Opine utilization by *Agrobacterium*. Pages 451–459 in: Molecular Biology of Plant Tumors. G. Kahl and J.S. Schell, eds. Academic Press, New York, NY.

107. Thies, J.E., Bohlool, B.B., and Singleton, P.W. 1991. Subgroups of the cowpea miscellany: symbiotic specificity within *Bradyrhizobium* spp. for *Vigna unguiculata*, *Phaseolus lunatus*, *Arachis hypogaea*, and *Macroptilium atropurpureum*. Appl. Environ. Microbiol. 57:1540–1545.

108. Thomashow, M.F., Panagopoulos, C.G., Gordon, M.P., and Nester, E.W. 1980. Host range of *Agrobacterium tumefaciens* is determined by the Ti plasmid. Nature (Lond.) 283:794–796.

109. Tiedje, J.M. 1994. Microbial diversity: of value to whom? ASM News. 60:524–525.

110. Tremblay, G., Lambert, R., Lebeuf, H., and Dion, P. 1987. Isolation of bacteria from soil and crown-gall tumors on the basis of their capacity for opine utilization. Phytoprotection 68:35–42.

111. Utkhede, R.S., and Smith, E.M. 1990. Effect of fumigants and *Agrobacterium radiobacter* strain 84 in controlling crown gall of apple seedlings. J. Phytopathol. 128:265–270.

112. Vicedo, B., Peñalver, R., Asins, M.J., and López, M.M. 1993. Biological control of *Agrobacterium tumefaciens*, colonization, and pAgK84 transfer with *Agrobacterium radiobacter* K84 and the Tra⁻ mutant strain K1026. Appl. Environ. Microbiol. 59:309–315.

113. Winans, S.C. 1992. Two-way chemical signaling in *Agrobacterium*-plant interactions. Microbiol. Rev. 56:12–31.

114. Yamada, T., Lee, P.-D., and Kosuge, T. 1986. Insertion sequence elements of *Pseudomonas savastanoi*: nucleotide sequence and homology with *Agrobacterium tumefaciens* transfer DNA. Proc. Natl. Acad. Sci. USA 83:8263–8267.

115. Zhang, L., Murphy, P.J., Kerr, A., and Tate, M.E. 1993. *Agrobacterium* conjugation and gene regulation by N-acyl-L-homoserine lactones. Nature (Lond.) 362:446–448.

Chapter 9

BIOLOGICAL CONTROL OF FUSARIUM WILTS: OPPORTUNITIES FOR DEVELOPING A COMMERCIAL PRODUCT

C. Alabouvette, P. Lemanceau, and C. Steinberg

Fusarium wilts induced by formae speciales of *Fusarium oxysporum*, and Fusarium root rot induced by more or less specific strains of *F. oxysporum*, are among the more severe plant diseases in the world. Fusarium wilts affect many plant species belonging to all botanical families except the Graminaceae.

The available control methods are not efficient or are difficult to apply. Although several molecules are effective in vitro against *F. oxysporum*, they fail to control the disease under field conditions where it is not possible to apply the fungicide directly to the roots. Soil fumigation may control Fusarium diseases for several months, but usually the pathogens rapidly recolonize the disinfested soil, and, at the end of the cropping period, damage may be as severe in the fumigated soil as in the untreated soil. Therefore, the best way to control Fusarium diseases is the use of resistant cultivars. For example, all the varieties of tomato grown in greenhouses for fresh fruit production are resistant to the two most common races of *F. oxysporum* f. sp. *lycopersici* and a gene conferring resistance to the race 2 recently identified in different countries is already available. But breeding for resistance can be very difficult when, for example, no dominant gene is known (e.g. carnation), or the plant species is dioecious (e.g. palm trees).

These difficulties in controlling the disease explain why studies of biological control of Fusarium wilts have been developed for a long time, independently of the recent concern for environmental protection.

FUSARIUM WILT-SUPPRESSIVE SOILS

The existence of soils that naturally limit the incidence of Fusarium wilts has been recognized since the end of last century and has stimulated research on biological control of Fusarium wilts during the last 20 years. During the sixties, studies by Stover (53), Stotzky and Martin (52), and their coworkers were devoted to the role of abiotic factors, mainly clays, in relation to the reduction of disease incidence in the so-called "long-life soils" in Central America. Then, studies by Smith and Snyder (50,51) and coworkers indicated that biological factors were involved in the mechanisms of soil suppressiveness.

Initiated in 1975, studies of the suppressive soils from Châteaurenard (France), clearly demonstrated that the saprophytic microflora is basically responsible for soil suppressiveness. Indeed, suppressiveness is destroyed by applying, to the suppressive soils, treatments that kill most of the microorganisms, but is restored by reintroduction of the microflora in the previously disinfested suppressive soil (34). The microbial nature of suppressiveness has been established for all the soils suppressive to Fusarium wilts now described in the world (47,54,56). However, the microbial interactions responsible for soil suppressiveness are always under the control of abiotic factors, as shown by Amir and Alabouvette (5).

Recognition of the involvement of saprophytic microflora in the mechanisms of soil suppressiveness led to the idea of making conducive soils suppressive by manipulating their microbial balance. Two approaches can be followed to transform a conducive soil into a suppressive soil: (i) enhance the low level of natural suppressiveness that exists in every soil; and (ii) isolate antagonistic microorganisms that can be mass produced and introduced in conducive soils to control Fusarium wilts. Although not the more ecological, the second approach has been followed by many teams that have been able to isolate antagonistic microorganisms from suppressive soils.

A review of the literature shows that many different types of organisms have been proposed to control Fusarium diseases (1). Most of them have been selected in vitro for their ability to inhibit the growth of *F. oxysporum*, and then have been applied to a conducive or a disinfested soil where they were able to reduce disease incidence or severity after artificial infestation with the pathogen. This type of approach does not prove that the selected microorganisms are responsible for the suppressiveness of the suppressive soil, nor that the single introduction of the antagonist makes the conducive soil suppressive to the disease. It just demonstrates that under certain conditions the selected antagonist could be used to control Fusarium diseases. Indeed, the natural phenomenon of soil suppressiveness is based on several mechanisms, involving several microbial populations acting alone or together to limit the activity of the pathogen. Therefore, soil suppressiveness will never be achieved by the introduction of a single strain of antagonist.

The two main models studied, the suppressive soils from Châteaurenard (France), and the suppressive soils from the Salinas Valley (California) led to the hypotheses that either nonpathogenic strains of *F. oxysporum* (46) or fluorescent pseudomonads were responsible for soil suppressiveness (48). Recently, it has been demonstrated that these two hypotheses are not mutually exclusive; on the contrary, the two types of antagonistic microorganisms can act together to control Fusarium wilts efficiently.

This paper will review the most pertinent facts indicating that nonpathogenic strains of *F. oxysporum*, and strains of *Pseudomonas fluorescens* and *P. putida* can be used as biological control agents.

NONPATHOGENIC *FUSARIUM OXYPORUM* AS BIOLOGICAL CONTROL AGENTS

Fusarium oxysporum is one of the more common soil fungi in cultivated soil, all over the world. It includes a large diversity of strains, all saprophytic, most parasitic (i.e. able to colonize to some extent plant tissues without inducing symptoms), and some pathogenic, inducing either root rot or tracheomycosis. There are no phenotypic traits enabling the recognition of these different types of *F. oxysporum*; the only way to identify the pathogenic *F. oxysporum* is to inoculate the host plant. But the wilt-inducing *F. oxyporum* present a high host specificity which led to the concept of forma specialis (specific to a plant species) and race (specific to a cultivar of the host plant). Because more than one hundred formae speciales have been described by plant pathologists (8), no one can test the strains studied against the entire range of susceptible plant species.

More recently, many studies have been devoted to the characterization of pathogenic versus nonpathogenic strains of *F. oxysporum* based on vegetative compatibility groups (VCG), isoenzyme patterns, and molecular traits (10,12,36,37). Pathogenic and nonpathogenic strains can belong to a same VCG (16), but molecular methods such as PCR-RFLP or RAPD offer the possibility of rapid identification of *F. oxysporum* strains at the subspecific level (15). Although the advantages and disadvantages of these different methods are still controversial, there is no doubt about the availability in the near future of new techniques for distinguishing between pathogenic and nonpathogenic *F. oxysporum*. But in this paper, the term "nonpathogenic strain" refers to strains not able to induce disease on the model plant used to conduct the study.

Involvement Of Nonpathogenic *F. oxysporum* In Soil Suppressiveness

The first evidence for a possible role of nonpathogenic *Fusarium* spp. in suppressive soils, provided by Smith and Snyder (50) and Toussoun (57), indicated that soil suppressive to Fusarium wilts supported a large population of nonpathogenic *Fusarium* spp. But the involvement of the nonpathogenic species of *Fusarium* was confirmed experimentally by a form of "Koch's postulates", i.e. by demonstration that the suppressiveness disappeared after elimination of *Fusarium* by heat treatment, and reappeared after reintroduction of the fungus into the heat-treated soil (46). Since then, numerous results point clearly to a role for nonpathogenic *Fusarium* spp. in the suppression of Fusarium wilts in suppressive soils from different areas of the world (6,44,49,54,56). Strains of *F. oxysporum* and *F. solani* were much more efficient in establishing suppressiveness in soil than *F. roseum* (54). Although isolation of *Fusarium* spp. from suppressive soils is an effective procedure for detecting strains able to control Fusarium diseases, the nonpathogenic strains of *F. oxysporum* selected by Ogawa and Komada (41) and Postma and Rattink (45) were obtained from the stem of healthy plants. Whatever their origin, the strains of *F. oxysporum* did not have the same ability to control Fusarium wilts (Figure 1). Therefore, screening procedures based on biotests have been developed to select the most efficient strains to control the disease (11). In an attempt to simplify the screening procedures, Amir and Mahdi (7) compared several properties of efficient and nonefficient strains in vitro and in disinfested soil. Unfortunately, there is no clear relationship between growth rate, conidial production, antagonism in vitro, colonization rate of sterilized soil, and ability to control the disease. The screening procedure proposed is based on too many tests and the results are too variable to be used routinely. Bioassays followed by an appropriate statistical test remain the most suitable method of screening for strains effective in controlling Fusarium wilts.

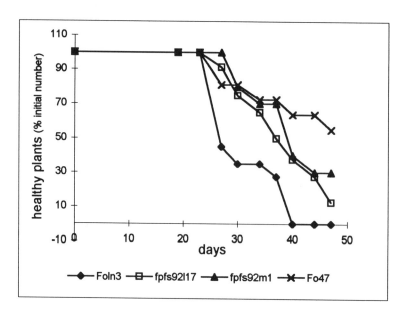

Figure 1. Ability of three different strains of nonpathogenic *F. oxysporum* to control Fusarium wilt of flax. The nonpathogenic strains were introduced at 1 × 10⁵ colony-forming units (CFU) ml⁻¹ and the pathogenic strain *F. oxysporum* f. sp. *lini* (Foln3) at 1 × 10³ CFU ml⁻¹ in 9-ml plugs of rockwool in which one flax plant was grown. There were three replicates of 11 plants.

Modes Of Action Of Nonpathogenic Strains Of *F. oxysporum*

Control of Fusarium wilts by nonpathogenic *F. oxysporum* can be achieved through several modes of action that are not mutually exclusive. Three modes of action have been investigated: (i) competition for nutrient in the rhizosphere, (ii) competition for infection sites on the rhizoplane, and (iii) induced resistance.

During the past years, progress made in biochemistry and molecular biology stimulated research on resistance mechanisms that were induced in the host plant inoculated with different types of antagonistic microorganisms, and that resulted in some reduction of disease severity (19). In the case of Fusarium wilts, it was established that preinoculation of a plant with an incompatible strain results in the mitigation of symptoms when the plant is later inoculated with a compatible strain (38). This phenomenon was described as premunition or cross-protection. Only recently has induced resistance been considered as a mechanism that could be responsible for disease control induced by nonpathogenic *F. oxysporum*. Mandeel and Baker (35), Kroon et al. (24), Fuchs and Défago (20), and Olivain et al. (42) have shown that a nonpathogenic strain applied on some roots of the plant can delay disease symptoms induced by a pathogen applied to other roots or in the stem of the plant. All these results tend to show that the plant reacts to inoculation with a nonpathogenic strain and produces a signal that stimulates the defense reactions. There is little information on the nature of the signal. However, Tamietti et al. (55) were able to demonstrate that tomato plants grown in suppressive soils showed a higher content in several hydrolytic enzymes related to the PR proteins. The same enzymatic activities were increased in plants grown in sterilized soil artificially infested with strains of nonpathogenic *F. oxysporum* that induced control of the disease. Although these results showed that induction of resistance could be one of the modes of action of the nonpathogenic *F. oxysporum*, much more research is needed to clearly identify the molecules responsible for induced resistance. Moreover, the protection offered by induced resistance, i.e. when the nonpathogenic strain is physically separated from the pathogenic strain, is less effective than control achieved when the two strains are introduced together in the soil. This observation indicates a role for other mechanisms, especially mechanisms of competition between pathogenic and nonpathogenic *F. oxysporum*.

Because the addition of glucose to a suppressive soil made it conducive, it was hypothesized that competition for carbon was a mechanism for soil suppressiveness. Indeed, it was shown that the total biomass was generally greater and more active in suppressive than in conducive soils, leading to a greater intensity of competition for nutrients in suppressive than in conducive soils (3). It was then demonstrated that strains of *F. oxysporum* differed in their ability to utilize glucose efficiently (13). Lemanceau et al. (29) were able to establish in vitro that competition for carbon was one of the mechanisms by which the nonpathogenic strain Fo47 inhibited the growth of *F. oxysporum* f. sp. *dianthi*, strain WCS816. These results are in accordance with those obtained by Couteaudier and Alabouvette (13) demonstrating that different strains of nonpathogenic *F. oxysporum* did not have the same ability to grow in soil enriched with glucose nor the same ability to compete with a pathogenic strain to colonize a disinfested soil. A mathematical model was used to calculate an index characteristic of the competitive ability of several strains of *F. oxysporum* growing together in a steamed soil. There was a strong correlation between the competitive index, the growth yield in disinfested soil enriched with glucose, and the ability of the strains to control Fusarium wilt of flax (2).

Finally, there is also evidence for the role of competition for infection sites at the root surface. Mandeel and Baker (35) stated that there are a finite number of infection sites that could be protected by increasing the inoculum density of the nonpathogenic strain. Indeed, control of the disease required not necessarily a high inoculum density of the nonpathogen but a high ratio of nonpathogen to pathogen. More recently, Eparvier and Alabouvette (17) demonstrated that strain Fo47 was effectively competing with the pathogen at the apex of the flax root, Fo47 reducing both the colonization rate by the pathogen and the activity of the pathogen in the root tissues.

FLUORESCENT *PSEUDOMONAS* SPECIES AS BIOLOGICAL CONTROL AGENTS

The fluorescent pseudomonads, mainly *Pseudomonas fluorescens* and *P. putida*, are among the most abundant bacteria in the rhizosphere, and on the root surface of plants. These bacteria have received considerable attention since the end of the seventies (23) because they are able to promote the growth of cultivated plants. The mechanisms by which they stimulate the growth of plants are not totally understood but it has been established that some strains are able to inhibit the growth or the activity of fungal pathogens including *F. oxysporum*. Indeed, Kloepper et al. (23) attributed to siderophores produced by fluorescent *Pseudomonas* spp. the suppressiveness of soil to Fusarium wilts. Scher and Baker (47) did not study the role of fluorescent pseudomonads in the suppressive soils from the Salinas Valley themselves, but they isolated strains of fluorescent *Pseudomonas* spp. from mycelial mats buried in these suppressive soils and demonstrated that some of these bacterial strains could induce suppressiveness in a conducive soil. This approach contrasts with that of Rouxel et al. (46), who showed that resident nonpathogenic *F. oxysporum* were involved in the mechanisms of suppression in the suppressive soils from Châteaurenard.

Modes Of Action Of Fluorescent *Pseudomonas* Species

Many studies have been devoted to the modes of action of fluorescent *Pseudomonas* spp. (9,14,33,60) but there are only a limited number of papers reporting significant control of Fusarium wilts by application of these bacteria (27). However, it has been well established that fluorescent *Pseudomonas* spp. producing the siderophores pseudobactin or pyoverdine are very efficient competitors for iron, and that competition for iron is one of the mechanisms responsible for soil suppressiveness to Fusarium wilts. Indeed, reducing the availability of iron by addition of EDDHA increased the level of soil suppressiveness; in contrast, soil suppressiveness was decreased by addition of FeEDTA, thereby providing iron for the pathogenic *F. oxysporum* (28,48). It was then assumed that fluorescent *Pseudomonas* spp. were responsible for the suppressiveness of the soils. In vitro, and in sterilized soil, the pyoverdine produced by fluorescent *Pseudomonas* spp. reduced the rate of chlamydospore germination and the growth of the germ tubes arising from the chlamydospores. Finally, the use of isogenic mutants affected in their capacity to produce siderophores provided further evidence implicating competition for iron as a mechanism of the antagonism expressed by fluorescent *Pseudomonas* spp. against *F. oxysporum* (30). The intensity of competition for iron depends on several environmental factors; in particular, the availability of iron decreases as pH increases. It is therefore noteworthy that most of the soils suppressive to Fusarium wilts exhibit a pH at or above 7.

Fluorescent *Pseudomonas* spp. produce different types of metabolites, many of them having antifungal activities. But the antifungal phenazines and glucinol have not been implicated in control of Fusarium diseases (27).

Although competition for iron seems to be the main mode of action of fluorescent *Pseudomonas* spp. against *F. oxysporum*, Van Peer et al. (59) observed that competition could not account for all the experimental data related to control of Fusarium wilt of carnation. They considered the hypothesis that the fluorescent strain WCS417 of *P. fluorescens* they were using could enhance the resistance of the host plant. Indeed, Van Peer et al. (58) obtained good evidence of induced resistance in the host after inoculation of the fluorescent *Pseudomonas* strain on the root, the pathogenic *F. oxysporum* being inoculated into the stem. Because both microorganisms were spatially separated, it was not possible to attribute the beneficial effect observed to microbial interaction. Moreover, Van Peer et al. (58) observed an increased accumulation of phytoalexins in the stem of carnation preinoculated with the bacteria. More recently, Hoffland et al. (22) demonstrated that both *P. fluorescens* strain WCS417r and salicylic acid induced resistance against *F. oxysporum* in *Arabidopsis thaliana* and radish but without accumulation of PR proteins.

As already stated for nonpathogenic strains of *F. oxysporum*, these different modes of action of the fluorescent *Pseudomonas* spp. are not mutually exclusive and they can probably act together to control Fusarium wilts.

ASSOCIATION OF FLUORESCENT *PSEUDOMONAS* WITH NONPATHOGENIC *F. OXYSPORUM*

From a theoretical point of view, competition for carbon, which occurs in the suppressive soils from Châteaurenard, does not exclude the possibility of competition for iron, demonstrated in the suppressive soils from the Salinas Valley. Therefore, considering both hypotheses, Lemanceau et al. (28) established that competition for both carbon and iron existed in the suppressive soils from Châteaurenard, even though the populations of fluorescent *Pseudomonas* spp. isolated from the suppressive soil were not more competitive for iron than the populations isolated from a conducive soil (32). Addition of carbon with EDDHA in a conducive soil resulted in a level of receptivity intermediate between the high conduciveness observed after addition of carbon and strong suppressiveness noted after addition of EDDHA (25). These observations prompted the hypothesis of a complementary effect of nonpathogenic *F. oxysporum* with fluorescent *Pseudomonas* spp. Indeed, following a specific screening procedure, Lemanceau and Alabouvette (26) isolated from the suppressive soil strains of fluorescent *Pseudomonas* spp. able to improve the efficacy of biological control achieved by the application of a strain of nonpathogenic *F. oxysporum*. Following another approach, Park et al. (43) also showed that interactions between *P. putida* and strains of nonpathogenic *F. oxysporum* could achieve biological control of Fusarium wilts.

The mechanisms of this beneficial interaction remained obscure until Lemanceau et al. (29,30), using a siderophore-deficient mutant of *P. fluorescens*, demonstrated that competition for iron resulting from the activity of the bacterial strain controlled the efficacy of competition for carbon between strains of *F. oxysporum*. Indeed, the growth yield of a strain of *F. oxysporum* growing on a single source of carbon was greatly reduced in the presence of the bacterial siderophore. Moreover, it was shown that the nonpathogenic strain Fo47 was less sensitive to pseudobactin-mediated iron competition than the pathogenic *F. oxysporum* f. sp. *dianthi* strain WCS816. These data together demonstrated that competition for iron, which results from siderophore production by *Pseudomonas* spp., increases the severity of the competition for carbon that results from the activity of both the total biomass and the nonpathogenic *F. oxysporum*. These mechanisms, which exist in naturally suppressive soils, may be used to achieve biological control of Fusarium wilts by introduction of selected strains of nonpathogenic *F. oxysporum* associated with fluorescent *Pseudomonas* spp. into conducive substrates.

CONDITIONS REQUIRED TO ACHIEVE BIOLOGICAL CONTROL

Obviously, the first condition to achieve biological control is to apply effective strains of nonpathogenic *F. oxysporum* and fluorescent *Pseudomonas* spp. However, strains differ in biological control efficacy. For example, Lemanceau and Alabouvette (26) found that only 10.8% of strains of fluorescent *Pseudomonas* spp. were able to improve biological control achieved by the strain Fo47. Screening procedures based on bioassays were developed to select effective strains of nonpathogenic *F. oxysporum* and *P. fluorescens*. These procedures resulted in the selection of strain Fo47 of *F. oxysporum* and strain C7 of *P. fluorescens*. Fo47 controlled Fusarium wilts effectively when applied alone (20,21,39,45) but its efficiency was always improved by association with the bacterial strain C7.

The efficacy of biological control depends on the population density of the antagonistic microorganisms. But, in the case of *F. oxysporum*, the ratio of the population density of the nonpathogen to the population density of the pathogen is more important than the absolute value of the population density itself. Under experimental conditions this ratio has to be greater than 10 (Figure 2) and depends on the plant pathogen model studied. But under commercial conditions, inoculum density of the pathogen is ignored. Therefore, the easiest solution is to introduce the nonpathogenic strain at a concentration close to the capacity of the soil or substrate. In the case of Fo47, this capacity is in the range 5×10^4 to 5×10^5 colony-forming units (CFU) ml^{-1} of substrate.

All the studies of population dynamics indicated that this strain survives very well in soil and in rockwool (18). For example, introduced into a disinfested soil at the rate of 5×10^4 CFU ml^{-1} to control Fusarium disease of asparagus, Fo47 was detected at a concentration greater than 5×10^3 CFU ml^{-1} 1 year later, during which time it prevented recolonization of soil by the pathogenic *F. oxysporum* that had survived in the deeper layers of the soil (C. Alabouvette and D. Didelot, unpublished).

In order to control the disease, the biological control agents have not

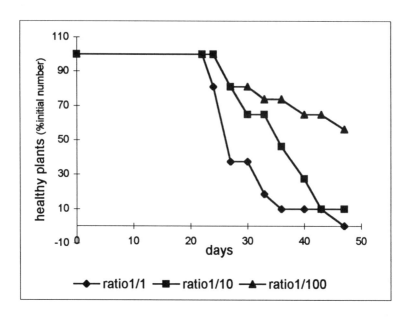

Figure 2. Influence of the ratio of the population density of the pathogen to that of the nonpathogen on biological control of Fusarium wilt of flax. The pathogenic strain Foln3 was introduced at 1×10^3 CFU ml^{-1} and the nonpathogen at 1×10^3, 1×10^4, and 1×10^5 CFU ml^{-1}. Both strains were introduced together at the time of sowing in 9-ml plugs of rockwool in which one flax plant was grown. There were three replicates of 11 plants.

only to be present at the right concentration but also to be active, i.e. expressing their antagonistic activity. Considering that the main mode of action for the nonpathogenic *F. oxysporum* and fluorescent *Pseudomonas* spp. is competition for carbon and iron, the efficacy of these species will be greater when the availability of carbon and iron is limited. It has been demonstrated in soilless culture that providing iron to the plant through FeEDDHA improves disease control due to the activity of *P. fluorescens* (59).

Competition for nutrients occurs only among microorganisms having the same requirements and sharing the same ecological niches. Therefore, the biological control agents must be present and active in the rhizosphere where the pathogenic *F. oxysporum* has to grow to infect the host plant.

It is well known that the rhizosphere is a zone of particularly intense microbial activity because root exudation and rhizodeposition provide nutrients for microorganisms. An effective biological control agent must be adapted to the specific conditions of the rhizosphere in which it has to be active. Rhizosphere competence is one criterion for selecting antagonistic microorganisms. However, depending on the species or cultivar considered, the host plant may select different types of microbial populations. The selection pressure may determine subtle changes affecting the diversity among a given population of microorganisms. Recently, Lemanceau et al. (31) assessed the diversity of fluorescent *Pseudomonas* spp. in an uncultivated soil, and in the rhizoplane and the root tissues of two plant species (flax and tomato) grown in the same soil. It was found that: (i) the plant selected certain strains of fluorescent pseudomonads, (ii) this selection was more or less intense depending on the plant species, and (iii) this selection was plant specific. These results demonstrate the need to study the influence of the plant on the occurrence of strains of fluorescent *Pseudomonas* spp. adapted to colonization of the host roots. One can assume that the strains antagonistic to *F. oxysporum* and adapted to root colonization will be the best biological control agents. The nonpathogenic strain Fo47 was isolated from a suppressive soil and, on average, it has always shown a high ability to control Fusarium wilts of different plant species. However, it seems that nonpathogenic strains isolated from healthy carnation plants had a higher capacity than Fo47 to protect carnation against wilt (45). Therefore, the question of the selection pressure exerted by the host plant on the population of nonpathogenic *F. oxysporum* must be addressed.

OPPORTUNITIES TO DEVELOP A COMMERCIAL PRODUCT

As stated above, much more knowledge is needed to understand the complex modes of action of the antagonistic strains and to apply them in the best conditions to achieve biological control of Fusarium diseases. However, biological control may be applied very soon in some specific cases; for example, in soilless cultures of vegetables or flowers grown in greenhouses where there are fewer constraints than in open fields.

Paradoxically, diseases caused by soilborne fungi are severe in soilless cultures, because the disinfested potting mixtures, or the sterile artificial substrates (e.g. rockwool, fiberglass), are rapidly colonized by fungi introduced by water or small particles of soil dispersed by air. In the absence of the microbial buffer produced in soil by the activity of the saprophytic microflora, the introduced pathogens proliferate rapidly and induce severe disease. Fortunately, under such conditions it is quite easy to introduce a biological control agent at the beginning of the cropping period to prevent colonization of the substrate by the pathogenic fungi. Because the volume of substrate is limited, it is possible to introduce enough propagules of the biological control agent to ensure an even distribution of the antagonist in the substrate in which the roots are growing. Under such conditions, the protective agent will precede the pathogen in the substrate and on the root surface, and thereby obtain a competitive advantage, increasing the probability of successful disease control. For example, Fo47 can be introduced into rockwool with the first watering of the substrate in the greenhouse, or it can be introduced during the processing of peat mixtures at the same time as mineral amendment.

Large-scale experiments of biological control require the production of large amounts of the biological control agent; therefore, the problems of mass-production and formulation must be resolved first. Fortunately, nonpathogenic *F. oxysporum* are easy to produce in fermenters. Under the right conditions, *F. oxysporum* will not produce mycelium but almost exclusively microconidia or budcells that can be harvested by centrifugation or filtration. The conidial suspension in water is mixed with talc and left to dry under a recirculating stream of air at 20°C. The walls of the propagules become thicker, and the so-called microchlamydospores can be stored for several months at room temperature, and for more than 1 year at 4°C, without losing their ability to germinate (2).

To be registered, a biological product must satisfy specific requirements of nontoxicity for humans and animals, and of safety for the environment. The nonpathogenic strain Fo47 has been recognized as nontoxic and it has been demonstrated that it is self-incompatible. Therefore, it would be very difficult for this strain to exchange genetic information with other strains of *F. oxysporum*. A French company called Natural Plant Protection recently started the process of registration for "Fusaclean", a product made from Fo47, and designed to control Fusarium diseases in vegetable and flower crops grown in greenhouses. Experiments are in progress in order to demonstrate the efficacy of the biological control under commercial greenhouse conditions. Obviously these results have to be positive to obtain registration of the product from the French administration. Examples of biological control experiments indicating success but also failures have been recently published (4).

One of the main problems encountered during evaluation of biological control agents concerns the necessity to have both a diseased control and a healthy control to assess the efficacy of the biological control. It is usually very difficult to convince growers to allow an untreated control, even if there are no fungicides registered to control Fusarium wilts. The strain Fo47 has been distributed to several colleagues who have been using it to control Fusarium wilts of cyclamen, carnation, tomato, and basil, with satisfactory results. Therefore, we are confident in its future registration and utilization to control Fusarium wilts under greenhouse conditions. But, as stated above, association of a selected strain of *P. fluorescens* with Fo47 always improves the efficacy of the control achieved by Fo47 alone and also improves the consistency of the control.

Until now, under experimental conditions, strain Fo47 has always been introduced into the substrate before planting and the bacterial strain C7 added after plantation by means of the drip irrigation system. The bacteria were grown on agar medium and harvested by washing the surface of the culture with sterile distilled water. Recent studies have been devoted to improving the process of production and application of the bacterial inoculum. Whether a commercial product based on two different microorganisms will be possible and accepted is still an open debate, but one may expect the commercialization of two products that can be used in association to improve biological control of Fusarium wilts.

Association of several microorganisms will be needed to control different diseases that affect the same crop. Indeed, most of the biological control agents are target specific; they are effective only against a given type of pathogen. Although this property represents an advantage from the environmental point of view, it is a great disadvantage for the growers who have to control several plant pathogens in the same crop. For example, cucumber and tomato grown in soilless cultures are severely affected by *Pythium* spp. responsible for root decay and yield losses (40). Given that some strains of fluorescent *Pseudomonas* spp. are able to control *Pythium*-induced diseases, it would be interesting to associate strains of fluorescent *Pseudomonas* spp. and nonpathogenic *F. oxysporum* in order to control simultaneously two of the most important root diseases that affect vegetables grown in soilless cultures.

LITERATURE CITED

1. Alabouvette, C. 1990. Biological control of Fusarium wilt pathogens in suppressive soils. Pages 27–43 in: Biological Control of Soil-borne Plant Pathogens. D. Hornby, ed. CAB International, Wallingford, UK.

2. Alabouvette, C., and Couteaudier, Y. 1992. Biological control of Fusarium wilts with nonpathogenic Fusaria. Pages 415–426 in: Biological Control of Plant Diseases. E.C. Tjamos, G.C. Papavizas, and R.J. Cook, eds. Plenum Press, New York, NY.

3. Alabouvette, C., Couteaudier, Y., and Louvet, J. 1985. Soils suppressive to Fusarium wilt: Mechanisms and management of suppressiveness. Pages 101–106 in: Ecology and Management of Soilborne Plant Pathogens. C.A. Parker, A.D. Rovira, K.J. Moore, P.T.W. Wong, and J.F. Kollmorgen, eds. The American Phytopathological Society, St. Paul, MN.

4. Alabouvette, C., Lemanceau, P., and Steinberg, C. 1993. Recent advances in the biological control of Fusarium wilts. Pestic. Sci. 37:365–373.

5. Amir, H., and Alabouvette, C. 1993. Involvement of soil abiotic factors in the mechanisms of soil suppressiveness to Fusarium wilts. Soil Biol. Biochem. 25:157–164.

6. Amir, H., and Amir, A. 1988. Le Palmier dattier et la fusariose. XIV. Antagonisme dans le sol de souches de *Fusarium solani* vis-à-vis de *Fusarium oxysporum* f. sp. *albedinis*. Rev. Ecol. Biol. Sol 25:161–174.

7. Amir, H., and Mahdi, N. 1992. Corrélations entre quelques caractéristiques écologiques de différentes souches de *Fusarium* avec référence particulière à leur persistance dans le sol. Soil Biol. Biochem. 24:249–258.

8. Armstrong, G.M., and Armstrong, J.K. 1981. Formae speciales and races of *Fusarium oxysporum* causing wilt diseases. Pages 391–399 in: Fusarium: Diseases, Biology, and Taxonomy. P.E. Nelson, T.A. Toussoun, and R.J. Cook, eds. Pennsylvania State University Press, University Park, PA.

9. Bakker, P.A.H.M., Van Peer, R., and Schippers, B. 1991. Suppression of soil-borne plant pathogens by fluorescent pseudomonads: mechanisms and prospects. Pages 217–230 in: Development in Agriculturally Managed-Forest Ecology. A.B.R. Beemster, G.J. Bollen, M. Gerlach, M.A. Ruissen, B. Schippers, and A. Tempel, eds. Elsevier, Amsterdam, The Netherlands.

10. Bosland, P.W., and Williams, P.H. 1987. An evaluation of *Fusarium oxysporum* from crucifers based on pathogenicity, isozyme polymorphism, vegetative compatibility and geographic origin. Can. J. Bot. 65:2067-2073.

11. Corman, A., Couteaudier, Y., Zegerman, M., and Alabouvette, C. 1986. Réceptivité des sols aux fusarioses vasculaires: méthode statistique d'analyse des résultats. Agronomie 6:751-757.

12. Correll, J.C. 1992. Genetic, biochemical and molecular techniques for the identification and detection of soil-borne plant pathogenic fungi. Pages 7-16 in: Methods for Research on Soilborne Phytopathogenic Fungi. L.L. Singleton, J.D. Mihail, and C.M. Rush, eds. The American Phytopathological Society, St. Paul, MN.

13. Couteaudier, Y., and Alabouvette, C. 1990. Quantitative comparison of *Fusarium oxysporum* competitiveness in relation to carbon utilization. FEMS Microbiol. Ecol. 74:261-268.

14. Défago, G., and Haas, D. 1990. Pseudomonads as antagonists of soilborne plant pathogens: mode of action and genetic analysis. Pages 249-291 in: Soil Biochemistry, Vol. 6. J.M. Bollag and G. Stotzky, eds. Marcel Dekker, New York, NY.

15. Edel, V., Steinberg, C., Avelange, I., Laguerre, G., and Alabouvette, C. 1995. Comparison of three molecular methods for the characterization of *Fusarium oxysporum* strains. Phytopathology 85:579-585.

16. Elmer, W.H., and Stephens, C.T. 1989. Classification of *Fusarium oxysporum* f. sp. *asparagi* into vegetatively compatible groups. Phytopathology 79:88-93.

17. Eparvier, A., and Alabouvette, C. 1994. Use of ELISA and GUS-transformed strains to study competition between pathogenic and non-pathogenic *Fusarium oxysporum* for root colonization. Biocontrol Sci. Technol. 4:35-47.

18. Eparvier, A., Lemanceau, P., and Alabouvette, C. 1991. Population dynamics of non-pathogenic *Fusarium* and fluorescent *Pseudomonas* strains in rockwool, a substratum for soilless culture. FEMS Microbiol. Ecol. 86:177-184.

19. Fritig, B., and Legrand, M. 1993. Mechanisms of Plant Defense Responses. Kluwer Academic Publishers, Dordrecht, The Netherlands. 480 pp.

20. Fuchs, J., and Défago, G. 1994. Induction of resistance in tomato plants against Fusarium wilt by the non-pathogenic strain Fo47 of *Fusarium oxysporum*. Page 41 in: Environmental Biotic Factors in Integrated Plant Disease Control. 3rd EFPP Conference, The Polish Phytopathological Society, Poznan, Poland.

21. Jeannequin, B., Martin, C., and Alabouvette, C. 1991. Lutte contre la fusariose du collet de la tomate. Pages 321–329 in: Troisième Conférence Internationale sur les Maladies des Plantes, Vol. I/III, ANPP, Paris, France.

22. Hoffland, E. 1994. Induced systemic resistance in *Arabidopsis* and radish: Involvement of PR proteins. Page 42 in: Environmental Biotic Factors in Integrated Plant Disease Control. 3rd EFPP Conference, The Polish Phytopathological Society, Poznan, Poland.

23. Kloepper, J.W., Leong, J., Teintze, M., and Schroth, M.N. 1980. Enhanced plant growth by siderophores produced by plant growth-promoting rhizobacteria. Nature (Lond.) 286:885–886.

24. Kroon, B.A.M., Scheffer, R.J., and Elgersma, D.M. 1991. Induced resistance in tomato plants against Fusarium wilt involved by *Fusarium oxysporum* f. sp. *dianthi*. Neth. J. Plant Pathol. 97:401–408.

25. Lemanceau, P. 1989. Role of competition for carbon and iron in mechanisms of soil suppressiveness to Fusarium wilts. Pages 386–396 in: Vascular Wilt Diseases of Plants: Basic Studies and Control. E.C. Tjamos and C.H. Beckman, eds. NATO ASI Series H28. Springer-Verlag, Berlin, Germany.

26. Lemanceau, P., and Alabouvette, C. 1991. Biological control of fusarium diseases by fluorescent *Pseudomonas* and non-pathogenic *Fusarium*. Crop Prot. 10:279–286.

27. Lemanceau, P., and Alabouvette, C. 1993. Suppression of Fusarium-wilts by fluorescent pseudomonads: mechanisms and applications. Biocontrol Sci. Technol. 3:219–234.

28. Lemanceau, P., Alabouvette, C., and Couteaudier, Y. 1988. Recherches sur la résistance des sols aux maladies. XIV. Modification du niveau de réceptivité d'un sol résistant et d'un sol sensible aux fusarioses vasculaires en réponse à des apports de fer ou de glucose. Agronomie 8:155–162.

29. Lemanceau, P., Bakker, P.A.H.M., De Kogel, W.J., Alabouvette, C., and Schippers, B. 1992. Effect of Pseudobactin 358 production by *Pseudomonas putida* WCS358 on suppression of Fusarium wilt of carnations by nonpathogenic *Fusarium oxysporum* Fo47. Appl. Environ. Microbiol. 58:2978–2982.

30. Lemanceau, P., Bakker, P.A.H.M., De Kogel, W.J., Alabouvette, C., and Schippers, B. 1993. Antagonistic effect on nonpathogenic *Fusarium oxysporum* strain Fo47 and pseudobactin 358 upon pathogenic *Fusarium oxysporum* f. sp. *dianthi*. Appl. Environ. Microbiol. 59:74–82.

31. Lemanceau, P., Corberand, T., Gardan, L., Latour, X., Laguerre, G., Boeufgras, J.-M., and Alabouvette, C. 1995. Effect of two plant species, flax (*Linum usitatissimum* L.) and tomato (*Lycopersicon esculentum* Mill.), on the diversity of soilborne populations of fluorescent pseudomonads. Appl. Environ. Microbiol. 61:1004–1012.

32. Lemanceau, P., Samson, R., and Alabouvette, C. 1988. Recherches sur la résistance des sols aux maladies. XV. Comparaison des populations de *Pseudomonas* fluorescents dans un sol résistant et un sol sensible aux fusarioses vasculaires. Agronomie 8:243–249.

33. Loper, J.E., and Lindow, S.E. 1993. Roles of competition and antibiosis in suppression of plant diseases by bacterial biological control agents. Pages 144–155 in: Pest Management: Biologically Based Technologies. R.D. Lumsden and J.L. Vaughn, eds. American Chemical Society, Washington, DC.

34. Louvet, J., Rouxel, F., and Alabouvette, C. 1976. Recherches sur la résistance des sols aux maladies. I—Mise en évidence de la nature microbiologique de la résistance d'un sol au développement de la fusariose vasculaire du melon. Ann. Phytopathol. 8:425–436.

35. Mandeel, Q., and Baker, R. 1991. Mechanisms involved in biological control of Fusarium wilt of cucumber with strains of nonpathogenic *Fusarium oxysporum*. Phytopathology 81:462–469.

36. Manicom, B., Bar, J.M., and Kutze, J.M. 1990. Molecular methods of potential use in the identification and taxonomy of filamentous fungi, particularly *Fusarium oxysporum*. Phytopathologica 22:233–239.

37.   Manulis, S., Kogan, N., Reuven, M., and Ben-Yephet, Y. 1994.
        Use of the RAPD technique for identification of *Fusarium*
        oxysporum f. sp. *dianthi* from carnation. Phytopathology
        84:98–101.
38.   Matta, A. 1989. Induced resistance to Fusarium wilt diseases.
        Pages 175–196 in: Vascular Wilt Diseases of Plants: Basic Studies
        and Control. E.C. Tjamos and C.H. Beckman, eds. NATO ASI
        Series H28. Springer-Verlag, Berlin, Germany.
39.   Mattusch, P. 1990. Biologische Bekämpfung von *Fusarium*
        oxysporum an einigen gärtnerischen Kulturpflanzen.
        Nachrichtenbl. Deut. Planzenschutzd. 42:148–150.
40.   Moulin, F., Lemanceau, P., and Alabouvette, C. 1994.
        Pathogenicity of *Pythium* species on cucumber in peat-sand,
        rockwool and hydroponics. Eur. J. Plant Pathol. 100:3–17.
41.   Ogawa, K., and Komada, H. 1984. Biological control of
        Fusarium wilt of sweet potato by non-pathogenic *Fusarium*
        oxysporum. Ann. Phytopathol. Soc. Japan 50:1–9.
42.   Olivain, C., Steinberg, C., and Alabouvette, C. 1994. Evidence
        of induced resistance in tomato inoculated by nonpathogenic
        strains of *Fusarium oxysporum*. Abstract in: Environmental Biotic
        Factors in Integrated Plant Disease Control. 3rd Conference
        EFPP, The Polish Phytopathological Society, Poznan, Poland.
43.   Park, C.-S., Paulitz, T.C., and Baker, R. 1988. Biocontrol of
        Fusarium wilt of cucumber resulting from interactions between
        Pseudomonas putida and nonpathogenic isolates of *Fusarium*
        oxysporum. Phytopathology 78:190–194.
44.   Paulitz, T.C., Park, C.S., and Baker, R. 1987. Biological control
        of Fusarium wilt of cucumber with nonpathogenic isolates of
        Fusarium oxysporum. Can. J. Microbiol. 33:349–353.
45.   Postma, J., and Rattink, H. 1992. Biological control of Fusarium
        wilt of carnation with a nonpathogenic isolate of *Fusarium*
        oxysporum. Can. J. Bot. 70:1199–1205.
46.   Rouxel, F., Alabouvette, C., and Louvet, J. 1979. Recherches
        sur la résistance des sols aux maladies. IV. Mise en évidence du
        rôle des *Fusarium* autochtones dans la résistance d'un sol à la
        fusariose vasculaire du Melon. Ann. Phytopathol. 11:199–207.
47.   Scher, F.M., and Baker, R. 1980. Mechanism of biological
        control in a *Fusarium*-suppressive soil. Phytopathology
        70:412–417.

48.   Scher, F.M, and Baker, R. 1982. Effect of *Pseudomonas putida* and a synthetic iron chelator on induction of soil suppressiveness to Fusarium wilt pathogens. Phytopathology 72:1567–1573.

49.   Schneider, R.W. 1984. Effects of nonpathogenic strains of *Fusarium oxysporum* on celery root infection by *F. oxysporum* f. sp. *apii* and a novel use of the Lineweaver-Burk double reciprocal plot technique. Phytopathology 74:646–653.

50.   Smith, S.N., and Snyder, W.C. 1971. Relationship of inoculum density and soil types to severity of Fusarium wilt of sweet potato. Phytopathology 61:1049–1051.

51.   Smith, S.N., and Snyder, W.C. 1972. Germination of *Fusarium oxysporum* chlamydospores in soils favorable and unfavorable to wilt establishment. Phytopathology 62:273–277.

52.   Stotzky, G., and Martin, R.T. 1963. Soil mineralogy in relation to the spread of Fusarium wilt of banana in Central America. Plant Soil 18:317–337.

53.   Stover, R.H. 1962. Fusarial Wilt (Panama disease) of Bananas and Other *Musa* Species. Phytopathological Paper 4. Commonwealth Mycological Institute, Kew, UK.

54.   Tamietti, G., and Alabouvette, C. 1986. Résistance des sols aux maladies: XIII—Rôle des *Fusarium oxysporum* non pathogènes dans les mécanismes de résistance d'un sol de Noirmoutier aux fusarioses vasculaires. Agronomie 6:541–548.

55.   Tamietti, G., Ferraris, L., Matta, A., and Gentile, I.A. 1993. Physiological responses of tomato plants grown in *Fusarium* suppressive soil. J. Phytopathol. 138:66–76.

56.   Tamietti, G., and Pramotton, R. 1990. La réceptivité des sols aux fusarioses vasculaires: rapports entre résistance et microflore autochtone avec référence particulière aux *Fusarium* non pathogènes. Agronomie 10:69–76.

57.   Toussoun, T.A. 1975. *Fusarium*-suppressive soils. Pages 145–151 in: Biology and Control of Soil-Borne Plant Pathogens. G.W. Bruehl, ed. The American Phytopathological Society, St. Paul, MN.

58.   Van Peer, R., Nieman, G.J., and Schippers, B. 1991. Induced resistance and phytoalexin accumulation in biological control of Fusarium wilt of carnation by *Pseudomonas* sp. strain WCS417r. Phytopathology 81:728–734.

59. Van Peer, R., Van Kuik, A.J., Rattink, H., and Schippers, B. 1990. Control of Fusarium wilt in carnation grown on rockwool by *Pseudomonas* sp. strain WCS417r and by Fe-EDDHA. Neth. J. Plant Pathol. 96:119–132.

60. Weller, D.M. 1988. Biological control of soilborne plant pathogens in the rhizosphere with bacteria. Annu. Rev. Phytopathol. 26:379–407.

Chapter 10

TRICHODERMA HARZIANUM IN BIOLOGICAL CONTROL OF FUNGAL DISEASES

A. Tronsmo

All plant disease problems seemed to be solved in the sixties when the new highly effective agrochemicals appeared on the market. Dark clouds, however, appeared in the blue sky when some of these "super" chemicals failed to work as the disease-causing organisms adapted to their new threat by becoming resistant. At the same time, there was a growing concern that the environment was being threatened by the modern industrial society. An especially strong movement against the exaggerated use of chemicals in the environment followed the eye-opening book "Silent Spring" by Rachel Carson (11). Many scientists started to search for alternatives to chemicals to control pests and diseases. This was not new, but more a continuation of earlier practices that had been overshadowed by the "chemical period."

The scientific community has shown increasing interest in alternative ways to control plant diseases using nature's defence weapons. Diseases have been controlled effectively by applying or stimulating antagonists on the plant surface or in soil. However, there is a gap between the small-scale experiment and the commercial use of such methods in agriculture. This is partly because biological control is not yet fully developed and it has been difficult to obtain consistent results over years.

To address this problem, much of the current research on biological control has moved from small-scale trials designed to demonstrate control of different diseases, to attempts at understanding the mechanisms by which biological agents reduce the impact of pathogens (57,88). As formulation of the biological control agent has proved to markedly affect the degree of control attained, research in this area has also increased (18,44,51). The development of principles and practices for biological control of soilborne plant pathogens benefits from studies on biological control of plant pathogens in other habitats. Therefore, this paper provides a comparative analysis of biological and integrated control of plants diseases, with emphasis on the use of *Trichoderma harzianum* in control of fungal diseases on subterranean parts of the plant, on cold stored products, and in the phyllosphere.

WHAT IS BIOLOGICAL CONTROL?

For more than 60 years the term biological control has been used by both entomologists and plant pathologists to mean control of one organism by another. In biological control of plant diseases, three antagonistic mechanisms are recognized: **antibiosis**, which means production of metabolites toxic or inhibitory to the plant pathogen; **competition** for nutrients and space; and **parasitism**, where the antagonist directly extracts nutrients from the pathogen.

In entomology, a current definition of biological control is "the action of parasites, predators, or pathogens in maintaining another organism's population density at a lower average than would occur in their absence" (23). Plant pathologists, on the other hand, have emphasized not only the reduction of population (inoculum) density, but also biological protection of plant surfaces and biological control inside the host plant. This includes the host as a biological system acting alone (resistance) or together with other organisms that can either be antagonistic to the pathogen or can induce host-plant resistance to the pathogen. A modern definition of biological control of plant pathogens has been proposed by Cook and Baker: "Biological control is the reduction of the amount of inoculum or disease-producing activity of a pathogen accomplished by or through one or more organisms other than man" (20). According to this broad definition, biological control includes cultural practices that create an environment favorable to antagonists, host-plant resistance, mass introduction of antagonists, nonpathogenic strains, or other beneficial organisms or agents as compost (20). In this chapter, only the part of biological control concerned with the action of microbial antagonists is considered.

NATURAL BIOLOGICAL CONTROL

Plants are attacked by a vast number of pests and pathogens. In the United States, it is calculated that the crops are attacked by at least 160 species of bacteria, 250 kinds of viruses, 8,000 species of pathogenic fungi, 8,000 species of insects, and 2,000 species of weeds (35). All these organisms harmful to plants are attacked by a complex guild of predators, parasites, pathogens, and antagonists. Cate (12) claims that these biotic antagonists of pests are largely responsible for the fact that less than 15% of the pest species constitute a serious threat to our well being and agricultural productivity.

That natural biological control exists is accepted in the scientific community, but its role and efficacy are hard to demonstrate. Fokkema et al. (36), however, were able to prove the occurrence of natural biological control by saprophytic yeasts on rye leaves. They sprayed the plants with benomyl, which did not have any effect on the pathogen *Cochliobolus sativus*, but nearly eliminated the natural microflora of yeasts. Inoculation experiments with the pathogen under field conditions showed that there was 60% less necrosis on the water-sprayed leaves than the benomyl-sprayed leaves.

There will always be nutrients available for microbes on the plant surface, either from plant exudates or from external sources such as pollen or honeydew. Naturally occurring saprophytes are important in reducing the amount of nutrients that could otherwise stimulate the pathogen. By interfering with the growth of the saprophytes, fungicide treatment could result in an increase in nutrient level on the plant surface, which may lead to greater infection by pathogens when the fungicide treatment ceases or when pathogens that resist or tolerate the fungicide appear. Artificially increasing the population of specific groups of naturally occurring antagonists to provide biological control in the field is a possible control method. This method is most likely to be successful at the time of the year when such groups form a dominant component of the phylloplane microflora. For example, on cereal leaves it would be appropriate to apply bacteria early in the season, yeasts in the middle of the season, and filamentous fungi later in the season (7).

Mechanisms Of Biological Control By *Trichoderma*

Antagonistic *Trichoderma* isolates possess all of the following antagonistic activities: competition, antibiosis, and mycoparasitism (24–26,81).

Competition occurs when there is a demand by two or more microorganisms for the same resource in excess of the immediate supply. Competition between the biological control agent and the pathogen may lead to disease control. However, there is another competition not often recognized and less understood—the competition between the antagonist and the indigenous microflora on the plant surface. Carbon, nitrogen, iron, vitamins, infection sites, and oxygen are the most important factors in competition (66). An example in which competition is important is the early colonization of fresh wounds, such as pruning wounds on fruit trees and stumps of forest trees. In this case, competition for nutrients has been demonstrated as the primary mechanism in the control of *Chondrostereum purpureum*, the silver leaf pathogen, by *Trichoderma viride* on plum trees (21). Competition also seems to be the most potent mechanism in the control of *Fusarium oxysporum* f. sp. *vasinfectum* and *F. oxysporum* f. sp. *melonis* by *T. harzianum* T-35 (72).

Antibiosis occurs when the production of toxic metabolites or antibiotics by one organism has a direct effect on another organism. Such compounds may be volatile or nonvolatile. Even though many antagonists are able to produce antibiotics or toxins in pure culture, there is little proof of the effect of such compounds in biological control (65). Thomashow and Weller (78) were, however, able to provide indirect proof by comparing the biological control effect of two *Pseudomonas fluorescens* isolates, one that produced the antibiotic phenazine, and the other a phenazine-negative mutant. Very little suppressiveness was shown by the mutants that did not produce phenazine, but replacing the phenazine gene restored the biological control activity. The importance of the toxin gliotoxin has also been proven in the antagonism of *Gliocladium virens* against *Pythium ultimum* (70).

Mycoparasitism, hyperparasitism, direct parasitism, and interfungal parasitism are terms used to describe the phenomenon of one fungus parasitizing another. Mycoparasitism covers a multitude of different interactions. Four stages can be distinguished in mycoparasitism (15). The first stage is chemotropic growth, where a chemical stimulus from the pathogen directs the growth of the parasite. This was detected in *Trichoderma* as early as 1981 (16). The mycoparasite starts to branch in an atypical way and these branches grow towards the pathogenic fungus. The next step is recognition. In most cases, the *Trichoderma* antagonist is rather specific and attacks only a few fungi. In this specific interaction, lectins may play an important role (3,32,48,49). The third step is attachment. The *Trichoderma* hypha can either grow along the host hypha or coil around it (26,43,86). The final step is the degradation of the host wall by production of lytic enzymes, chitinases and β-1,3-glucanases (13,33). Lysed sites and penetration holes were found in hyphae of the host fungus following the removal of the parasitic hyphae (33,69).

It is very difficult to estimate the relative importance of each of the three mechanisms, because the importance of the different biological control mechanisms is dependent on the isolates used, the target organism, and the environmental conditions. In conferences on biological control it has often been debated which mechanism is most important, and an agreement is never reached. All three mechanisms are found in some biological control agents, and it seems likely that cooperation between different mechanisms must be an advantage for the antagonist. Strong support for the importance of cell wall-degrading enzymes produced by *Trichoderma* and *Gliocladium* against fungal diseases has been given by Harman and coworkers. They have demonstrated the antifungal activity of purified endochitinase, chitobiosidase, N-acetyl-β-1,4-glucosaminidase, and β-1,3-glucanases, or combinations thereof, against nine different fungal species. Spore germination (or cell replication) and germ tube elongation were inhibited in all chitin-containing fungi except *T. harzianum* P1, the producer of the enzymes. The degree of inhibition was proportional to the level of chitin in the cell wall of the target fungi. Combination of the purified enzymes resulted in a synergistic increase in antifungal activity (27,56–58,88). Especially high synergistic effects are found when β-1,3-glucanases and chitinolytic enzymes are combined (88). This is in accordance with earlier findings concerning a combination of glucanase and endochitinase isolated from plants (60). Synergism between chitinolytic enzymes and fungicides is also described. Sensitivity to agrochemicals such as fusilazole and miconazol, which attack the cytoplasmic membrane of the pathogen, was increased up to a 100-fold when these were combined with small amounts of chitinolytic enzymes (57). The mechanism of action is probably that the hydrolytic enzymes weakened the cell walls, thereby facilitating penetration by the pesticide. This observation has a profound commercial potential, as combination with chitinolytic enzymes can markedly reduce the amount of pesticide needed for effective control of plant diseases.

SELECTION OF BIOLOGICAL CONTROL AGENTS

Screening potentially useful crop protection agents, both chemical and biological, has been going on for many years (10). There are numerous well developed and published screening systems with plants for chemical agents, but many fewer for biological agents, even though there may be advanced unpublished screening methods used in industry. In developing a screening system, an important question is whether there is a correlation between the test for antagonism in dual culture on a synthetic agar medium and the antagonist's effect under natural field and greenhouse conditions. This question has been reviewed by Fravel (37), who documents that there are numerous examples of lack of correlation between in vitro tests and biological control, but also some examples of correlation. Rather than screening thousands of possible biological control agents, one can accelerate the process by taking environmental conditions for the antagonist, pathogen, and plant to be protected into consideration.

With *Trichoderma* spp. as the biological control agent, both successful and unsuccessful screenings on agar plates have been reported. A correlation between the lytic activity of several strains of *T. harzianum* on cell walls of *Sclerotium rolfsii*, *Rhizoctonia solani*, and *Pythium aphanidermatum* and the degree of biological control of those pathogens in vivo has been found (34). On the other hand, no correlation was found between the penetration of *S. rolfsii* sclerotia on agar and antagonistic activity in soil by *T. harzianum* (46). However, by modifying the test to study the inhibition of sclerotial germination on plates with natural soil, good correlation was found with the capacity to control damping-off of bean in the greenhouse (47). No correlation was found between the production of lytic enzymes in liquid culture by three isolates of *Trichoderma* and the control capability against Fusarium wilts of muskmelon and cotton (63). Similarly, Ridout et al. (68) reported that the most effective isolate for biological control of *R. solani* on lettuce produced only small amounts of extracellular protein with low β-1,3-D-glucanase and cellulase activity. Conversely, *T. harzianum* isolates with high in vitro enzyme activity gave poor disease control. This indicates that production of these enzymes under natural conditions is unlikely to be a major factor for these isolates in disease control (17). Sivan and Chet (71) have proposed an explanation of the lack of correlation between in vitro chitinase production by *Trichoderma* spp. and biological control of *Fusarium solani*. They postulated that *F. solani* is protected from chitinases and β-1,3-D-glucanases by a protein layer that could mask the carbohydrate polymers. Therefore, enzymes in addition to chitinases and β-1,3-D-glucanases are probably important for successful antagonism against this pathogen.

Today there is no simple in vitro assay that can replace test plants for evaluating antagonistic isolates, but even with test plants, it is important that the culture conditions mimic natural conditions. If not, one can overestimate the potential of the antagonist. This was illustrated in control experiments using different isolates of *Trichoderma* against *Botrytis cinerea* on strawberry in the greenhouse and in the field. In greenhouse trials, nearly 100% control was obtained with all isolates under constant temperature and humidity conditions, whereas in the field test, less effective and variable control with the different isolates was obtained (A. Tronsmo, unpublished).

There may also be other criteria to consider to make the screening more effective. If the aim is to develop a commercial biological control agent, one has to screen for organisms that grow fast on cheap nutrients and produce propagules with an acceptable shelf life. But even with targeted screens, the success of the selection can only be proven under natural conditions. For example, Powell and Faull (67) estimated that only 5% of the proposed biological control agents from a well-planned screening test actually worked in the field trials.

PRODUCTION OF *TRICHODERMA* FOR BIOLOGICAL CONTROL

In biological control with *Trichoderma*, the biological control agent is usually applied to the plants or soil as chlamydospores or conidia. Because of the difficulties most scientists experienced in producing *Trichoderma* spores in liquid fermentation, conidia have usually been produced on agar plates. This is convenient for small-scale experiments but not when large amounts of conidia are needed for field trials. Backman and Rodriguez-Kabana (2) developed a system using molasses-enriched clay granules as a food base for growing the antagonist for protection against soilborne diseases. This carrier also facilitated dispersal of the fungus into the soil. They obtained good control of *S. rolfsii* in a peanut field by applying 140 kg ha^{-1} of the *Trichoderma* granules. Hadar et al. (40,41) used wheat bran as a substrate for growth and sporulation of *T. harzianum*. Sivan et al. (73) further improved the growth medium by mixing peat and wheat bran. This method has been successfully used in biological control trials against a range of soilborne pathogens (14,74).

Another semisolid fermentation method makes use of barley autoclaved in plastic bags as the solid support and nutrient (79,83). The grain creates a large surface and under the right humidity conditions 5 × 10^8 spores can be produced per gram of barley grains after 14 days (A. Tronsmo, unpublished). This production method has been used with success for *T. harzianum* P1 and *T. harzianum* 107 (80,83,84). This method is, however, not suitable for another biological control agent, *T. harzianum* 1295-22, which does not sporulate under these conditions (A. Tronsmo, unpublished). This illustrates that particular growth conditions have to be provided for each individual biological control agent.

Papavizas et al. (64) have developed a liquid fermentation system that does not need sophisticated fermentation tanks. They have obtained a high biomass yield with a large proportion of chlamydospores of *Trichoderma hamatum*, *T. harzianum*, *T. viride*, and *G. virens*. After the fermentation, the biomass is mixed with alginate and dropped into $CaCl_2$ to make pellets (55). This system works well for biological control of soilborne pathogens. A product named Gliogard (W.R. Grace & Co.), based on *G. virens* strain G20, has now been registered by the United States Environmental Protection Agency for control of soilborne diseases of greenhouse plants (59). This fermentation method does not, however, work with other *Trichoderma* biological control agents such as *T. harzianum* 107, *T. harzianum* P1, and *T. harzianum* 1295-22, because only very low numbers of conidia and chlamydospores are produced by these organisms under the described conditions.

For commercial production of biological control agents in general, liquid fermentation in large commercial fermenters is probably the most realistic method. With regard to *Trichoderma* spp., it is possible to choose hyphae, chlamydospores, or conidia as the basis of a biological control preparation. Hyphae are unlikely to be useful because they will not withstand drying. Papavizas et al. (54) have chosen to produce a biomass dominated by chlamydospores. Under favorable conditions, however, conidia are usually produced more abundantly than chlamydospores. Since conidia are usually produced in aerial environments and not in submerged culture, media must be found that encourage production of conidia in liquid culture. Harman et al. (44) were able to develop a liquid fermentation system that worked well for production of conidia by *T. harzianum* 1295-22. However, although conidia of this strain were produced in abundance, only a small percentage of these propagules survived drying (44). This problem was alleviated by the inclusion of appropriate osmoticants in the medium to a level of about -2 MPa (51). For commercial spore production, good spore yield and good shelf life at ambient temperature are necessary. Harman et al. (44) have succeeded in developing a semisolid fermentation technology now used for commercial production of their registered biological control agent *T. harzianum* 1295-22.

It is not sufficient to have viable propagules when treating the plants or soil; the propagules also have to be properly formulated. Formulation is the answer to the question: "What can transform a potential biological control agent from a laboratory curiosity to a commercially successful product?" (18). A successful biological control formulation is one that is economically produced, safe, stable in the environment, easily applied using conventional agricultural equipment, and gives effective and consistent results under a variety of environmental conditions. A reasonable goal, but difficult to reach, is to make a product that is stable for at least 18 months at 40°C. Such a product would have 2 years in the distribution system (18).

One important factor in the formulation is the nutrient status of the support or additives to the liquid spray. Biological control agents usually must rapidly proliferate and become established where they are applied. Extra nutrients in *Trichoderma* spray solutions are in most cases necessary for the conidia to germinate (22). On the other hand, added nutrients may have the negative effect of stimulating the pathogens (42). Chitin is a nutrient that may be accessible to a limited range of fungi, and it was shown that chitin and *Bacillus cereus* in combination gave better control than *B. cereus* alone against *Cercospora arachidicola* on peanut leaves (52). For application in a solid support or in a liquid spray on seeds, a nutrient addition stimulatory to the antagonist and not available for the pathogens is the goal.

Specificity Of *Trichoderma* Isolates In Biological Control

In biological control, the specificity of the antagonist has been debated. The ideal situation would be that the antagonist would control all plant pathogens and not affect other saprophytic or mycorrhizal fungi. Although this is unrealistic, biological control agents are usually considered to control only a narrow range of pathogens. Bell et al. (4) found, in paired cultures on agar, that an isolate highly effective against one isolate of a pathogen could have only minimal effect on other isolates of the same species. Many of the antagonists tested in the field contain only a single strain and thus may be effective only in localized areas. Consequently, Cook (19) feels that many strains have to be formulated together to increase the range of control. However, Harman et al. (45) have shown that such complex mixtures are not necessary. In numerous trials in several different systems (soil, seed treatment, treatment of the phyllosphere), and against a range of pathogens, including *P. ultimum*, *R. solani*, *Fusarium* spp., *S. rolfsii*, and *B. cinerea*, the antagonist *T. harzianum* 1295-22 made by protoplast fusion (76) has provided good control if properly formulated (42). This is very promising for the future of commercial biological control.

INTEGRATED CONTROL

Toxicants may be employed as a supplement to different biological control systems. The addition of chemicals may allow disease control under unfavorable edaphic conditions. For example, some strains of *Trichoderma* may be ineffective as seed treatments in cold soils, but addition of a compatible protectant such as metalaxyl may provide an effective treatment (77). The use of integrated biological and chemical treatments may also permit control of pathogens outside the range of the bioprotectant itself. Biocides can stress and weaken pathogen propagules and render them more susceptible to attack by antagonists. Sublethal doses of soil fumigants may interfere with the defence mechanisms of resting structures such as sclerotia so that they are more rapidly biologically degraded in soil (61,62). The use of pesticide-resistant *Trichoderma* biological control strains is possible because it is easy to create pesticide-resistant isolates of this fungus (1,81,84). This permits integrated control employing both biological protectants and chemical pesticides (84). Like nonspecific resistance, integrated control is seldom complete, but it seldom fails (9). Integrated control has the advantage over exclusively chemical control, in that there is less risk of development of fungicide resistance when the combined effect of a biological control agent and a fungicide is used to manage the disease.

BIOLOGICAL CONTROL OF POSTHARVEST DISEASES

Biological control of diseases on cold stored products has received little attention compared to biological control in other environments. However, according to Wilson and Pusey (92), there should be good possibilities for control of postharvest diseases for the following reasons: (i) it is possible to control the storage environment; (ii) it is easy to target the application to the product to be protected; (iii) because of the high value of the harvested crop, rather elaborate control procedures, that may not be economically feasible under field conditions, are cost effective for harvested food.

However, even if the potential for biological control of postharvest diseases is good, few results so far have been published where realistic storage conditions and storage times have been applied (93). One of the main reasons is that the climatic conditions best suited for storage of the crop, often close to 0°C, are not suitable for the growth and antagonistic activity of most biological control agents. Changing the conditions to favor the antagonist instead of the crop is not a solution because that will lead to new and often severe problems (75). The demand for cold-tolerant antagonists limits the amount of possible biological control agents, as most isolated antagonists do not have any significant activity below 10°C. Work with antagonistic isolates of *Trichoderma* spp., however, has shown that it is possible to select isolates that act antagonistically at low temperatures (53,86). One of these isolates has been used to control storage rot on carrots and apples.

Control Of Storage Rot Of Carrot

Carrots (*Daucus carota* L.) were harvested from two fields in Norway, one naturally infested by *Mycocentrospora acerina*, and another naturally infested by *Rhizoctonia carotae*. At harvest, the carrots were placed in polyethylene nets and the biological control treatment was performed by soaking the nets in a conidial suspension (10^7 conidia ml^{-1}) of the cold-tolerant isolate *T. harzianum* P1 (ATCC 74058) (83). The carrots were stored in 400-kg bins in which the bottom, sides, and top were covered with polyethylene to prevent the carrots from drying out (87). Room temperature during the storage period was 0 to 0.5°C, but it took 2 months for the temperature in the middle of the produce to sink to 2°C, where it stabilized for the rest of the storage period. The biological treatment significantly reduced the amount of infected roots from both the *M. acerina*-infested and *R. carotae*-infested fields. On average, the amount of marketable crop had increased by 47% after 6 months and 75% after 8.5 months in cold storage with the *Trichoderma* treatment. There was a significant reduction in the attack by *B. cinerea*, *R. carotae* and *Sclerotinia sclerotiorum* in both the *R. carotae*-infested and *M. acerina*-infested fields. The percentage of rot caused by *M. acerina* and *R. carotae* on the untreated and *Trichoderma*-treated carrots is shown in Figure 1.

Control Of Storage Rot Of Apple

Botrytis cinerea can be a serious cause of storage rot on apple. Preharvest treatment against dry eye rot in the flowering period has not had any effect on the storage rot (84), but Janisiewicz et al. (50) have obtained very good results with the yeast *Sporobolomyces roseus* against rot caused by *B. cinerea* and *Penicillium expansum* on wound-inoculated apples stored at 1°C for 3 months.

Gloeosporium or bull's eye rot (*Pezicula malicorticis*, anamorph *Cryptosporiopsis curvispora*) is another serious storage rot on many apple

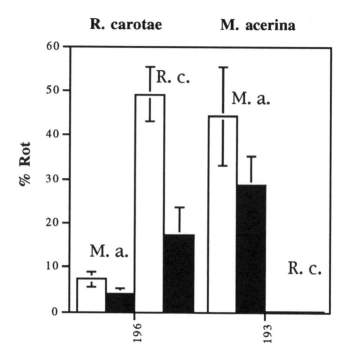

Figure 1. Percentage of carrots with visible symptoms of disease caused by *Mycocentrospora acerina* (M.a.) and *Rhizoctonia carotae* (R.c.) after 196 days (*R. carotae*-infested field) and 193 days (*M. acerina*-infested field) in cold storage. The first four bars from the left refer to the field infested with *R. carotae*; the remaining bars refer to the field infested with *M. acerina*. Open bars = untreated, solid bars = treated with *Trichoderma harzianum* P1 before storage. Standard errors are indicated by vertical lines.

cultivars in Norway. Control has been attempted by chemical spray in the field 14 days before harvest, but has often failed. In 1994, a small-scale experiment was performed on the apple cultivar Aroma, a very popular cultivar but unfortunately highly susceptible to bull's eye rot. The apple trees were sprayed with a conidial suspension of three different *T. harzianum* isolates, P11, Th P1a, or Th-22, to which 0.5% chitosan or 0.01% Triton-X-100 was added. The apples were stored at 4°C for up to 100 days. Figure 2 shows that two of the *T. harzianum* treatments significantly reduced the rot caused by *P. malicorticis*, and this indicates that preharvest biological treatment can be a possible control method against this serious postharvest disease.

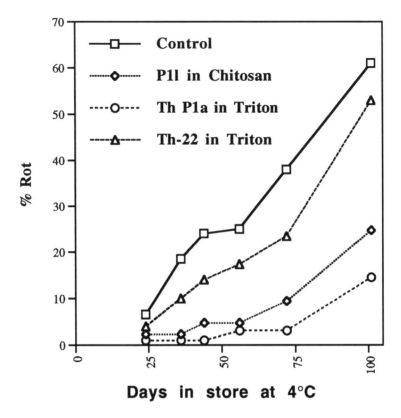

Figure 2. Percentage of apples stored at 4°C for up to 100 days showing visible symptoms of bull's eye rot caused by *Pezicula malicorticis*. Two isolates of *Trichoderma harzianum* (P11 and Th P1a) reduced the rot; a third isolate (Th-22) did not.

BIOLOGICAL CONTROL IN THE PHYLLOSPHERE

The surface of flowers and leaves is called the phyllosphere or phylloplane. This environment is more hostile for microorganisms than the soil. The microorganisms are frequently affected by catastrophic disturbances caused by low nutrient availability, extreme temperatures, drought, and intense radiation. As a consequence, the microbial communities and the interactions within the two systems may change rapidly over time and it may be difficult for an antagonist to establish itself in this environment. However, even if the environment is hostile for most of the microbial community, there is always some microorganism on all plant surfaces, and if an aggressive colonist is able to establish itself on the plant surface in advance of a pathogen, it may be able to inhibit the pathogens from successful infection of the plant.

Botrytis cinerea is a widespread parasite with good saprophytic ability on a great number of plants of considerable economic importance. R.K.S. Wood demonstrated biological control of this pathogen as early as 1950, and suggested that the activity of the saprophytic organisms on dead lettuce tissue was largely responsible for the control of *B. cinerea* under natural conditions (94). Bhatt and Vaughan (5) were able to increase the yield of strawberry artificially inoculated with *B. cinerea* in the greenhouse by spraying with *Cladosporium herbarum*. However, attempts to repeat these results have been unsuccessful (8). Better results have been obtained using *Trichoderma* spp. as antagonists. Using *T. harzianum* and *T. viride*, Tronsmo and Dennis (85) were able to reduce natural infection by *B. cinerea* in the field and in storage to the same level as with chemical treatment with dichlofluanid. Further attempts to control the disease have, however, been less successful. The control has been acceptable in some regions, but no control has been obtained under other geographic and climatic conditions (80).

Botrytis cinerea is one of the more serious pathogens on grape vines. The disease has been successfully controlled under field conditions with *Trichoderma* spp. in Italy (6,38,39), France (28,29), and Israel (31). Systematic trials to control this disease showed that the time of treatment is important for control. Most effective protection was obtained with treatment from flowering to 3 weeks before harvest, but partial protection was also obtained with spraying during the flowering period (30).

Another disease caused by *B. cinerea*, and controllable by *Trichoderma*, is dry eye rot of apple. Under humid conditions, *B. cinerea* may also attack apple flowers, grow from them into the sepals, and cause the dry eye rot (90). Wilted petals are an important nutrient base for the pathogen, and the disease has to be controlled at this stage to prevent the establishment of the pathogen. The first attempt at biological control showed that *Trichoderma pseudokoningii* could reduce infection caused by artificial inoculation, but not the natural infection of *B. cinerea* in the field (89). The antagonistic isolate used was unable to grow below 9°C, and because the temperature during the flowering period often was below 9°C, new trials were performed with an isolate selected for antagonistic properties at low temperature (86). With *T. harzianum* isolates, the natural infection by *B. cinerea* was significantly reduced (80,91).

Table 1. Effect of biological, chemical and integrated control on the frequency of dry eye rot (*Botrytis cinerea*) of apple at harvest. *Trichoderma harzianum* P1 (Th P1) was used as the antagonist, 0.5% carboxymethylcellulose (CMC) as a nutrient amendment, and Ronilan (50% a.i. vinclozolin) or Euparen M (50% a.i. tolyfluanid) as fungicides.

Treatment	Fruits with rot (%)		
	1983	1986	1987
Control	7.3	5.7	7.1
Th P1 + CMC	3.6	1.4	5.3
Th P1 + CMC + 0.02% Ronilan	1.7	1.0	4.8
0.02% Ronilan	3.0	2.7	3.9
0.10% Ronilan	1.5	2.5	3.2
0.15% Euparen	2.8	10.6	8.0
LSD 0.05	3.2	4.2	2.2

Integrated control of *B. cinerea* on apple is also reported. In integrated control, fungicide-resistant or fungicide-tolerant biological control agents are needed. From an experiment on fungicide resistance (82), one isolate, *T. harzianum* P1, was selected. This isolate had the same growth characteristics and fitness as the parent strain *T. harzianum* 107, and it was tolerant to the dicarboximide fungicides vinclozolin and iprodione (84). The *T. harzianum* P1 isolate gave better control than the parent strain, either alone or in combination with reduced concentrations of vinclozolin, and was able to significantly reduce the rot caused by *B. cinerea* on apple by natural infection in the field (Table 1).

CONCLUSION AND NEW APPROACHES

Biological control of plant pathogens has come a long way since Weindling in 1932 described the concept of using fungal antagonists against plant pathogenic fungi (20). Due to numerous examples of effective control in small experiments in the greenhouse and field, there is no doubt about the potential of the methods, but the problem has so far been to bring a commercial product on the market. We already have a number of superior biological control agents. What is missing in most cases, however, is the step from production of the biological control agent for small field trials to large-scale production methods and optimal formulation of the biological control agent. To develop these, a fruitful cooperation between industry and university or research institutes is necessary.

Recent research in biological control has also focused on approaches other than the direct use of antagonists. Among such approaches is the possible use of metabolites produced by antagonists, such as chitinase. Cell wall-degrading enzymes have a potential used alone or in combination with other chemicals such as fungicides. Our research has proved that adding purified chitinases markedly increased the sensitivity of fungal pathogens to agrochemicals (56,57,88). However, to be able to use this technique on a commercial scale, more research using genetic and protein engineering, together with carefully designed biological experiments, still needs to be done.

LITERATURE CITED

1. Abd-El Moity, T.H., Papavizas, G.C., and Shatla, M.N. 1982. Induction of new isolates of *Trichoderma harzianum* tolerant to fungicides and their experimental use for control of white rot on onion. Phytopathology 72:396–400.

2. Backman, P.A., and Rodriguez-Kabana, R. 1975. A system for the growth and delivery of biological control agents to the soil. Phytopathology 65:819–821.

3. Barak, R., Elad, Y., Mirelman, D., and Chet, I. 1985. Lectins: a possible basis for specific recognition in the interaction of *Trichoderma* and *Sclerotium rolfsii*. Phytopathology 75:458–462.

4. Bell, D.K., Wells, H.D., and Markham, C.R. 1982. In vitro antagonism of *Trichoderma* species against six fungal plant pathogens. Phytopathology 72:379–382.

5. Bhatt, D.D., and Vaughan, E.K. 1962. Preliminary investigations on biological control of gray mold (*Botrytis cinerea*) of strawberries. Plant Dis. Rep. 46:342–345.

6. Bisiach, M., Minervini, G., Vercesi, A., and Zerbetto, F. 1985. Six years of experimental trials on biological control against grapevine grey mould. Quad. Vitic. Enol. Univ. Torino 9:285–297.

7.   Blakeman, J.P. 1985. Ecological succession of leaf surface microorganisms in relation to biological control. Pages 6–30 in: Biological Control on the Phylloplane. C.E. Windels and S.E. Lindow, eds. The American Phytopathological Society, St. Paul, MN.

8.   Blakeman, J.P., and Fokkema, N.J. 1982. Potential for biological control of plant diseases on the phylloplane. Annu. Rev. Phytopathol. 20:167–192.

9.   Bruehl, G.W. 1989. Integrated control of soil-borne plant pathogens: an overview. Can. J. Plant Pathol. 11:153–157.

10.   Campbell, R. 1994. Biological control of soil-borne diseases: some present problems and different approaches. Crop Prot. 13:4–13.

11.   Carson, R.L. 1962. Silent Spring. Houghton Mifflin Company, New York, NY. 368 pp.

12.   Cate, J.R. 1990. Biological control of pests and diseases: integrating a diverse heritage. Pages 23–43 in: New Directions in Biological Control: Alternatives for Suppressing Agricultural Pests and Diseases. R.R. Baker and P.E. Dunn, eds. Alan R. Liss Inc., New York, NY.

13.   Cherif, M., and Benhamou, N. 1990. Cytochemical aspects of chitin breakdown during the parasitic action of a *Trichoderma* sp. on *Fusarium oxysporum* f. sp. *radicis-lycopersici*. Phytopathology 80:1406–1414.

14.   Chet, I. 1987. *Trichoderma*—application, mode of action, and potential as a biocontrol agent of soil-borne plant pathogenic fungi. Pages 137–160 in: Innovative Approaches to Plant Disease Control. I. Chet, ed. John Wiley & Sons, New York, NY.

15.   Chet, I. 1990. Mycoparasitism—recognition, physiology and ecology. Pages 725–733 in: New Directions in Biological Control: Alternatives for Suppressing Agricultural Pests and Diseases. R.R. Baker and P.E. Dunn, eds. Alan R. Liss Inc., New York, NY.

16.   Chet, I., Harman, G.E., and Baker, R. 1981. *Trichoderma hamatum*: its hyphal interactions with *Rhizoctonia solani* and *Pythium* spp. Microb. Ecol. 7:29–38.

17.   Coley-Smith, J.R., Ridout, C.J., Mitchell, C.M., and Lynch, J.M. 1991. Control of bottom rot disease of lettuce (*Rhizoctonia solani*) using preparations of *Trichoderma viride*, *T. harzianum* or tolclofos-methyl. Plant Pathol. 40:359–366.

18.   Connick Jr., W.J., Lewis, J.A., and Quimby Jr., P.C. 1990. Formulation of biocontrol agents for use in plant pathology. Pages 345–372 in: New Directions in Biological Control: Alternatives for Suppressing Agricultural Pests and Diseases. R.R. Baker and P.E. Dunn, eds. Alan R. Liss Inc., New York, NY.

19.   Cook, R.J. 1993. The role of biological control in pest management in the 21st century. Pages 10–20 in: Pest Management: Biologically Based Technologies. R.D. Lumsden and J.L. Vaughn, eds. American Chemical Society, Washington, DC.

20.   Cook, R.J., and Baker, K.F. 1983. The Nature and Practice of Biological Control of Plant Pathogens. The American Phytopathological Society, St. Paul, MN. 539 pp.

21.   Corke, A.T.K., and Hunter, T. 1979. Biocontrol of *Nectria galligena* infection of pruning wounds on apple shoots. J. Hort. Sci. 54:47–55.

22.   Danielson, R.M., and Davey, C.B. 1973. Effect of nutrients and acidity on phialospore germination of *Trichoderma in vitro*. Soil Biol. Biochem. 5:517–524.

23.   DeBach, P., ed. 1964. Biological Control of Insect Pests and Weeds. Chapman and Hall, London, UK. 844 pp.

24.   Dennis, C., and Webster, J. 1971. Antagonistic properties of species-groups of *Trichoderma*. I. Production of non-volatile antibiotics. Trans. Br. Mycol. Soc. 57:25–39.

25.   Dennis, C., and Webster, J. 1971. Antagonistic properties of species-groups of *Trichoderma*. II. Production of volatile antibiotics. Trans. Br. Mycol. Soc. 57:41–48.

26.   Dennis, C., and Webster, J. 1971. Antagonistic properties of species-groups of *Trichoderma*. III. Hyphal interaction. Trans. Br. Mycol. Soc. 57:363–369.

27.   Di Pietro, A., Lorito, M., Hayes, C.K., Broadway, R.M., and Harman, G.E. 1993. Endochitinase from *Gliocladium virens*: Isolation, characterization, and synergistic antifungal activity in combination with gliotoxin. Phytopathology 83:308–313.

28.   Dubos, B., and Bulit, J. 1981. Filamentous fungi as biological control agents on aerial plant surfaces. Pages 353–367 in: Microbial Ecology of the Phylloplane. J.P. Blakeman, ed. Academic Press, London, UK.

29.   Dubos, B., Bulit, J., Bugaret, Y., and Verdu, D. 1978. Possibilités d'utilisation du *Trichoderma viride* Pers. comme moyen biologique de lutte contre la Pourriture grise (*Botrytis cinerea* Pers.) et l'Excoriose (*Phomopsis viticola* Sacc.) de la Vigne. C.R. Acad. Agric. Fr. 64:1159–1168.

30.   Dubos, B., Jailloux, F., and Bulit, J. 1982. Protection du vignoble contre la pourriture grise: les propriétés antagonistes du *Trichoderma* à l'égard du *Botrytis cinerea*. Les Colloques de l'INRA. 11:205–219.

31.   Elad, Y. 1994. Biological control of grape gray mould by *Trichoderma harzianum*. Crop Prot. 13:35–38.

32.   Elad, Y., Barak, R., and Chet, I. 1983. Possible role of lectins in mycoparasitism. J. Bacteriol. 154:1431–1435.

33.   Elad, Y., Chet, I., Boyle, P., and Henis, Y. 1983. Parasitism of *Trichoderma* spp. on *Rhizoctonia solani* and *Sclerotium rolfsii*—Scanning electron microscopy and fluorescence microscopy. Phytopathology 73:85–88.

34.   Elad, Y., Chet, I., and Henis, Y. 1982. Degradation of plant pathogenic fungi by *Trichoderma harzianum*. Can. J. Microbiol. 28:719–725.

35.   Ennis Jr., W.B., Dowler, W.M., and Klassen, W. 1975. Crop protection to increase food supplies. Science. 188:593–598.

36.   Fokkema, N.J., Van de Laar, J.A.J., Nelis-Blomberg, A.L., and Schippers, B. 1975. The buffering capacity of the natural mycoflora of rye leaves to infection by *Cochliobolus sativus*, and its susceptibility to benomyl. Neth. J. Plant Pathol. 81:176–186.

37.   Fravel, D.R. 1988. Role of antibiosis in the biocontrol of plant diseases. Annu. Rev. Phytopathol. 26:75–91.

38.   Gullino, M.L. 1992. Control of Botrytis rot of grapes and vegetables with *Trichoderma* spp. Pages 125–132 in: Biological Control of Plant Diseases. E.C. Tjamos, G.C. Papavizas, and R.J. Cook, eds. Plenum Press, New York, NY.

39.   Gullino, M.L., and Garibaldi, A. 1983. Situation actuelle et perspectives d'avenir de la lutte biologique intégrée contre la pourriture grise de la vigne en Italie. Les Colloques de l'INRA 18:91–97.

40.   Hadar, E., Elad, Y., Ovadia, S., Hadar, Y., and Chet, I. 1979. Biological and chemical control of *Rhizoctonia solani* in carnation. Phytoparasitica 7:55. (Abstr.).

41. Hadar, Y., Chet, I., and Henis, Y. 1979. Biological control of *Rhizoctonia solani* damping-off with wheat bran culture of *Trichoderma harzianum*. Phytopathology 69:64–68.

42. Harman, G.E. 1990. Deployment tactics for biocontrol agents in plant pathology. Pages 779–792 in: New Directions in Biological Control: Alternatives for Suppressing Agricultural Pests and Diseases. R.R. Baker and P.E. Dunn, eds. Alan R. Liss Inc., New York, NY.

43. Harman, G.E., Chet, I., and Baker, R. 1981. Factors affecting *Trichoderma hamatum* applied to seeds as a biocontrol agent. Phytopathology 71:569–572.

44. Harman, G.E., Jin, X., Stasz, T.E., Peruzzotti, G., Leopold, A.C., and Taylor, A.G. 1991. Production of conidial biomass of *Trichoderma harzianum* for biological control. Biol. Control 1:23–28.

45. Harman, G.E., Taylor, A.G., and Stasz, T.E. 1989. Combining effective strains of *Trichoderma harzianum* and solid matrix priming to improve biological seed treatments. Plant Dis. 73:631–637.

46. Henis, Y., Adams, P.B., Lewis, J.A., and Papavizas, G.C. 1983. Penetration of sclerotia of *Sclerotium rolfsii* by *Trichoderma* spp. Phytopathology 73:1043–1046.

47. Henis, Y., Lewis, J.A., and Papavizas, G.C. 1984. Interactions between *Sclerotium rolfsii* and *Trichoderma* spp: relationship between antagonism and disease control. Soil Biol. Biochem. 16:391–395.

48. Inbar, J., and Chet, I. 1992. Biomimics of fungal cell-cell recognition by use of lectin-coated nylon fibers. J. Bacteriol. 174:1055–1059.

49. Inbar, J., and Chet, I. 1994. A newly isolated lectin from the plant pathogenic fungus *Sclerotium rolfsii*: purification, characterization and role in mycoparasitism. Microbiology 140:651–657.

50. Janisiewicz, W.J., Peterson, D.L., and Bors, R. 1994. Control of storage decay of apples with *Sporobolomyces roseus*. Plant Dis. 78:466–470.

51. Jin, X., Harman, G.E., and Taylor, A.G. 1991. Conidial biomass and desiccation tolerance of *Trichoderma harzianum* produced at different medium water potentials. Biol. Control 1:237–243.

52.   Kokalis-Burelle, N., Backman, P.A., Rodriguez-Kabana, R., and Ploper, L.D. 1992. Potential for biological control of early leafspot of peanut using *Bacillus cereus* and chitin as foliar amendments. Biol. Control 2:321–328.

53.   Köhl, J., and Schlösser, E. 1989. Decay of sclerotia of *Botrytis cinerea* by *Trichoderma* spp. at low temperatures. J. Phytopathol. 125:320–326.

54.   Lewis, J.A., and Papavizas, G.C. 1983. Production of chlamydospores and conidia by *Trichoderma* spp. in liquid and solid growth media. Soil Biol. Biochem. 15:351–357.

55.   Lewis, J.A., and Papavizas, G.C. 1985. Characteristics of alginate pellets formulated with *Trichoderma* and *Gliocladium* and their effect on the proliferation of the fungi in soil. Plant Pathol. 34:571–577.

56.   Lorito, M., Hays, C.K., Di Pietro, A., Woo, S.L., and Harman, G.E. 1994. Purification, characterization, and synergistic activity of a glucan 1,3-β-glucosidase and an N-acetyl-β-glucosaminidase from *Trichoderma harzianum*. Phytopathology 84:398–405.

57.   Lorito, M., Peterbauer, C., Hays, C.K., and Harman, G.E. 1994. Synergistic interaction between fungal cell wall degrading enzymes and different antifungal compounds enhances inhibition of spore germination. Microbiology 140:623–629.

58.   Lorito, M., Di Pietro, A., Hayes, C.K., Woo, S.L., and Harman, G.E. 1993. Antifungal, synergistic interaction between chitinolytic enzymes from *Trichoderma harzianum* and *Enterobacter cloacae*. Phytopathology 83:721–728.

59.   Lumsden, R.D., Locke, J.C., and Walter, J.F. 1991. Approval of *Gliocladium virens* by the U.S. Environmental Protection Agency for biological control of Pythium and Rhizoctonia damping-off. Petria 1:138.

60.   Mauch, F., Mauch-Mani, B., and Boller, T. 1988. Antifungal hydrolases in pea tissue. II. Inhibition of fungal growth by combinations of chitinase and β-1,3-glucanase. Plant Physiol. 88:936–942.

61.   Merriman, P.R. 1976. Survival of sclerotia of *Sclerotinia sclerotiorum* in soil. Soil Biol. Biochem. 8:385–389.

62.   Ohr, H.D., Munnecke, D.E., and Bricker, J.L. 1973. The interaction of *Armillaria mellea* and *Trichoderma* spp. as modified by methyl bromide. Phytopathology 63:965–973.

63.   Ordentlich, A., Migheli, Q., and Chet, I. 1991. Biological control
        activity of three *Trichoderma* isolates against Fusarium wilts of
        cotton and muskmelon and lack of correlation with their lytic
        enzymes. J. Phytopathol. 133:177–186.

64.   Papavizas, G.C., Dunn, M.T., Lewis, J.A., and Beagle-Ristaino,
        J. 1984. Liquid fermentation technology for experimental
        production of biocontrol fungi. Phytopathology 74:1171–1175.

65.   Papavizas, G.C., and Lumsden, R.D. 1980. Biological control of
        soilborne fungal propagules. Annu. Rev. Phytopathol.
        18:389–413.

66.   Paulitz, T.C. 1990. Biochemical and ecological aspects of
        competition in biological control. Pages 713–724 in: New
        Directions in Biological Control: Alternatives for Suppressing
        Agricultural Pests and Diseases. R.R. Baker and P.E. Dunn, eds.
        Alan R. Liss Inc., New York, NY.

67.   Powell, K.A., and Faull, J.L. 1989. Commercial approaches to
        the use of biological control agents. Pages 259–275 in:
        Biotechnology of Fungi for Improving Plant Growth. J.M. Wipps
        and R.D. Lumsden, eds. Cambridge University Press,
        Cambridge, UK.

68.   Ridout, C.J., Coley-Smith, J.R., and Lynch, J.M. 1986. Enzyme
        activity and electrophoretic profile of extracellular protein induced
        in *Trichoderma* spp. by cell walls of *Rhizoctonia solani*. J. Gen.
        Microbiol. 132:2345–2352.

69.   Ridout, C.J., Coley-Smith, J.R., and Lynch, J.M. 1988.
        Fractionation of extracellular enzymes from a mycoparasitic strain
        of *Trichoderma harzianum*. Enzyme Microb. Technol.
        10:180–187.

70.   Roberts, D.P., and Lumsden, R.D. 1990. Effect of extracellular
        metabolites from *Gliocladium virens* on germination of sporangia
        and mycelial growth of *Pythium ultimum*. Phytopathology
        80:461–465.

71.   Sivan, A., and Chet, I. 1989. Degradation of fungal cell walls by
        lytic enzymes of *Trichoderma harzianum*. J. Gen. Microbiol.
        135:675–682.

72.   Sivan, A., and Chet, I. 1989. The possible role of competition
        between *Trichoderma harzianum* and *Fusarium oxysporum* on
        rhizosphere colonization. Phytopathology 79:198–203.

73.   Sivan, A., Elad, Y., and Chet, I. 1984. Biological control effects
        of a new isolate of *Trichoderma harzianum* on *Pythium
        aphanidermatum*. Phytopathology 74:498–501.

74.   Smith, V.L., Wilcox, W.F., and Harman, G.E. 1990. Potential for biological control of Phytophthora root and crown rots of apple by *Trichoderma* and *Gliocladium* spp. Phytopathology 80:880–885.

75.   Sommer, N.F. 1982. Postharvest handling practices and postharvest diseases of fruit. Plant Dis. 66:357–364.

76.   Stasz, T.E., Harman, G.E., and Weeden, N.F. 1988. Protoplast preparation and fusion in two biocontrol strains of *Trichoderma harzianum*. Mycologia 80:141–150.

77.   Taylor, A.G., and Harman, G.E. 1990. Concepts and technologies of selected seed treatments. Annu. Rev. Phytopathol. 28:321–339.

78.   Thomashow, L.S., and Weller, D.M. 1988. Role of a phenazine antibiotic from *Pseudomonas fluorescens* in biological control of *Gaeumannomyces graminis* var. *tritici*. J. Bacteriol. 170:3499–3508.

79.   Tribe, H.T., and Ahmed, A.H.M. 1975. Use of autoclavable plastic bags in fungal culture work. Trans. Br. Mycol. Soc. 64:362–363.

80.   Tronsmo, A. 1986. *Trichoderma* used as a biocontrol agent against *Botrytis cinerea* rots on strawberry and apple. Sci. Rep. Agric. Univ. Norway. 65(17):1–22.

81.   Tronsmo, A. 1986. Use of *Trichoderma* spp. in biological control of necrotrophic pathogens. Pages 348–362 in: Microbiology of the Phyllosphere. N.J. Fokkema and J. Van den Heuvel, eds. Cambridge University Press, Cambridge, UK.

82.   Tronsmo, A. 1989. Effect of fungicides and insecticides on growth of *Botrytis cinerea*, *Trichoderma viride* and *T. harzianum*. Norw. J. Agric. Sci. 3:151–156.

83.   Tronsmo, A. 1989. *Trichoderma harzianum* used for biological control of storage rot on carrots. Norw. J. Agric. Sci. 3:157–161.

84.   Tronsmo, A. 1991. Biological and integrated controls of *Botrytis cinerea* on apple with *Trichoderma harzianum*. Biol. Control 1:59–62.

85.   Tronsmo, A., and Dennis, C. 1977. The use of *Trichoderma* species to control strawberry fruit rots. Neth. J. Plant Pathol. 83 (Suppl. 1):449–455.

86.   Tronsmo, A., and Dennis, C. 1978. Effect of temperature on antagonistic properties of *Trichoderma* species. Trans. Br. Mycol. Soc. 71:469–474.

87.   Tronsmo, A., and Hoftun, H. 1984. Storage and distribution of carrots. Effect on quality of long term storage in ice bank cooler and cold room, and of different packing materials during distribution. Acta Hortic. 163:143–150.

88.   Tronsmo, A., Klemsdal, S.S., Hayes, C.K., Lorito, M., and Harman, G.E. 1993. The role of hydrolytic enzymes produced by *Trichoderma harzianum* in biological control of plant diseases. Pages 159–168 in: *Trichoderma reesei* Cellulases and Other Hydrolases: Enzyme Structures, Biochemistry, Genetics and Applications, Vol. 8. P. Suominen and T. Reinikainen, eds. Foundation for Biotechnical and Industrial Fermentation Research, Helsinki, Finland.

89.   Tronsmo, A., and Raa, J. 1977. Antagonistic action of *Trichoderma pseudokoningii* against the apple pathogen *Botrytis cinerea*. Phytopathol. Z. 89:216–220.

90.   Tronsmo, A., and Raa, J. 1977. Life cycle of the dry eye rot pathogen *Botrytis cinerea* Pers. on apple. Phytopathol. Z. 89:203–207.

91.   Tronsmo, A., and Ystaas, J. 1980. Biological control of *Botrytis cinerea* on apple. Plant Dis. 64:1009.

92.   Wilson, C.L., and Pusey, P.L. 1985. Potential for biological control of postharvest plant diseases. Plant Dis. 69:375–378.

93.   Wilson, C.L., and Wisniewski, M.E. 1989. Biological control of postharvest diseases of fruits and vegetables: an emerging technology. Annu. Rev. Phytopathol. 27:425–441.

94.   Wood, R.K.S. 1950. The control of diseases of lettuce by use of antagonistic organisms. 1. The control of *Botrytis cinerea* Pers. Ann. Appl. Biol. 38:203–216.

Chapter 11

RELATIONSHIPS AMONG ORGANIC MATTER DECOMPOSITION LEVEL, MICROBIAL SPECIES DIVERSITY, AND SOILBORNE DISEASE SEVERITY

H.A.J. Hoitink, L.V. Madden, and M.J. Boehm

INTRODUCTION

Agriculturalists have recognized for decades that the level of decomposition of soil organic matter plays an important role in soilborne disease severity (9). Management of diseases with soil amendments remains an art rather than a science, however. Lack of progress in this field has resulted largely from our inability to characterize adequately the availability of biological energy in soil or the amendments. Recent breakthroughs in instrumentation, coupled with advances made in the analysis of microbial populations in soil, promise to solve this problem. Increased acceptance of composting as the preferred strategy for treatment of solid wastes has provided additional impetus for support of research on quality control of soil organic matter (composts). As a result, procedures based on respirometry are now available to monitor the stability of composts (24). Furthermore, nondestructive direct spectroscopic procedures are available to characterize stability or biological energy availability in organic matter (6,25). Much of the quantitative information available in this field was developed for composts. This review emphasizes this literature on composting, therefore.

During the 1960s, nurserymen across the United States explored the possibility of using wood industry wastes as peat substitutes to reduce potting mix costs. Since that time, procedures for composting of tree bark have been developed that avoid nitrogen immobilization and allelopathic toxin problems associated with fresh bark (23). During the past decade, similar procedures have been developed for composts produced from food industry and agricultural wastes, as well as from municipal wastes (22).

Early during the utilization of bark composts, improved plant growth and decreased losses caused by Phytophthora root rots were observed as side benefits by the nursery industry in the United States. In practice, control of such root rots with composts can be as effective as that obtained with fungicides (17,21,37). In many parts of the world, therefore, the nursery industry relies heavily on composts for control of diseases caused by these soilborne plant pathogens.

Composts must be of consistent quality and maturity to be used successfully in biological control of diseases of horticultural crops, particularly if used in container media (26). Variability in this quality parameter is one of the main factors limiting compost utilization for this purpose. In ground bed or field agriculture, maturity is less important as long as adequate heating has occurred to kill pathogens, and the compost is applied well ahead of planting to allow for additional stabilization. Lack of maturity frequently causes problems here as well, however.

Predicting maturity of composts (6), as related to the potential for improved plant health, is now possible. In addition, as a result of increased research on biological control during the 1980s, information is available that eventually will allow commercial-scale formulation of composts for suppression of soilborne diseases caused by *Fusarium* spp., *Phytophthora* spp., *Pythium* spp., *Rhizoctonia solani*, and other pathogens (14,19). To maintain quality related to both plant growth and disease control, compost producers must develop a basic understanding of the processes involved in this method of biological control. Therefore, a brief review of the principles involved is presented here.

FATE OF BIOLOGICAL CONTROL AGENTS DURING COMPOSTING

The composting process often is divided into three phases. The first phase occurs during the first 24 to 48 hours as temperatures gradually rise to between 40 and 50°C and when sugars and other easily biodegradable substances are destroyed. During the second phase, when temperatures of 40 to 65°C prevail, cellulosic substances that are less biodegradable are destroyed. Lignins, the darker components in plant tissues, break down even more slowly but the rate varies among plant species. Thermophilic microorganisms predominate during this part of the process. Plant pathogens and seeds are killed by the heat generated during this high temperature phase (3,11). Compost piles must be turned frequently to expose all parts to high temperature to produce a homogeneous product free of pathogens and weed seeds.

The third or curing phase of composting begins as the concentration of readily biodegradable components in wastes declines. As a result, temperatures and rates of decomposition and heat output decline. At this time, mesophilic microorganisms recolonize the compost from the outer cool layer into the pile. Humic substances accumulate in increasing quantities by this time. Mature composts consist largely of lignins, humic substances and microbial biomass, and have a dark color (6).

Beneficial as well as detrimental microorganisms are killed during the high temperature phase of composting. Therefore, suppression of pathogens, or disease, or both, is largely induced during curing as biological control agents recolonize composts after peak heating. *Bacillus* spp., *Enterobacter* spp., *Flavobacterium balustinum, Pseudomonas* spp., other bacterial genera and *Streptomyces* spp., as well as *Penicillium* spp., *Trichoderma* spp., *Gliocladium virens* and other fungi have been identified as biological control agents in composts (8,16,17,20,36,38).

Variability in suppression of Rhizoctonia damping-off encountered in substrates amended with mature composts results, in part, from random recolonization after peak heating of compost by a microflora with efficacy against *R. solani*. Compost produced in the open near a forest, an environment that is high in microbial species diversity, is more consistently suppressive to Rhizoctonia diseases than the same produced in a partially enclosed facility where few microbial species survive heat treatment (30). Composts prepared from municipal sewage sludges and other materials such as manures and municipal solid wastes are consistently conducive to *R. solani* because care is taken to kill fecal pathogens and parasites with heat exposure. As mentioned earlier, this process kills most beneficial microorganisms as well. These composts, although naturally suppressive to Pythium diseases, have to be incubated for a month or more before they become naturally colonized by chance by the specific microflora that render them suppressive to *R. solani* (29). In field soil, several months may pass before suppression is induced (32).

To solve the problem of variability in suppressiveness of compost to *R. solani*, a specific inoculant containing a fungus (*Trichoderma hamatum* 382) and a bacterium (*F. balustinum* 299) has been developed that when introduced into compost after peak heating, but before significant levels of recolonization have occurred, induces consistent levels of suppression. Patents have been issued to the Ohio State University for this process (18). In Japan, Phae et al. (38) isolated a *Bacillus* strain that induces predictable biological control in composts. It survives peak heating.

MECHANISMS OF BIOLOGICAL CONTROL IN COMPOSTS

Two mechanisms of biological control, based on competition, antibiosis, hyperparasitism, and systemic acquired resistance in plants, have been described for compost-amended substrates. Propagules of plant pathogens, including *Pythium* and *Phytophthora* spp., are suppressed through a mechanism known as "general suppression" (2,4,5,9,17,33). Many types of microorganisms present in compost-amended container media function as biological control agents for diseases caused by *Phytophthora* and *Pythium* spp. (2,17). Propagules of these pathogens, if inadvertently introduced into composts, do not germinate in response to nutrients released in the form of seed or root exudates. The high microbial activity and biomass, caused by the "general soil microflora" in compost-amended substrates, prevents germination of spores of the pathogen and infection of the host, presumably through microbiostasis (4,33). Propagules of these pathogens remain dormant and are typically not killed if introduced into compost-amended soil (4,33). An enzyme assay that determines microbial activity based on the rate of hydrolysis of fluorescein diacetate is used to predict suppressiveness of potting mixes to Pythium diseases (1,4,33). Similar information has been developed for soils on "organic farms" where soilborne diseases are less prevalent (46).

The mechanism of biological control of *R. solani* in compost-amended substrates is different from that of *Pythium* and *Phytophthora* spp. A narrow group of microorganisms is capable of eradicating sclerotia of *R. solani*. This type of suppression we refer to as "specific suppression" (21). *Trichoderma* spp, including *T. hamatum* and *T. harzianum*, are the predominant hyperparasites recovered from composts prepared from lignocellulosic wastes (30,36). These fungi interact with various bacterial isolates in biological control of Rhizoctonia damping-off (31). It is of interest that *Penicillium* spp. are the predominant hyperparasites recovered from sclerotia of *Sclerotium rolfsii* in composted grape pomace, a waste high in sugar content (16). *Trichoderma* spp. were not recovered from this compost. The composition of the parent product, as expected, thus appears to have an impact on the microflora in composts active in biological control. Further work may show that this is a major factor affecting biological control efficacy in field agriculture as well.

ROLE OF ORGANIC MATTER DECOMPOSITION LEVEL

The decomposition level of organic matter in compost-amended substrates has a major impact on disease suppression. *Rhizoctonia solani* is highly competitive as a saprophyte (12). It can utilize cellulose and therefore colonize cellulose-rich fresh wastes but not cellulose-poor mature compost (7). On the other hand, *Trichoderma* isolates that function as biological control agents of *R. solani* are capable of colonizing fresh as well as mature compost (36). In fresh, undecomposed organic matter, biological control does not occur because both fungi grow as saprophytes and *R. solani* remains capable of causing disease. Presumably, synthesis of lytic enzymes involved in hyperparasitism (10) is repressed here because of high glucose concentrations in fresh wastes. The same may apply to antibiotic production. In mature compost, where concentrations of free nutrients are low (4), sclerotia of *R. solani* are killed by the hyperparasite and biological control prevails (36).

Because organic matter decomposition level is so important, composts must be stabilized adequately to reach that decomposition level where biological control is feasible. In practice, this occurs in composts (tree barks, yard wastes, etc.) that have been stabilized far enough to not induce phytotoxicity (due to lack of stability) nor immobilize nitrogen during plant growth, but are colonized by the appropriate specific microflora. Practical guidelines that define this critical stage of decomposition in terms of biological control are not yet available. Industry presently controls decomposition level by maintaining constant conditions during the entire process and adhering to a given time schedule.

Excessively stabilized organic matter does not support adequate activity of biological control agents. In a highly mineralized soil, where humic substances are the predominant forms of organic matter, suppression is lacking and soilborne diseases are severe (46). It has not yet been determined how long composts incorporated in soil support adequate levels of activity for biological control. Presumably, the length of time varies with soil temperature and the type of organic matter from which the compost was prepared. Loading rates and farming practices, of course also play a role.

We have studied the "carrying capacity" of soil organic matter in potting mixes prepared with sphagnum peat to bring a partial solution to this problem (1,2). Sphagnum peat typically competes with compost as a source of organic matter in horticulture. Both the microflora and the organic matter in peat itself can affect suppression of soilborne diseases. The literature on that effect is reviewed briefly here, therefore.

Because of its stability (resistance to decomposition), organic matter in sphagnum peat generally does not support high microbial activity. Dark, more decomposed sphagnum peat, harvested from a foot or greater depths in peat bogs, is low in activity and consistently conducive to Pythium and Phytophthora root rots (1). On the other hand, light sphagnum peats harvested from the surface of peat bogs, are less decomposed and have a higher microbial activity. The suppressive effect of light peat to Pythium root rots is temporary, however (1,42,45). Light peats, therefore, can be used most effectively for short production cycles, such as in plug and flat mixes used in the ornamentals industry. Composts have longer lasting effects (1,2).

The rate of hydrolysis of fluorescein diacetate (FDA) predicts suppressiveness of compost-amended substrate and of peat mixes to Pythium root rot (1). As FDA activity in suppressive light sphagnum peat substrates declines to < 3.2 μg FDA hydrolyzed per minute per gram dry weight of mix, the population of *Pythium ultimum* increases and root rot develops. During this collapse in suppressiveness, bacterial species diversity in the rhizosphere or the edaphic substrate does not change. However, species composition is affected (2). A microflora typical of suppressive soils, which includes *Pseudomonas* spp. and other rod-shaped gram negative bacteria as the predominant rhizosphere colonizers, is replaced by pleomorphic gram positive bacteria and putative oligotrophs (47). Thus, the microflora of the conducive substrate resembles that of highly mineralized niches in soil (28).

Nondestructive analysis of soil organic matter, utilizing Fourier Transform Infra Red spectroscopy (FT-IR) and Cross Polarization Magic Angle Spinning-[13]Carbon Nuclear Magnetic Resonance Spectroscopy (CPMAS-^{13}CNMR), allows characterization of biodegradable components of soil organic fractions (6,25). CPMAS-^{13}CNMR allows quantitative analysis of concentrations of readily biodegradable substances such as "carbohydrates" (hemicellulose, cellulose, etc.) versus lignins and humic substances in soil organic matter (reviewed in 6). In a preliminary report, Wu et al. (47) have shown that the concentration of "carbohydrates" in a suppressive light peat mix declines as suppressiveness is lost. When the peat is no longer suppressive, the concentration of carbohydrate is not different from that in the conducive, dark, more decomposed peat. They also have shown that biological control agents inoculated into the more decomposed peat are not able to induce sustained biological control of Pythium root rot. Thus, biological control of this disease is determined by the "carrying capacity" of the substrate which regulates species composition and activity and, in turn, the potential for sustained biological control.

COMPOST STEEPAGES FOR DISEASE CONTROL

During the past decade, a series of projects have been published on the control of plant diseases of above-ground plant parts with water extracts of compost (44). The steepages are prepared by soaking specific mature manure composts in water (still culture; 1:1, w/w) for 7 to 10 days. The steepage is filtered and then sprayed on plants. Efficacy varies with crops and the disease under question. Sackenheim (40), using plate-counting procedures, has shown that aerobic microorganisms predominate in these steepages. The microflora includes strains of bacteria and isolates of fungi already known as biological control agents. A number of enrichment strategies that employed nutrients as well as microorganisms were developed to improve efficacy of the steepages.

Disease control induced by compost steepages was formerly attributed in part to systemic acquired resistance induced in plants by microbes present in the extracts. The recent work by Sackenheim (40) on grape does not support this assumption, however. A factor that has been entirely overlooked but could play a role in efficacy of steepages is the condition of organic matter and the associated microflora in the soil in which treated plants are produced. Soils naturally suppressive to soilborne plant pathogens (e.g. compost-amended soils) harbor active populations of biological control agents (2) that could induce protection in the leaves of plants to foliar pathogens (35,43). Thus, further research may show that plants produced in suppressive soils not only have less root rot but could also be less prone to attack by vascular wilt and foliar pathogens and respond better to compost steepages.

DISEASE SUPPRESSION: CONCLUSIONS AND OUTLOOK

It has been known for centuries that composts may provide biological control of diseases caused by soilborne plant pathogens. Many factors affect the potential for composts to provide this control. Heat during composting kills or inactivates pathogens and weed seeds if the process is monitored properly. Unfortunately, most beneficial microorganisms also are killed by this heat treatment. Thus, biological control agents must recolonize composts after peak heating. The raw feedstock, the environment in which the compost is produced, and conditions during curing and utilization determine the potential for recolonization by this microflora and the induction of disease suppression. Controlled inoculation of compost with biological control agents is necessary to induce consistent levels of suppression on a commercial scale.

The decomposition level (stability) of composts affects suppressiveness. Immature composts serve as food for pathogens and increase disease even when biological control agents are present. *Trichoderma* strains effective as biological control agents in mature compost grow as saprophytes in fresh organic matter because the high concentrations of free nutrients repress hyperparasitism. On the other hand, excessively stabilized organic amendments such as highly decomposed peats do not support the activity of biological control agents. Biological control agents introduced into these sources of organic matter decline in population density and do not induce sustained biological control.

Success in biological control of diseases caused by *Fusarium, Pythium* and *Phytophthora* spp. is possible only if all factors involved in the production of composts are defined and kept constant. Most composts cannot meet these criteria because emphasis on quality control in general still is inadequate. Composted bark and light sphagnum peat, therefore, remain the principal organic components used for the preparation of potting mixes or soils naturally suppressive to soilborne plant pathogens. Composted manures, yard and food wastes steadily are gaining in popularity, and offer the same potential (13,15,27,34,41).

The general suppression phenomenon, based on present concepts, at best covers those diseases caused by pathogens suppressed through microbiostasis. Microbial activity in a mix, based on the rate of hydrolysis of fluorescein diacetate, is one procedure that now can be used effectively to determine suppressiveness to Pythium and Phytophthora root rots. This procedure by itself, however, does not predict how long the effect will last. The potential for biodegradable carbon in a mix to support an active and effective microbial biomass determines that phenomenon. CPMAS-^{13}CNMR spectroscopy predicts this property for peat. The same basic information will have to be developed for composts and other sources of organic amendments. Predictive utilization of composts for disease control will remain an "open loop" production system until more quantitative guidelines are developed.

Controlled inoculation of composts with biological control agents is a procedure that must be developed on a commercial scale to induce consistent levels of suppression to pathogens such as *R. solani* (14,21,38). This new field of biotechnology is in its infancy, however. Major research and development efforts must be directed into this approach to disease control. Recycling through composting increasingly is being chosen as the preferred strategy for waste treatment. This also applies to farm manures (39). For this reason, composts are becoming available in greater quantities. Peat, on the other hand, is a limited resource that cannot be recycled. Future opportunities for natural or controlled induced suppression of soilborne plant pathogens, using composts as the foodstuff for biological control agents, therefore appear bright.

ACKNOWLEDGEMENT

Salaries and research support were provided by State and Federal Funds appropriated to the Ohio Agricultural Research and Development Center, The Ohio State University, and by grant no. US2196-22 from BARD, the United States-Israel Binational Research and Development Fund. Manuscript number 107–94.

LITERATURE CITED

1. Boehm, M.J., and Hoitink, H.A.J.. 1992. Sustenance of microbial activity in potting mixes and its impact on severity of Pythium root rot of poinsettia. Phytopathology 82:259–264.

2. Boehm, M.J., Madden, L.V., and Hoitink, H.A.J. 1993. Effect of organic matter decomposition level on bacterial species diversity and composition in relationship to Pythium damping-off severity. Appl. Environ. Microbiol. 59:4171–4179.

3. Bollen, G.J. 1993. Factors involved in inactivation of plant pathogens during composting of crop residues. Pages 301–318 in: Science and Engineering of Composting: Design, Environmental, Microbiological and Utilization Aspects. H.A.J. Hoitink and H.M. Keener, eds. Renaissance Publications, Worthington, OH.

4. Chen, W., Hoitink, H.A.J., Schmitthenner, A.F., and Tuovinen, O.H. 1988. The role of microbial activity in suppression of damping-off caused by *Pythium ultimum*. Phytopathology 78:314–322.

5. Chen, W., Hoitink, H.A.J., and Madden, L.V. 1988. Microbial activity and biomass in container media for predicting suppressiveness to damping-off caused by *Pythium ultimum*. Phytopathology 78:1447–1450.

6. Chen, Y., and Inbar, Y. 1993. Chemical and spectroscopical analyses of organic matter transformations during composting in relation to compost maturity. Pages 551–600 in: Science and Engineering of Composting: Design, Environmental, Microbiological and Utilization Aspects. H.A.J. Hoitink and H.M. Keener, eds. Renaissance Publications, Worthington, OH.

7.   Chung, Y.R., Hoitink, H.A.J., and Lipps, P.E. 1988. Interactions between organic-matter decomposition level and soilborne disease severity. Agric. Ecosyst. Environ. 24:183–193.

8.   Chung, Y.R., and Hoitink, H.A.J. 1990. Interactions between thermophilic fungi and *Trichoderma hamatum* in suppression of Rhizoctonia damping-off in a bark compost-amended container medium. Phytopathology 80:73–77.

9.   Cook, R.J., and Baker, K.F. 1983. The Nature and Practice of Biological Control of Plant Pathogens. The American Phytopathological Society, St. Paul, MN. 539 pp.

10.   Cruz, J. de la, Rey, M., Lora, J.M., Hidalgo-Gallego, A., Domínguez, F., Pintor-Toro, J.A., Llobell, A., and Benítez, T. 1993. Carbon source control on β-glucanases, chitobiase and chitinase from *Trichoderma harzianum*. Arch. Microbiol. 159:316–322.

11.   Farrell, J.B. 1993. Fecal pathogen control during composting. Pages 282–300 in: Science and Engineering of Composting: Design, Environmental, Microbiological and Utilization Aspects. H.A.J. Hoitink and H.M. Keener, eds. Renaissance Publications, Worthington, OH.

12.   Garrett, S.D. 1962. Decomposition of cellulose in soil by *Rhizoctonia solani* Kuhn. Trans. Br. Mycol. Soc. 45:114–120.

13.   Gorodecki B., and Hadar, Y. 1990. Suppression of *Rhizoctonia solani* and *Sclerotium rolfsii* diseases in container media containing composted separated cattle manure and composted grape marc. Crop Prot. 9:271–274.

14.   Grebus, M.E., Feldman, K.A., Musselman, C.A., and Hoitink, H.A.J. 1993. Production of biocontrol agent-fortified compost-amended potting mixes for predictable disease suppression. Phytopathology 83:1406. (Abstr.).

15.   Grebus, M.E., Watson, M.E., and Hoitink, H.A.J. 1994. Biological, chemical and physical properties of composted yard trimmings as indicators of maturity and plant disease suppression. Compost Sci. Util. 2:57–71.

16.   Hadar, Y., and Gorodecki, B. 1991. Suppression of germination of sclerotia of *Sclerotium rolfsii* in compost. Soil Biol. Biochem. 23:303–306.

17.   Hardy, G.E.St.J., and Sivasithamparam, K. 1991. Suppression of *Phytophthora* root rot by a composted *Eucalyptus* bark mix. Aust. J. Bot. 39:153–159.

18.   Hoitink, H.A.J. 1990. Production of disease suppressive compost and container media, and microorganism culture for use therein. U.S. Patent 4960348. 13 February 1990.

19.   Hoitink, H.A.J., Boehm, M.J., and Hadar, Y. 1993. Mechanisms of suppression of soilborne plant pathogens in compost-amended substrates. Pages 601–621 in: Science and Engineering of Composting: Design, Environmental, Microbiological and Utilization Aspects. H.A.J. Hoitink and H.M. Keener, eds. Renaissance Publications, Worthington, OH.

20.   Hoitink, H.A.J., and Fahy, P.C. 1986. Basis for the control of soilborne plant pathogens with composts. Annu. Rev. Phytopathol. 24:93–114.

21.   Hoitink, H.A.J., Inhar, Y., and Boehm, M.J. 1991. Status of compost-amended potting mixes naturally suppressive to soilborne diseases of floricultural crops. Plant Dis. 75:869–873.

22.   Hoitink, H.A.J., and Keener, H.M., eds. 1993. Science and Engineering of Composting: Design, Environmental, Microbiological and Utilization Aspects. Renaissance Publications, Worthington, OH.

23.   Hoitink, H.A.J., and Kuter, G.A. 1986. Effects of composts in growth media on soilborne pathogens. Pages 289–306 in: The Role of Organic Matter in Modern Agriculture. Y. Chen and Y. Avnimelech, eds. Martinus Nijhoff Publishers, Dordrecht, The Netherlands.

24.   Iannotti, D.A., Pang, T., Toth, B.L., Elwell, D.L., Keener, H.M., and Hoitink, H.A.J. 1993. A quantitative respirometric method for monitoring compost stability. Compost Sci. Util. 1:52–65.

25.   Inbar, Y., Chen, Y., and Hadar, Y. 1989. Solid-state carbon-13 nuclear magnetic resonance and infrared spectroscopy of composted organic matter. Soil Sci. Soc. Am. J. 53:1695–1701.

26.   Inbar, Y., Chen, Y., and Hoitink, H.A.J. 1993. Properties for establishing standards for utilization of composts in container media. Pages 668–694 in: Science and Engineering of Composting: Design, Environmental, Microbiological and Utilization Aspects. H.A.J. Hoitink and H.M. Keener, eds. Renaissance Publications, Worthington, OH.

27.   Inbar, Y., Hadar, Y., and Chen, Y. 1993. Recycling of cattle manure: the composting process and characterization of maturity. J. Environ. Qual. 22:857–863.

28. Kanazawa S., and Filip, Z. 1986. Distribution of microorganisms, total biomass, and enzyme activities in different particles of brown soil. Microb. Ecol. 12:205–215.

29. Kuter, G.A., Hoitink, H.A.J., and Chen, W. 1988. Effects of municipal sludge compost curing time on suppression of *Pythium* and *Rhizoctonia* diseases of ornamental plants. Plant Dis. 72:751–756.

30. Kuter, G.A., Nelson, E.B., Hoitink, H.A.J., and Madden, L.V. 1983. Fungal populations in container media amended with composted hardwood bark suppressive and conducive to Rhizoctonia damping-off. Phytopathology 73:1450–1456.

31. Kwok, O.C.H., Fahy, P.C., Hoitink, H.A.J., and Kuter, G.A. 1987. Interactions between bacteria and *Trichoderma hamatum* in suppression of Rhizoctonia damping-off in bark compost media. Phytopathology 77:1206-1212.

32. Lumsden, R.D., Lewis, J.A., and Millner, P.D. 1983. Effect of composted sewage sludge on several soilborne pathogens and diseases. Phytopathology 73:1543–1548.

33. Mandelbaum, R., and Hadar, Y. 1990. Effects of available carbon source on microbial activity and suppression of *Pythium aphanidermatum* in compost and peat container media. Phytopathology 80:794–804.

34. Marugg C., Grebus, M.E., Hansen, R.C., Keener, H.M., and Hoitink, H.A.J. 1993. A kinetic model of the yard waste composting process. Compost Sci. Util. 1:38–51.

35. Maurhofer, M., Hase, C., Meuwly P., Métraux, J.-P., and Défago, G. 1994. Induction of systemic resistance of tobacco to tobacco necrosis virus by the root-colonizing *Pseudomonas fluorescens* strain CHA0: influence of the *gacA* gene and of pyoverdine production. Phytopathology 84:139–146.

36. Nelson, E. B., Kuter, G.A., and Hoitink, H.A.J. 1983. Effects of fungal antagonists and compost age on suppression of Rhizoctonia damping-off in container media amended with composted hardwood bark. Phytopathology 73:1457-1462.

37. Ownley, B.H., and Benson, D.M. 1991. Relationship of matric water potential and air-filled porosity of container media to development of Phytophthora root rot of rhododendron. Phytopathology 81:936–941.

38.   Phae, C.-G., Sasaki, M., Shoda, M., and Kubota, H. 1990. Characteristics of *Bacillus subtilis* isolated from composts suppressing phytopathogenic microorganisms. Soil Sci. Plant Nutr. 36:575–586.

39.   Rynk, R., ed. 1992. On-Farm Composting Handbook. Publication NRAES-54, Cooperative Extension, Northeast Regional Agricultural Engineering Service, Ithaca, NY. 186 pp.

40.   Sackenheim, R. 1993. Untersuchungen über Wirkungen von wässerigen, mikrobiologisch aktiven Extracten aus kompostierten Substraten auf den Befall der Weinrebe *(Vitis vinifera)* mit *Plasmopora viticola, Uncinula necator, Botrytis cinerea* und *Pseudopezicula tracheiphila*. Ph.D. thesis, Rheinische Friedrich-Wilhelms Universität, Bonn, Germany. 157 pp.

41.   Schüler, C., Pikny, J., Nasir, M., and Vogtmann, H. 1993. Effects of composted organic kitchen and garden waste on *Mycosphaerella pinodes* (Berk. et Blox) Vestergr., causal organism of foot rot on peas *(Pisum sativum* L.). Biol. Agric. Hort. 9:353–360.

42.   Tahvonen, R. 1982. The suppressiveness of Finnish light coloured *Sphagnum* peat. J. Sci. Agric. Soc. Finl. 54:345–356.

43.   Wei, G., Kloepper, J.W., and Tuzun, S. 1991. Induction of systemic resistance of cucumber to *Colletotrichum orbiculare* by select strains of plant growth-promoting rhizobacteria. Phytopathology 81:1508–1512.

44.   Weltzhien, H.C. 1992. Biocontrol of foliar fungal diseases with compost extracts. Pages 430–450 in: Microbial Ecology of Leaves. J.H. Andrews and S. Hirano, eds. Springer-Verlag, New York, NY.

45.   Wolffhechel, H. 1988. The suppressiveness of sphagnum peat to *Pythium* spp. Acta Hortic. 221:217–222.

46.   Workneh, F., Van Bruggen, A.H.C., Drinkwater, L.E., and Shennan, C. 1993. Variables associated with corky root and Phytophthora root rot of tomatoes in organic and conventional farms. Phytopathology 83:581–589.

47.   Wu, T., Boehm, M.J., Madden, L.V., and Hoitink, H.A.J. 1993. Sustained suppression of Pythium root rot: a function of microbial carrying capacity and bacterial species composition. Phytopathology 83:1365. (Abstr.).

Chapter 12

SOIL SOLARIZATION: INTEGRATED CONTROL ASPECTS

J. Katan

SOILBORNE PATHOGENS AND CONCEPTS RELATED TO THEIR CONTROL

Soilborne pathogens, such as bacteria, fungi, nematodes, parasitic plants and other organisms frequently cause heavy losses to major crops, by affecting both yield and quality. In severe cases, they may totally destroy the crop, forcing the farmer to either abandon the land or shift to less susceptible, but often less profitable crops. There is, therefore, a need to develop effective control methods to ensure crop productivity and yield stability. These methods have to be economically, environmentally, and technologically sound.

All agricultural practices disturb the biological equilibrium in the plant ecosystem. Even in the most primitive agricultural system, the population of a single plant species is more crowded and uniform than in a natural plant community and it is frequently grown repeatedly on the same land. Thus, the buildup of pest populations resulting from frequent cropping began in the early days of agriculture. Certain modern agricultural practices, such as monoculture of highly profitable crops, aggravate this buildup, while others, e.g. effective pest control, delay it. Of the latter, soil disinfestation is extreme; when properly applied it very effectively eradicates pests, but concomitantly, may disturb the biological balance in the soil (66).

Many methods—chemical, cultural, and biological—have been developed for the control of plant diseases caused by soilborne pathogens. The existence of many methods of control reflects a weakness rather than an abundance of options, since a general rule is that no method is perfect nor can it be used in all instances. Thus, any new method of control, even if restricted in its use, is of value because it adds to our rather limited arsenal of control methods. This is especially true of novel nonchemical methods of control, which are needed to replace hazardous pesticides.

The concept of control of soilborne pathogens has changed over the last few decades. In the early days, the control of soilborne pathogens focused on eradicating the pathogen population, while ignoring or underestimating the consequences to nontarget biotic and abiotic components of the environment. Although this approach can lead to effective control, it may also lead to harmful consequences. Later, it was realized that effective control can be achieved, not only by eradicating the pathogen, but also by interrupting the disease cycle via the manipulation of any of the three biotic components involved in disease—pathogen population, plant resistance, or the microbial balance—with the ultimate goal being economic reduction of disease level, not necessarily absolute control. The integrated control and integrated pest management (IPM) concepts encompass many elements, including: pest control programs for the crop itself; crop husbandry; pesticide reduction (not necessarily total elimination); integration of all the available methods of control; monitoring of pest populations; and economic, environmental, legal, and sociological parameters. IPM is a holistic and sophisticated approach, an ultimate goal to be achieved, in sequential steps. These concepts were first established by entomologists and later adopted and adapted by plant pathologists. Entomologists pioneered the use of biological control on a large scale while plant pathologists were more active in developing resistant cultivars. How to determine optimal combinations of control methods is one of the key questions in developing strategies of control.

Soil disinfestation is one of the approaches for the control of root diseases, and is especially common with intensive and high-value crops, e.g. greenhouse and horticultural crops. It is sophisticated, expensive, and an effective method of control, which has great advantages, but also limitations. The basic principle is to *eradicate* a wide spectrum of harmful agents in the soil *before* planting, usually using *drastic* chemical or physical means. Although the main goal of soil disinfestation is to eliminate the major pathogens when an economically intolerable level is reached, additional (initially unintended) benefits may also be obtained. The elimination of accompanying detrimental biotic and abiotic agents, which have been accumulating in the soil throughout its long history of cropping, is an example of such a benefit. Such detrimental agents are the minor pathogens (also referred to as weak parasites, subclinical, or deleterious soil microorganisms) and toxic substances, produced directly by soil microorganisms, or originating from decomposed plant residues

(17,35,36). Such phenomena are typical of "soil sickness" in monoculture and replanting disease in orchards. On the other hand, because of the drastic nature of soil disinfestation, a harmful effect on the nontarget biotic and abiotic components of the soil is also possible.

There are three approaches to soil disinfestation. The first two, steaming and fumigation, were developed about 120 years ago, and, until recently, were the only approaches (17,82,106). The third, relatively new, approach is soil solarization (also called solar heating) (66,70).

Numerous studies on soil solarization have been published since its inception in 1976 (70). During the first decade (1976–86), at least 173 reports were published (71). By 1995, the number of publications probably exceeded 350. By 1991 (66), research on solarization had been carried out in at least 38 countries, mostly in arid regions, but with some important exceptions, such as the UK. Since then, solarization research has been carried out in additional countries including Canada, Taiwan, Guadeloupe, Guatemala, Cuba, Czechoslovakia, Trinidad, Southern Germany and Holland. Solarization is frequently discussed at international meetings, including those dedicated solely to this subject, such as the international conference held in Amman, Jordan, in 1990 (24). Attempts to use solar energy in plant protection, by heating plant material or soil, have been known for centuries, for example, in the ancient civilization of India for seed treatment (87). This was recently verified (61). As early as 1933, Hagan (57) heated soils in Hawaii by mulching them with cellophane. He assumed that this heating could lead to the control of nematodes such as *Heterodera radicicola*. Moreover, he suggested that plowing fields in the appropriate season (i.e. exposing the soil to natural solar heating) would be beneficial for the control of nematodes. The practice of deep plowing to control soilborne pathogens has long been known in many countries, but has unfortunately been neglected. In 1939, Groshevoy (54) demonstrated that exposing soils to natural solar heating raises their temperature and controls pathogens. Thus, the present version of soil solarization, first presented in 1976 and based on covering (mulching) the soil with transparent polyethylene, is a modern modification of known principles, using technologies which were not previously available. These technologies enable more intensive and controlled heating, and hence a more effective pest control. Recently, as a result of environmental concerns, there has been a renewed and increasing interest in using heat as a control agent in plant protection, e.g. for treating products at the postharvest stage (9).

Many aspects of soil solarization for pest control have been described in numerous reviews, books, and articles in English and other languages

(13,17,21,23,24,30,44,51,60,63,64,66–68,96,100,111, and many others). Therefore, in this chapter I shall briefly review the major current developments and recent findings in soil solarization and discuss possible future developments. I shall refer to the integration of solarization with other methods and to questions related to its implementation and novel uses.

A BRIEF DESCRIPTION OF THE PRINCIPLES OF SOIL SOLARIZATION

The aim of soil solarization is to harness solar energy to raise the temperature of moistened soil. Mulching (covering) the soil with transparent polyethylene or polyvinyl sheets is, at present, the most common means of achieving this. Future technologies may provide simpler and cheaper tools and within short periods to capture solar energy for plant protection. The use of sprayable plastic material, instead of laying plastic films, may revolutionize soil solarization. Encouraging results have been reported from California (97). Ghini (46), in Brazil, developed a very efficient solar collector for the disinfestation of small volumes of substrate for seedling production. One day of treatment (as compared to the 20 to 40 days needed for conventional solarization) was sufficient for controlling *Sclerotium rolfsii*, *Sclerotinia sclerotiorum*, *Fusarium solani* f. sp. *phaseoli*, and *Pythium aphanidermatum*.

Soil solarization is a hydrothermal process that brings about thermal and other physical, chemical, and biological changes in the moist soil during, and even after, mulching (100). The term soil solarization is now very common, although other terms such as solar heating of the soil, plastic mulching (tarping), solar pasteurization and solar disinfestation are also in use. I prefer "soil solarization", which was introduced by American plant pathologists (86), because it is both widely accepted and concise. In addition, "soil solarization" implies an active process of solar heating of the soil, rather than the usual passive solar heating of a soil exposed to the sunlight. In indexes in plant protection journals, both solarization and solar heating (of soil) serve as key words.

The principles of soil solarization are summarized as follows.

1. Soil mulching should be carried out during the period of high temperatures and intense solar irradiation.
2. The soil should be kept moist to increase the thermal sensitivity of resting structures and to improve heat conduction.
3. The thinnest polyethylene tarp which can be used (25 to 50 μm thickness) is recommended, since it is both cheaper and somewhat

more effective than thicker ones. Because the upper soil layer is heated more intensely than the lower ones, the mulching period should be sufficiently long—usually 4 weeks or more—to achieve pest control at all desired depths. The longer the mulching period, the greater the depth of effective activity, and the higher the pathogen-killing rates (e.g. 49,64,70).

4.   Solarization heats the soil through repeated daily cycles. At increasing soil depths, maximal temperatures decrease, are reached later in the day, and are maintained for longer periods of time. In solarized plots where effective disease and weed control were obtained, typical maximal temperatures were within the range of 45 to 50°C and 38 to 45°C at depths of 10 and 20 cm, respectively, although temperatures that were 5 to 10°C higher were also recorded. The temperatures in the solarized soil are 5 to 15°C higher than those in comparable nonsolarized ones.

5.   The best time for soil solarization, when climatic conditions are the most favorable, can be determined experimentally by tarping the soil, measuring the resulting temperatures and assessing pathogen population. Meteorological data from previous years and predictive models (see below) further facilitate this task. Currently, a wealth of data on temperature regimes of solarized soils in different regions in the world exists. Any research on solarization in a new area should make use of this existing knowledge. A comprehensive year-long study for the determination of the best time for solarization has recently been carried out in Brazil (48).

A convenient method that facilitates solarization research is the use of small plots in which the effect of solarization on soil temperatures and on pathogen populations is evaluated. Suitable inocula (e.g. sclerotia, chlamydospores) of selected pathogens are incorporated into the soil to various depths and the decline in their viability is quantified with time (70). Such a study, if done properly, enables a first evaluation of the effectiveness of solarization with respect to optimal time of year, length of treatment, sensitivity of different pathogens, the depth to which each pathogen is controlled, weed control, and other aspects. However, the results from small-plot studies should be interpreted with caution, because control effectiveness of solarization in these plots is expressed as a reduction of pathogen population or inoculum density. To what extent this reduction reflects comparable disease control and yield increase will depend on a variety of factors, some of which are difficult to assess or predict. For example, Davis and Sorenson (22) found that solarization did not affect population levels of *Verticillium dahliae* and of the nematode *Pratylenchus*. Yet, when potato was planted in the solarized plots, disease incidence was significantly reduced, as compared to nonsolarized plots, and both yield and quality were improved. The opposite, namely, a pronounced decrease in pathogen population accompanied by only a partial disease control, may also occur. The lack of a correlation between inoculum density and disease incidence could be attributed to suppressiveness induced in the soil (50,58,62,68), to weakening of the propagules (32,33,83,109), or the recontamination of the previously solarized soil by pathogens. These effects are not necessarily reflected in initial changes in inoculum density and may lead to underestimation or overestimation of solarization effectiveness. Hence, data based solely on inoculum density may not properly predict the level of disease control by solarization. The small plots should not be smaller than 3 to 4 m × 3 to 4 m and the data should be collected from the center of the plot to avoid the "border effect", i.e. lower soil temperatures near the edges of the plastic mulch (53,79). The use of inocula produced under laboratory conditions (e.g. *Fusarium* conidia), may lead to incorrect conclusions. Since solarization is a climate-dependent technique, the experiments need to be repeated to evaluate their reproducibility, as should be done when developing any control method.

6.   As with any disinfestation method, recontamination of the soil after the termination of solarization, e.g. by infected propagation material, or infested soil or water, should be absolutely avoided.

REDUCTION OF PEST POPULATIONS, DISEASE CONTROL, AND CROP IMPROVEMENT

Many studies describing the control by solarization of a variety of pathogens such as fungi, nematodes, bacteria, weeds, arthropod pests, as described below, and some unidentified agents (73), have been published. In most cases, the reduction in pest population (inoculum density) was accompanied by disease reduction, or yield increase, or both. Pathogenic fungi, pathogenic nematodes, and weeds have been more extensively studied than other pests. Davis (21) lists 35 pathogens and 32 weeds that have been controlled by solarization. Some pathogens (and weeds) are not controlled, or are only partially or inconsistently controlled, by solarization, including the fungi *Macrophomina*, *Monosporascus*, *Fusarium oxysporum* f. sp. *pisi*, and the nematode *Meloidogyne*. Pathogenic bacteria such as *Pseudomonas tomato* pv. *tomato*,

Streptomyces scabies and *Clavibacter*, and arthropods such as the mite *Rhizoglyphus thornei* and certain soil mites and insects, are controlled by solarization (5,16,22,45,47). Some pathogens that are highly sensitive to solarization, e.g. *V. dahliae* and species of *Phytophthora*, were effectively controlled in most studies, whereas the level of control of more heat-tolerant pathogens such as *S. rolfsii* varied from site to site (52,80,91). Effective control of *V. dahliae* and *Pratylenchus* was also obtained in temperate regions such as Idaho in the United States (22), and central Ontario in Canada (75). In the latter study, pathogen populations in the upper soil layers, and incidence of Verticillium wilt in potato planted the following year, were reduced. Although the effectiveness of control was not as high as obtained in the Middle East or California, it is nevertheless significant and impressive and has many implications. It is consistent with other studies carried out in relatively cool climates, with sensitive pathogens (20,22). The effective control obtained under marginal climatic conditions is attributed, among other things, to the enhancement of beneficial biological processes in the soil. The lesson to be learned is that it is worth trying new and less-known control methods (preferably with certain modifications and adjustments) even under conditions which appear less than promising. However, in these cases, success also depends on the type of pathogen involved—*V. dahliae* is especially sensitive to heat treatment (70,85,86). Certain technologies can further improve solarization (see below) and may increase its effectiveness in regions which at this stage are considered less suitable for this method.

The incidence of some foliar diseases has been surprisingly reduced by solarization in certain cases (63,103). This might be attributed to the eradication of primary inoculum, as shown with *Botrytis* (79), or to changes in the mineral content of the soil (17), which may increase plant resistance. On the other hand, solarization frequently improves plant growth, resulting in a larger plant canopy, which may favor the development of foliar diseases. An interesting aspect of the effect of solarization on primary inoculum was shown by Phillips (83) with *Sclerotinia minor*. Solarization reduced not only the populations of the sclerotia of this pathogen, but also the ability of the surviving sclerotia to form apothecia.

Yield increases following solarization vary and depend on the effectiveness of control, yield loss caused by the pest, the level of soil infestation, and compensation by neighboring plants, as shown with Fusarium wilt of cotton (69). In some cases, quality is also improved (22,31,35,52). Solarization is also effective in controlling diseases and in increasing yield and biomass of plants grown in artificial growth substrates (35).   Yield decline and growth retardation were evident in *Gypsophila* grown in either soil or artificial growth substrate under continuous cropping, although major pathogens were absent.   In this study, solarization (and fumigation) nullified these phenomena; yield was increased by 17 to 97%.   Solarization of tuff container medium increased the yield and improved the quality of the flowers and had a long-term effect.   Apparently, solarization controlled minor pathogens that are involved in yield decline in monoculture, as also seen in another study (36).   However, solarization was much less effective than chemical fumigation in treating replant disease of apple (56).

A long-term effect of solarization on disease control, yield increases, or both, for a second, or even a third season after solarization was observed in various regions and with a variety of crops (15,69,86,93,107).   The effect appears to be caused by the drastic reduction in inoculum density and by the induced suppressiveness (see below), which delayed reinfestation of the soil by the pathogens. Combining solarization with other methods, especially biological control agents, should enable better and longer-lasting disease control, and may require less frequent application.   The long-term effect has economic implications, whereby the cost per crop is reduced.   Economic analyses of soil solarization and other disinfestation methods are discussed elsewhere (111).

Weed control is one of the visible results of soil solarization and can be used as an indicator of its effectiveness (30,77,92).   Usually, annual weeds are more sensitive to solarization than perennial weeds.   Purple nutsedge is usually not controlled by solarization, and may even increase its population (27).   A comprehensive study on the effect of solarization on weeds was carried out in a Mediterranean climate at ICARDA, in Syria (77).   Of 57 weed species from 25 families, 46 were reduced by solarization, six were unaffected, and five were increased.   Most annual weeds were controlled by up to 100%.   Low to zero control, or even stimulation, occurred with weed species having bulbs, heat-tolerant seeds, a deep-lying root system, or other perennial organs.

Solarization for weed control has advantages over herbicides in that (i) it is a nonchemical method; (ii) there is no harmful residual effect, and (iii) the weed-killing effect of solarization may extend to deeper soil layers.   Other aspects of weed control by solarization are discussed below.

Postplant Solarization

Most solarization studies deal with preplanting applications of this technique, mostly for annual crops but also in some cases for perennial horticultural crops such as artichoke (108). Using solarization in existing plantings of trees, as a postplanting treatment for controlling soilborne pathogens, is a departure from the standard preplanting use of solarization for annual crops. This approach was pioneered by Ashworth and his associates and was tested for controlling Verticillium wilt in established pistachio nut groves (6). The success of this approach depends on a variety of factors, such as the extent of damage inflicted on the trees during solarization, and the effectiveness of solarization in shaded areas. Thus, postplanting solarization reduced the incidence of Verticillium wilt in pistachio (6) and olives (107), and of *Rosellinia necatrix* in apple (34), and improved the growth of peach seedlings (99). The recovery of infested trees following the application of solarization is a very desirable goal. This approach was extensively studied with olive trees infected with *V. dahliae* (107); infected 10- to 15-year-old trees recovered after a single application. Recovery from infection with *R. necatrix* was demonstrated in apples (34).

A new approach for postplanting solarization with black polyethylene has recently been developed (26,101). Black polyethylene film (which is less effective in heating the soil) was used to avoid heat damage to the solarized trees. This type of solarization also significantly conserved water, increased the trunk diameter of peach trees (although that of almond was reduced), controlled weeds, and reduced the rate of nematode galling as well as populations of certain nematodes. In a study aimed at controlling Verticillium wilt in apricot and almond, black polyethylene mulching gave better overall results than transparent polyethylene (101). The intermediate temperatures produced did not chronically harm the trees, yet the prolonged period of soil heating provided control of *Verticillium* that was equivalent to that achieved with transparent polyethylene. In another innovative and successful use, postplant solarization was applied to planted tomato in order to avoid interfering with normal agronomic practices (81). This resulted in effective control of *V. dahliae* and increased yield in the subsequent year.

In conclusion, it seems that there are many potential unusual uses of soil solarization, which could be very valuable in uncommon or specific situations.

BIOTIC AND ABIOTIC CHANGES DURING AND AFTER SOLARIZATION: MECHANISMS OF SOLARIZATION

Physical, chemical and biological processes occur during and after solarization. The long-term effect indicates that inoculum density was drastically reduced and that a new biological equilibrium in the soil may have been created. This strongly suggests that biological control processes are induced or stimulated in the soil, thus contributing to pathogen control. Data from various regions and soils and with different pathogens have verified this concept. Although the positive effects of solarization, expressed as pathogen and disease control and yield increases, are more frequent than the negative ones (such as growth retardation due to harmful effects on *Rhizobium* or mycorrhizae), the emphasis in research should be placed on the negative effects and on developing means to detect and avoid them.

The reduction in disease incidence occurring in plants growing in solarized soils, as with any soil treatment, results from the effects exerted on each of the three living components involved in disease—host, pathogen, and surrounding soil microorganisms—as well as on the physical and chemical environment which, in turn, affects the activities and the interrelationships of these organisms (23,63,64). Although these microbial processes occur primarily during solarization, they may continue to various extents and in different ways after removal of the polyethylene sheets and planting. The most pronounced effect of soil mulching with polyethylene is physical, i.e. an increase in soil temperatures for several hours daily during the solarization period. However, other accompanying processes, such as shifts in the microbial populations, changes in the chemical composition and physical structure of the soil, high moisture levels maintained by the polyethylene mulch, and changes in the gas composition of the soil should also be considered when analyzing mechanisms of disease control. The effect of solarization on each component or on certain combinations will be described below.

Thermal Inactivation Of Pathogens

Thermal death or inactivation of the population of an organism depends on both the temperature and the exposure time, which are inversely related. In many cases, heat-versus-mortality curves are exponential. Some studies investigating the time-versus-temperature relationship of thermal killing have shown that straight lines are obtained by plotting the logarithm of the number of survivors against exposure time at a given temperature. Pullman et al. (85) obtained a linear relationship by plotting the logarithm of the time required to kill 90% of the propagules of various soilborne fungal pathogens against temperature. However, pathogen control cannot be attributed solely to soil heating. Thus, reduction in inoculum density might be also observed at relatively lower temperatures.

Processes Affecting Pathogen-Microbe Interactions

Microbial processes induced by solarization may contribute to disease control (in addition to the physical effect of heat-induced mortality), since the impact of any lethal agent in the soil extends beyond the target organisms. Such processes may be especially useful where the cumulative effect of heat may be insufficient for disease control, e.g. in the deeper soil layers or in climatically marginal seasons or regions. Biological control may operate at any stage of pathogen survival or disease development during or after solarization, through antibiosis, lysis, parasitism, or competition. Solarization may affect inoculum density, inoculum potential, or both.

Most studies of the mechanisms of conventional biological control, soil disinfestation, or other methods of disease control of soilborne pathogens emphasize reduction of density of inoculum existing in the soil. Although eradication of the primary inoculum is a prerequisite for disease control, studying the behavior and fate of inoculum introduced into previously treated soil is crucial for the evaluation of disease control during the growing season, because reinfestation is common under field conditions. Studies of induced suppressiveness may be especially relevant to the problem of reinfestation.

Populations of antagonistic flora may be stimulated to attack propagules of pathogens and reduce inoculum density following solarization. Populations of *Talaromyces flavus*, an antagonist of *V. dahliae*, increased in solarized soils (107). Soil solarization increased rate of infection of juveniles of the nematode *Meloidogyne javanica* by the bacterial antagonist *Pasteuria penetrans* for at least 10 months (110). Antagonism may be further improved if propagules of the pathogens are *weakened* by sublethal heating and become vulnerable to microbial activity. Reductions in populations of sclerotia of *S. minor* by solarization were attributed to microbial colonization and degradation of sclerotia weakened by sublethal temperatures (83). Their reduced capacity to produce sclerotia also reflects weakening. In another study with this pathogen (109), solarization enhanced microbial colonization of the sclerotia and accelerated their exponential decay. Sublethal heating of conidia and chlamydospores of *F. oxysporum* f. sp. *niveum* reduced propagule viability and weakened the surviving propagules (33). The weakening effect was shown as a delay in germination, a reduction in the growth of conidial and chlamydospore germ tubes, and enhanced decline of the population density of viable conidia in the soil. Populations of conidia that were heat-treated or exposed to solarized soil declined faster than unheated conidia in a soil suspension culture. Disease incidence in watermelon seedlings inoculated with the heat-treated conidia was reduced. Sublethally heated conidia of *Fusarium* respond by producing heat-shock proteins (32). This response may provide a tool for assessing and predicting the weakening of propagules upon sublethal heating.

Studies have frequently shown that solarized soils become hostile, i.e. less receptive, to pathogen reinfestation. This can be regarded as an induced suppressiveness (34,39,50,58,62), resembling the natural phenomenon. In solarized and subsequently inoculated soil, the incidences of disease caused by *Fusarium* and *S. rolfsii* were lower than in untreated inoculated soil, the population of lytic microorganisms was higher, and suppression of chlamydospore formation was frequently observed (50). Similar phenomena of induced suppressiveness in solarized soils were found with Fusarium wilt of carnation (58) and other pathogens.

Among the bacteria affected by solarization are *Pseudomonas* and *Bacillus* spp. (23). Oxidase-negative fluorescent pseudomonads are highly sensitive to soil solarization and their populations drop to low levels; however, they quickly recolonize solarized soils and their populations reach high densities. Some pseudomonads are rhizosphere-competent and, as such, afford not only a degree of protection against fungal pathogens but also cause an increased growth response in host plants. By coating radish and sugar beet seeds with isolates of *Pseudomonas fluorescens* selected for their beneficial effects on plant growth and yield, Stapleton and DeVay (98) found up to a sixfold increase in root colonization by these bacteria in solarized soil, as compared with plants in nonsolarized soil. The population densities of different types of fluorescent pseudomonads in the rhizosphere and roots of tomato plants growing in solarized soils also increased considerably, although they decreased initially in the solarized soil (36). *Bacillus* spp. are spore formers; they can often tolerate extreme temperatures and were found to be increased by solarization (40). The stimulation of fluorescent pseudomonads was connected with changes in the composition of root exudates from plants growing in solarized soil, as expressed by increased levels of amino acids and amino compounds (37). These exudates enhance chemotaxis of these bacteria towards the roots (38) and also affect the balance between various groups of microorganisms in soil. The signals that occur in solarized soils and trigger these changes in the root exudates are as yet unknown.

Processes Affecting Plant Growth

The phenomenon of increased growth response (IGR) denotes the improvement in plant growth when disinfestation is carried out in soils free of known pathogens. This phenomenon was detected in both artificially heated and fumigated soils several decades ago. It has also been found in solarized soils (1,15,17,18,36,55,91,98,102). Mechanisms that can explain IGR are either chemical (release of mineral nutrients or growth factors, nullification of toxins) or biological (elimination of minor or unknown pathogens, stimulation of beneficial microorganisms). Higher concentrations of mineral and organic substances were found in solutions of solarized soils. Since IGR has economic implications, methods for IGR prediction in various soils should be developed. In a recent study (55), IGR was recorded in shoots of tomato seedlings, 15 days after transplanting, and in roots only 2 weeks later. Detailed studies showed significantly higher levels of chlorophyll and protein contents in plants from solarized soil as compared to those from the control. In addition, the degradation of these compounds, and decrease of net photosynthesis at near-saturation light intensity and of photochemical yield with ageing were delayed in plants growing in solarized soil, as compared to the control. It was concluded that the initial IGR in solarized soil is independent of improved root growth. Delayed leaf senescence appeared to be a plant response contributing to IGR. It will be interesting to carry out similar studies also with plants growing in fumigated soils.

IMPROVING SOIL SOLARIZATION

Any control method undergoes constant modification and improvement at various stages of its development and implementation to realistic farm conditions. The aim is to adapt the method to a wide variety of conditions and to overcome difficulties, limitations, and existing negative side effects which have already been encountered or may potentially occur.

In improving solarization, we seek to achieve the following goals.

1. Improve the level of control.
2. Control pathogens that are not adequately controlled by

solarization because of heat tolerance, e.g. *Macrophomina*, or for other reasons. Such improvement would enable us to broaden the spectrum of pathogen control.

3. Achieve long-lasting effects, throughout a season, or over successive seasons.
4. Expand its use to climatically marginal regions or seasons.
5. Shorten the 20- to 40-day period usually needed to achieve effective control.
6. Avoid or curb negative side effects, e.g. a negative effect on mycorrhizae.
7. Reduce cost.
8. Increase the reliability and reproducibility of the method, which is climate-dependent.

The following approaches might be adopted to improve solarization.

1. Combine solarization with other methods of control: cultural, biological, or chemical (preferably at reduced dosages).
2. Improve the technology of application.

Combining Solarization With Other Methods: Integrated Control

The combining of solarization with other methods of control should be considered from its two different sides: (i) as a way of improving solarization (as detailed above), or (ii) as a way of improving the other method. From the latter point of view, a suitable combination of solarization with a pesticide at a *reduced* dosage makes the pesticide less harmful (and more acceptable), without reducing control effectiveness and maybe even increasing it. Combining control methods is at the heart of integrated control.

Many studies have combined control methods with solarization, frequently with good results. Such a combination may result in either a synergistic or an additive effect. The weakening phenomenon described above is such a mechanism, potentially leading to synergism, or at least improved control. In addition to reduction in fumigant dosage, the combining of methyl bromide fumigation with solarization confers an added benefit—the same plastic sheeting can be used for both purposes. Indeed, good results with such a combination have been reported in the control of yield decline in *Gypsophila* (35) and other diseases (64). Methyl bromide-chloropicrin mixture combined with solarization was effective in controlling *Pseudomonas solanacearum* in tomato (16).

Combining solarization with pesticides may also activate the pesticide upon heating. Improvements in disease control, yield, or both, have been demonstrated with a variety of combinations, such as dazomet with solarization for the control of pink root of onion (84), and metham-sodium with solarization for the control of delimited shell spot of peanut (31) and soilborne diseases of strawberry (59). The solarization-metham combination (10) killed more propagules of *F. oxysporum* f. sp. *vasinfectum*, and faster, than solarization alone. Thus, the period during which solarization is effective might be longer than previously thought, when a suitable improvement is used. On the other hand, solarization may increase phytotoxicity of herbicides by suppressing their degradation in soil (7). This has also been demonstrated with the combination of solarization and the herbicide dacthal used with collards (104). Combining biological control agents or other beneficial organisms with solarization is an especially attractive alternative. Its effectiveness has been studied with *Trichoderma harzianum* combined with solarization for the control of Fusarium crown and root rot of tomato (95) and of *Rhizoctonia* (19,29), and with *Gliocladium virens* combined with solarization for controlling *S. rolfsii* (91). The benefit of vesicular-arbuscular mycorrhizal inoculation appears greatest with solarization (3). Combining solarization with the use of tolerant cultivars improved the control of Verticillium wilt of potato (22). Combining solarization with a certain crop sequence improved the control of Fusarium wilt of cotton (69).

When organic amendments that release volatile components toxic to the pathogen are combined with solarization, the plastic tarp prevents or delays the escape of the volatiles and increases their effectiveness. This has been demonstrated by combining cruciferous amendments (which have a high content of sulfur-containing compounds) with solarization for controlling cabbage yellows (88,89), or *Pythium* and *S. rolfsii* (41). In the latter study, a variety of volatiles generated by this amendment were identified, including alcohols, aldehydes, sulfides and isothiocyanates. Combining solarization with chicken compost was very effective in controlling the southern root-knot nematode and increased the yield of lettuce over two successive seasons (40). The rhizosphere of plants in the solarized soil was intensely colonized by fluorescent pseudomonads and *Bacillus* spp. However, in certain situations, lettuce growth was inhibited. Thus, for each combination we have to determine the sequence of applications, or whether to combine the methods simultaneously or to alternate them.

An interesting approach for improving the results of solarization is to combine different plastic mulches, as was done in the Jordan Valley (2). Soil solarization with transparent polyethylene mulch reduced weed growth and improved yield of tomato and squash. However, additional mulching with black polyethylene further improved the results. To reduce costs, the soil was solarized with black polyethylene, which was kept in place and perforated for crop planting, and used as a mulch for the crop for the rest of the growing season.

Improving The Technology Of Application

The following approaches might be adopted to improve the technology of application.

1.   The use of plastic materials that are more efficient in heating the soil.

2.   A double layer of plastic heats the soil to higher levels than a single layer. A static layer of air above the plastic or between double layers of plastic apparently acts as an insulator between the plastic and the roof of the glasshouse. This reduces both moisture and convective heat losses from the transparent plastic sheeting covering the soil (14). The effectiveness of a double layer of film for solarization was studied by Raymundo and Alcazar (90) in Peru for controlling nematodes in potato-growing soils. They observed that the soil temperature at a 10-cm depth under two layers of polyethylene film separated by 50 cm of air was 12.5°C higher than that under a single layer of film (60.0°C versus 47.5°C); the temperature of noncovered soil was 32.5°C. Ben-Yephet et al. (11) found that solarization with two layers of 25 μm-thick polyethylene film separated by a 6-cm air layer caused soil temperatures at a depth of 15 cm to rise by 12.7°C and 3.6°C over those in noncovered soil or soil covered by one layer of film, respectively; at a depth of 30 cm, the respective differences in temperature were 11.2°C and 2.7°C. Viability of propagules (mainly chlamydospores) of *F. oxysporum* f. sp. *vasinfectum* that had been buried at a depth of 30 cm was reduced after 31 days of solarization by 97.5%, 58%, and 0% under a double-film layer, a single layer, and in noncovered soil, respectively.

Solarization with a double layer of film was effective in controlling *Pythium*, *Fusarium* spp. and *Rhizoctonia* in a forest nursery in Central Italy in a region that is considered marginal for this technique (4). This approach was also followed successfully in Australia with nursery potting mix (25). Temperatures at a depth of 25 cm reached maximum values of 51°C and 44.6°C under a double and single layer of clear plastic mulch, respectively. Three soilborne plant pathogens, *Pythium myriotylum*, *Phytophthora nicotianae* var. *nicotianae*, and *S.*

rolfsii, were eliminated under a double layer of plastic within 2 to 8 days during spring, summer and autumn experiments, whereas it took 4 to 20 days at the same time of the year to eliminate the same pathogens using a single layer of plastic. However, in a study carried out in Egypt, solarization with a single layer of polyethylene was as effective as a double layer, and more effective than fumigation with methyl bromide in controlling Fusarium wilt in tomato (28).

3.   An interesting approach developed independently in Italy, Japan and other countries, is mulching the soil with polyethylene or vinyl *inside* closed glasshouses or plastic houses, to intensify soil heating. This can be regarded as another version of the double layer approach. In northern Italy, solarization in the open field was not effective, but when it was carried out inside closed glasshouses it controlled corky root of tomato (42,43,105). Mulching inside plastic houses was less effective. This approach has been extensively developed in Japan, e.g. for the control of *Fusarium* in strawberry (60,72). Recently, this approach has also been developed for sanitation purposes, namely, to heat the greenhouse to over 60°C and eradicate the inoculum remaining in the greenhouse structure (94). Besri and Diop (12) successfully used solarization to control *Didymella lycopersici* in tomato supports in which this pathogen survives.

4.   Improving the technologies of application by improving the machinery (51), which can lead to improved control and cost reduction.

5.   Reducing the border effect (heat loss from the plastic edges) by broadcast solarization (53).

SIMULATIVE AND PREDICTIVE MODELS OF SOIL HEATING

The patterns of soil heating, both with and without plastic, are very typical. Numerical models can serve as a tool for predicting soil temperatures at each soil layer, thus facilitating our work. Models for predicting temperatures of mulches and nonmulched soils at various depths and under a variety of conditions, were developed (8,79). These models also enable a first evaluation of the effect of environmental conditions and agricultural practices (Figure 1). The physical models have to be coupled with biological ones in predicting the rate of heat mortality of pathogens under different climatic regimes and under constant or fluctuating temperatures (74,78,85,112).

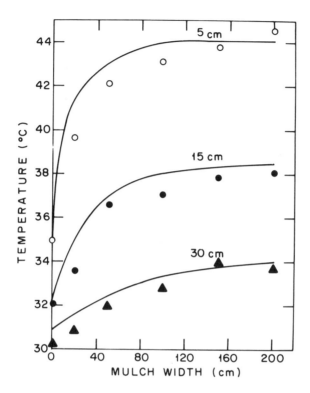

Figure 1. Observed and predicted average daily maximum soil temperatures for the period 2 to 15 October 1979 at soil depths of 5, 15, and 30 cm as a function of the mulch width. Observed values are indicated by ○, ●, and ▲, respectively. After Mahrer and Katan (79).

IMPLEMENTATION OF SOIL SOLARIZATION

As with the introduction of any nonchemical method of control, there are many steps to be followed in order to bring the method to the farmer's field (65), the major ones being: (i) testing the reproducibility and validity of the results under realistic conditions; (ii) running an upscaling program that evaluates the effectiveness of the method up to farmer plot, including the establishment of demonstration plots and the use of extension tools; (iii) performing an economic analysis (111); and (iv) developing the proper technologies for application (51).

Solarization is already being used by farmers in several countries, such as Israel, Greece, Italy, Iraq, Mexico and Japan. In some cases, this was achieved by following systematic research and development plans, whereas in others the adoption of the method was achieved by trial and error.

The development of pest control methods requires a continuous evaluation of its advantages and limitations, and the development of means to avoid or curb its negative side effects. The major advantages of soil solarization are that it is a nonchemical and simple method that appears to cause no major or frequent undesirable disturbances to the soil's biological balance. Difficulties stem from its restriction to certain climatic areas and seasons, and to the fact that it limits other activities on the land for several weeks. There are some reports of temporary negative effects of solarization on beneficial organisms, such as *Rhizobium* (1,15) and mycorrhizae; however, the effect is usually less detrimental than that of fumigation.

CONCLUSIONS

The control of soilborne pathogens in their natural habitat, be it the soil or inside plant tissue, is a very difficult task. After more than 120 years of soil disinfestation, the arsenal of chemical disinfestants is still very limited and the future of methyl bromide, the major agent, is unclear. The arsenal of nonchemical agents for soil disinfestation is even more limited. Therefore, the integration of control methods, rather than a reliance on one powerful control agent, is not only desirable but also the only feasible solution for coping with our need for methods of controlling soilborne pathogens in an atmosphere of environmental awareness, concerns, and pressure. To achieve this goal, we need to make a "commitment to long-term agricultural research, namely, the provision of a stable research capability that allows lessons to be learned from the past and encourages creative planning for the future" (76). There is increasing concern that contamination with chemical pollutants is degrading the quality of soil. By analogy, soilborne pathogens should be regarded as biological pollutants of agricultural soils. They may even be more dangerous than the chemical pollutants because of their capacity to reproduce and increase their population in the soil, and because of the ability of some of the biotic agents to survive in the soil even for decades. Hence, the proper control of soilborne pathogens should be considered a remediation of agricultural soils from the hazards of biopollutants.

LITERATURE CITED

1. Abdel-Rahim, M.F., Satour, M.M., Mickail, K.Y., El Eraki, S.A., Grinstein, A., Chen, Y., and Katan, J. 1988. Effectiveness of soil solarization in furrow-irrigated Egyptian soils. Plant Dis. 72:143–146.

2. Abu-Irmaileh, B.E. 1991. Weed control in squash and tomato fields by soil solarization in the Jordan Valley. Weed Res. 31:125–133.

3. Afek, V., Menge, J.A., and Johnson, E.L.V. 1991. Interaction among mycorrhizae, soil solarization, metalaxyl, and plants in the field. Plant Dis. 75:665–671.

4. Annesi, T., and Motta, E. 1994. Soil solarization in an Italian forest nursery. Eur. J. For. Pathol. 24:203–209.

5. Antoniou, P.P., Tjamos, E.C., and Panagopoulos, C.G. 1995. Use of soil solarization for controlling bacterial canker of tomato in plastic houses in Greece. Plant Pathol. 44:438–447.

6. Ashworth Jr., L.J., and Gaona, S.A. 1982. Evaluation of clear polyethylene mulch for controlling Verticillium wilt in established pistachio nut groves. Phytopathology 72:243–246.

7. Avidov, E., Aharonson, N., Katan, J., Rubin, B., and Yarden, O. 1985. Persistence of terbutryn and atrazine in soil as affected by soil disinfestation and fungicides. Weed Sci. 33:457–461.

8. Avissar, R., Naot, O., Mahrer, Y., and Katan, J. 1986. Field aging of transparent polyethylene mulches: II. Influence on the effectiveness of soil heating. Soil Sci. Soc. Am. J. 50:205–209.

9. Barkai-Golan, R., and Phillips, D.J. 1991. Postharvest heat treatment of fresh fruit and vegetables for decay and control. Plant Dis. 75:1085–1089.

10. Ben-Yephet, Y., Melero-Vera, J.M., and DeVay, J.E. 1988. Interaction of soil solarization and metham-sodium in the destruction of *Verticillium dahliae* and *Fusarium oxysporum* f. sp. *vasinfectum*. Crop Prot. 7:327–331.

11. Ben-Yephet, Y., Stapleton, J.J., Wakeman, R.J., and DeVay, J.E. 1987. Comparative effects of soil solarization with single and double layers of polyethylene film on survival of *Fusarium oxysporum* f. sp. *vasinfectum*. Phytoparasitica 15:181–185.

12. Besri, M., and Diop, M. 1985. Control of *Didymella lycopersici* in tomato by storing the supports in plastic tunnels: new application of the solar heating or solarization. Rev. Hort. 58:99–102.

13. Borges, L.V. 1982. Solarizacon do solo novo metodo de pasteurizacao de solo. Rev. Cienc. Agrar. 5:1–5.

14. Cenis, J.L. 1987. Double plastic sheet for improving soil solarization efficiency. Page 73 in: Proceedings, Seventh Conference of the Mediterranean Phytopathological Union. Sociedad Espanola de Fitopatologia, Granada, Spain.

15. Chauhan, Y.S., Nene, Y.L., Johansen, C., Haware, M.P., Saxena, N.P., Singh, S., Sharma, S.B., Sahrawat, K.L., Burford, J.R., Rupela, O.P., Kumar Rao, J.V.D.K., and Sithanantham, S. 1988. Effects of soil solarization on pigeonpea and chickpea. ICRISAT Bull. 11, Andhra Pradesh, India. 16 pp.

16. Chellemi, D.O., Olson, S.M., and Mitchell, D.J. 1994. Effects of soil solarization and fumigation on survival of soilborne pathogens of tomato in northern Florida. Plant Dis. 78:1167–1172.

17. Chen, Y., Gamliel, A., Stapleton, J.J., and Aviad, T. 1991. Chemical, physical, and microbial changes related to plant growth in disinfested soils. Pages 103–129 in: Soil Solarization. J. Katan and J.E. DeVay, eds. CRC Press, Boca Raton, FL.

18. Chen, Y., and Katan, J. 1980. Effect of solar heating of soils by transparent polyethylene mulching on their chemical properties. Soil Sci. 130:271–277.

19. Chet, I., Elad, Y., Kalfon, A., Hadar, Y., and Katan, J. 1982. Integrated control of soilborne and bulbborne pathogens in iris. Phytoparasitica 10:229–236.

20. Cook, R.J., Sittor, J.W., and Haglund, W.A. 1987. Influence of soil treatments on growth and yield of wheat and implications for control of Pythium root rot. Phytopathology 77:1192–1198.

21. Davis, J.R. 1991. Soil solarization: pathogen and disease control and increases in crop yield and quality: short- and long-term effects and integrated control. Pages 39–50 in: Soil Solarization. J. Katan and J.E. DeVay, eds. CRC Press, Boca Raton, FL.

22. Davis, J.R., and Sorensen, L.H. 1986. Influence of soil solarization at moderate temperatures on potato genotypes with differing resistance to *Verticillium dahliae*. Phytopathology 76:1021–1026.

23. DeVay, J.E., and Katan, J. 1991. Mechanisms of pathogen control in solarized soils. Pages 87–101 in: Soil Solarization. J. Katan and J.E. DeVay, eds. CRC Press, Boca Raton, FL.

24. DeVay, J.E., Stapleton, J.J., and Elmore, C.L., eds. 1991. Soil Solarization. Proceedings, First International Conference on Soil Solarization, Amman, Jordan. FAO Plant Production and Protection Paper 109, FAO, Rome, Italy. 396 pp.

25. Duff, J.D., and Connelly, M.I. 1993. Effect of solarization using single and double layers of clear plastic mulch in *Pythium*, *Phytophthora* and *Sclerotium* species in a nursery potting mix. Australas. Plant Pathol. 22:28–35.

26. Duncan, R.A., Stapleton, J.J., and McKenry, M.V. 1992. Establishment of orchards with black polyethylene film mulching: effect on nematode and fungal pathogens, water conservation, and tree growth. J. Nematol. Suppl. 24:681–687.

27. Egley, G.H. 1983. Weed seed and seedling reductions by soil solarization with transparent polyethylene sheets. Weed Sci. 31:404–409.

28. El-Shami, M., Salem, D.E., Fadl, F.A., Ashour, W.E., and El-Zayat, M.M. 1990. Soil solarization and plant disease management: II. Effect of soil solarization in comparison with soil fumigation on the management of Fusarium wilt of tomato. Agric. Res. Rev. 68:601–611.

29. Elad, Y., Katan, J., and Chet, I. 1980. Physical, biological, and chemical control integrated for soilborne diseases in potatoes. Phytopathology 70:418–422.

30. Elmore, C.L. 1991. Weed control by soil solarization. Pages 61–72 in: Soil Solarization. J. Katan and J.E. DeVay, eds. CRC Press, Boca Raton, FL.

31. Frank, Z.R., Ben-Yephet, Y., and Katan, J. 1986. Synergistic effect of metham and solarization in controlling delimited shell spots of peanut pods. Crop Prot. 5:199–202.

32. Freeman, S., Ginzburg, C., and Katan, J. 1989. Heat shock protein synthesis in propagules of *Fusarium oxysporum* f. sp. *niveum*. Phytopathology 79:1054–1058.

33. Freeman, S., and Katan, J. 1988. Weakening effect on propagules of *Fusarium* by sublethal heating. Phytopathology 78:1656–1661.

34. Freeman, S., Sztejnberg, A., Shabi, E., and Katan, J. 1990. Long- term effect of soil solarization for the control of *Rosellinia necatrix* in apple. Crop Prot. 9:312–316.

35. Gamliel, A., Hadar, E., and Katan, J. 1993. Improvement of growth and yield of *Gypsophila paniculata* by solarization or fumigation of soil or container medium in continuous cropping systems. Plant Dis. 77:933–938.

36.   Gamliel, A., and Katan, J. 1991. Involvement of fluorescent pseudomonads and other microorganisms in increased growth response of plants in solarized soils. Phytopathology 81:494–502.

37.   Gamliel, A., and Katan, J. 1992. Influence of seed and root exudates on fluorescent pseudomonads and fungi in solarized soil. Phytopathology 82:320–327.

38.   Gamliel, A., and Katan, J. 1992. Chemotaxis of fluorescent pseudomonads towards seed exudates and germinating seeds in solarized soil. Phytopathology 82:328–332.

39.   Gamliel, A., and Katan, J. 1993. Suppression of major and minor pathogens by fluorescent pseudomonads in solarized and nonsolarized soils. Phytopathology 83:68–75.

40.   Gamliel, A., and Stapleton, J.J. 1993. Effect of chicken compost or ammonium phosphate and solarization on pathogen control, rhizosphere microorganisms, and lettuce growth. Plant Dis. 77:886–891.

41.   Gamliel, A., and Stapleton, J.J. 1993. Characterization of antifungal volatile compounds evolved from solarized soil amended with cabbage residues. Phytopathology 83:899–905.

42.   Garibaldi, A., and Tamietti, G. 1983. Attempts to use soil solarization in closed glasshouses in northern Italy for controlling corky root of tomato. Acta Hortic. 152:237–243.

43.   Garibaldi, A., and Tamietti, G. 1989. Solar heating: recent results obtained in northern Italy. Acta Hortic. 255:125–129.

44.   Gaur, H.S., and Perry, R.N. 1991. The use of soil solarization for control of plant parasitic nematodes. Nematol. Abstr. 60:153–167.

45.   Gerson, U., Yathom, S., and Katan, J. 1981. A demonstration of bulb mite control by solar heating of the soil. Phytoparasitica 9:153–155.

46.   Ghini, R. 1993. A solar collector for soil disinfestation. Neth. J. Plant Pathol. 99:45–50.

47.   Ghini, R., Betiol, W., Spadotto, C.A., Moraes, G.J. de, Paraiba, L.C., Mineiro, J.L. de. 1993. Soil solarization for the control of tomato and eggplant Verticillium wilt and its effect on weed and minor arthropod communities. Summa Phytopathol. 19:183–189.

48.   Ghini, R., Paraiba, L.C., and Lima, M.W.P. 1994. Period determination for soil solarization in the region of Campinas SP. Summa Phytopathol. 20:131–133.

49. González-Torres, R., Meléro-Vara, J.M., Gómez-Vázquez, J., and Jiménez-Díaz, R.M. 1993. The effects of soil solarization and soil fumigation on fusarium wilt of watermelon grown in plastic houses in south-eastern Spain. Plant Pathol. 42:858–864.

50. Greenberger, A., Yogev, A., and Katan, J. 1987. Induced suppressiveness in solarized soils. Phytopathology 77:1663–1667.

51. Grinstein, A., and Hetzroni, A. 1991. The technology of soil solarization. Pages 159–170 in: Soil Solarization. J. Katan and J.E. DeVay, eds. CRC Press, Boca Raton, FL.

52. Grinstein, A., Katan, J., Abdul-Razik, A., Zeydan, O., and Elad, Y. 1979. Control of *Sclerotium rolfsii* and weeds in peanuts by solar heating of the soil. Plant Dis. Rep. 63:1056–1059.

53. Grinstein, A., Kritzman, G., Hetzroni, A., Gamliel, A., Mor, M., and Katan, J. 1995. The border effect of soil solarization. Crop Prot. 14:315–320.

54. Grooshevoy, S.E. 1939. Disinfection of seed-bed soil in cold frames by solar energy. The A.I. Mikoyan Pan-Soviet Sci. Res. Inst. Tob. and Indian Tob. Ind. (VITIM), Krasnodar, Publ. 137:51–56. Abstract in: Rev. Appl. Mycol. 18:635–636.

55. Gruenzweig, J.M., Rabinowitch, H.D., and Katan, J. 1993. Physiological and developmental aspects of increased plant growth in solarized soils. Ann. Appl. Biol. 122:579–591.

56. Gur, A., Cohen, Y., Katan, J., and Barkai, Z. 1991. Preplant application of soil fumigants and solarization for treating replant diseases of peaches and apples. Sci. Hort. 45:215–224.

57. Hagan, H.R. 1933. Hawaiian pineapple field soil temperatures in relation to the nematode *Heterodera radicicola* (Greef) Müller. Soil Sci. 36:83–95.

58. Hardy, G.E.St.J., and Sivasithamparam, K. 1985. Soil solarization: effects on Fusarium wilt of carnation and Verticillium wilt of eggplant. Pages 279–281 in: Ecology and Management of Soilborne Plant Pathogens. C.A. Parker, A.D. Rovira, K.J. Moore, P.T.W. Wong, and J.F. Kollmorgen, eds. The American Phytopathological Society, St. Paul, MN.

59. Hartz, T.K., DeVay, J.E., and Elmore, C.L. 1993. Solarization is an effective soil disinfestation technique for strawberry production. HortScience 28:104–106.

60. Horiuchi, S. 1984. Soil solarization for suppressing soilborne diseases in Japan. Pages 215–227 in: FFTC Book Series No. 26, Soilborne Crop Diseases in Asia. Food and Fertilizer Technology Center, Taiwan, ROC.

61. Jindal, K.K., Thind, B.S., and Soni, P.S. 1989. Physical and chemical agents for the control of *Xanthomonas campestris* pv. *vignicola* from cowpea seeds. Seed Sci. Technol. 17:371–382.

62. Kassaby, F.Y. 1985. Solar-heating soil for control of damping-off diseases. Soil Biol. Biochem. 17:429–434.

63. Katan, J. 1981. Solar heating (solarization) of soil for control of soilborne pests. Annu. Rev. Phytopathol. 19:211–236.

64. Katan, J. 1987. Soil solarization. Pages 77–105 in: Innovative Approaches to Plant Disease Control. I. Chet, ed. John Wiley & Sons, New York, NY.

65. Katan, J. 1993. Replacing pesticides with nonchemical tools for the control of soilborne pathogens—a realistic goal? Phytoparasitica 21:95–99.

66. Katan, J., and DeVay, J.E., eds. 1991. Soil Solarization. CRC Press, Boca Raton, FL. 267 pp.

67. Katan, J., and DeVay, J.E. 1991. Soil solarization: historical perspectives, principles, and uses. Pages 23–37 in: Soil Solarization. J. Katan and J.E. DeVay, eds. CRC Press, Boca Raton, FL.

68. Katan, J., DeVay, J.E., and Greenberger, A. 1989. The biological control induced by soil solarization. Pages 293–299 in: Vascular Wilt Diseases of Plants: Basic Studies and Control. E.C. Tjamos and C.H. Beckman, eds. NATO ASI Series H28. Springer-Verlag, Berlin, Germany.

69. Katan, J., Fishler, G., and Grinstein, A. 1983. Short- and long-term effects of soil solarization and crop sequence on Fusarium wilt and yield of cotton in Israel. Phytopathology 73:1215–1219.

70. Katan, J., Greenberger, A., Alon, H., and Grinstein, A. 1976. Solar heating by polyethylene mulching for the control of diseases caused by soil-borne pathogens. Phytopathology 66:683–688.

71. Katan, J., Grinstein, A., Greenberger, A., Yarden, O., and DeVay, J.E. 1987. The first decade (1976–1986) of soil solarization (solar heating): a chronological bibliography. Phytoparasitica 15:229–255.

72. Kodama, T., and Fukui, T. 1982. Solar heating in closed plastic house for control of soilborne diseases. V. Application for control of Fusarium wilt of strawberry. Ann. Phytopathol. Soc. Japan 48:570–577 (Japanese with English summary).

73. Laemlin, F.F. 1992. Control of tomato decline using solarization. Acta Hortic. 301:237–242.

74. Lazarovitz, G., Hawke, M.A., Olthof, Th.H.A., and Coutu-Sundy, J. 1991. Influence of temperature on survival of *Pratylenchus penetrans* and of microsclerotia of *Verticillium dahliae* in soil. Can. J. Plant Pathol. 13:106–111.

75. Lazarovitz, G., Hawke, M.A., Tomlin, A.D., Olthof, Th.H.A., and Squire, S. 1991. Soil solarization to control *Verticillium dahliae* and *Pratylenchus penetrans* on potatoes in central Ontario. Can. J. Plant Pathol. 13:116–123.

76. Lewis, T. 1994. Commitment to long-term agricultural research: a message for science, sponsors and industry. Pages 3–20 in: Brighton Crop Protection Conference—Pest and Diseases, Vol. 1. British Crop Protection Council, Farnham, UK.

77. Linke, K.H. 1994. Effects of soil solarization on arable weeds under Mediterranean conditions: control, lack of response or stimulation. Crop Prot. 13:115–120.

78. López-Herrera, C.J., Verdú-Valiente, B., and Melero-Vara, J.M. 1994. Eradication of primary inoculum of *Botrytis cinerea* by soil solarization. Plant Dis. 78:594–597.

79. Mahrer, Y., and Katan, J. 1981. Spatial soil temperature regime under transparent polyethylene mulch: numerical and experimental studies. Soil Sci. 131:82–87.

80. Mihail, J.D., and Alcorn, S.M. 1984. Effects of soil solarization on *Macrophomina phaseolina* and *Sclerotium rolfsii*. Plant Dis. 68:156–159.

81. Morgan, D.P., Liebman, J.A., Epstein, L., and Jimenez, M.J. 1991. Solarizing soil planted with cherry tomatoes vs. solarizing fallow ground for control of Verticillium wilt. Plant Dis. 75:148–151.

82. Newhall, A.G. 1955. Disinfestation of soil by heat, flooding and fumigation. Bot. Rev. 21:189–250.

83. Philips, A.J.L. 1990. The effects of soil solarization on sclerotial populations of *Sclerotinia sclerotiorum*. Plant Pathol. 39:38–43.

84. Porter, I.J., Merriman, P.R., and Keane, P.J. 1989. Integrated control of pink root (*Pyrenochaeta terrestris*) of onions by dazomet and soil solarization. Aust. J. Agric. Res. 40:861–869.

85. Pullman, G.S., DeVay, J.E., and Garber, R.H. 1981. Soil solarization and thermal death: a logarithmic relationship between time and temperature for four soilborne plant pathogens. Phytopathology 71:959–964.

86. Pullman, G.S., DeVay, J.E., Garber, R.H., and Weinhold, A.R. 1981. Soil solarization: effects on Verticillium wilt of cotton and soilborne populations of *Verticillium dahliae*, *Pythium* spp., *Rhizoctonia solani*, and *Thielaviopsis basicola*. Phytopathology 71:954–959.

87. Raghaven, D. 1964. Agriculture in Ancient India. ICAR Publ., New Delhi, India. 164 pp.

88. Ramirez-Villapudua, J., and Munnecke, D.E. 1987. Control of cabbage yellows (*Fusarium oxysporum* f. sp. *conglutinans*) by solar heating of field soils amended with dry cabbage residues. Plant Dis. 71:217–221.

89. Ramirez-Villapudua, J., and Munnecke, D.E. 1988. Effect of solar heating and soil amendments of cruciferous residues on *Fusarium oxysporum* f. sp. *conglutinans* and other organisms. Phytopathology 78:289–295.

90. Raymundo, S.A., and Alcazar, J. 1986. Increasing efficiency of soil solarization in controlling root-knot nematodes by using two layers of plastic mulch. J. Nematol. 18:628. (Abstr.).

91. Ristaino, J.B., Perry, K.B., and Lumsden, R.D. 1991. Effect of solarization and *Gliocladium virens* on sclerotia of *Sclerotium rolfsii*, soil microbiota, and the incidence of southern blight of tomato. Phytopathology 81:1117–1124.

92. Rubin, B., and Benjamin, A. 1983. Solar heating of the soil: effect on weed control and on soil-incorporated herbicides. Weed Sci. 31:819–825.

93. Satour, M.M., Abdel-Rahim, M.F., Grinstein, A., Rabinowitch, H.D., and Katan, J. 1989. Soil solarization in onion fields in Egypt and Israel: short- and long-term effects. Acta Hortic. 255:151–159.

94. Shlevin, E., Katan, J., Mahrer, Y., and Kritzman, G. 1994. Space solarization of the greenhouse structure for the elimination of surviving soilborne pathogens. Phytoparasitica 22:82. (Abstr.).

95. Sivan, A., and Chet, I. 1993. Integrated control of fusarium crown and root rot of tomato with *Trichoderma harzianum* in combination with methyl bromide or soil solarization. Crop Prot. 12:380–386.

96. Souza, N.L. 1994. Solarizacao do solo. Summa Phytopathol. 20:3–15.

97. Stapleton, J.J. 1991. Physical characteristics of spray-applied polymer mulches as related to potential management of soilborne plant diseases in the San Joaquin Valley. Phytopathology 81:1185. (Abstr.).

98. Stapleton, J.J., and DeVay, J.E. 1984. Thermal components of soil solarization as related to changes in soil and root microflora and increased plant growth response. Phytopathology 74:255–259.

99. Stapleton, J.J., and DeVay, J.E. 1985. Soil solarization as a post-plant treatment to increase growth of nursery trees. Phytopathology 75:1179. (Abstr.).

100. Stapleton, J.J., and DeVay, J.E. 1986. Soil solarization: a non-chemical approach for management of plant pathogens and pests. Crop Prot. 5:190–198.

101. Stapleton, J.J., Paplomatas, E.J., Wakeman, R.J., and DeVay, J.E. 1993. Establishment of apricot and almond trees using soil mulching with transparent (solarization) and black polyethylene film: effects on Verticillium wilt and tree health. Plant Pathol. 42:333–338.

102. Stapleton, J.J., Quick, J., and DeVay, J.E. 1985. Soil solarization: effects on soil properties, crop fertilization and plant growth. Soil Biol. Biochem. 17:369–373.

103. Stevens, C., Khan, V.A., Collins, D., Rodriguez-Kabana, R., Ploper, L.D, Adeyeye, O., Brown, J., and Backman, P. 1992. Use of soil solarization to reduce the severity of early blight, southern blight and root-knot in tomatoes. Phytopathology 82:500. (Abstr.).

104. Stevens, C., Khan, V.A., Okoronkuro, T., Tang, A.-Y., Wilson, M.A., Lu, J., and Brown, J.E. 1990. Soil solarization and dachtal: influence on weeds, growth, and root microflora of collards. HortScience 25:1260–1262.

105. Tamietti, G., and Garibaldi, A. 1981. Control of corky root in tomato by solar heating of the soil in greenhouse in Riviera Ligure. La Difesa Della Piente 2:143–150 (In Italian).

106. Tietz, H. 1970. One centennium of soil fumigation: its first years. Pages 203–207 in: Root Diseases and Soil-Borne Pathogens. T.A. Toussoun, R.V. Bega, and P.E. Nelson, eds. University of California Press, Berkeley, CA.

107. Tjamos, E.C., Biris, D.A., and Paplomatas, E.J. 1991. Recovery of olive trees with Verticillium wilt after individual application of soil solarization in established olive orchards. Plant Dis. 75:557–562.

108. Tjamos, E.C., and Paplomatas, E.J. 1988. Long-term effect of soil solarization in controlling verticillium wilt of globe artichokes in Greece. Plant Pathol. 37:507–515.

109. Vannacci, G., Triolo, E., and Materrazi, A. 1988. Survival of *Sclerotinia minor* Jagger sclerotia in solarized soil. Plant Soil 109:49–55.

110. Walker, G.E., and Wachtel, M.F. 1988. The influence of soil solarization and non-fumigant nematicides on infection of *Meloidogyne javanica* by *Pasteuria penetrans*. Nematologica 34:477–483.

111. Yaron, D., Regev, A., and Spector, R. 1991. Economic evaluation of soil solarization and disinfestation. Pages 171–190 in: Soil Solarization. J. Katan and J.E. DeVay, eds. CRC Press, Boca Raton, FL.

112. Yarwood, C.E. 1975. Temperature coefficients in plant pathology. Phytopathology 65:1198–1201.

Chapter 13

INOCULUM DYNAMICS OF *FUSARIUM SOLANI* F. SP. *PHASEOLI* AND MANAGEMENT OF FUSARIUM ROOT ROT OF BEAN

R. Hall

Critical observations and methodology open the door to understanding in scientific enquiry. In 1961, Nash et al. (65) reported that chlamydospores function as the survival and infectious units of *Fusarium solani* (Mart.) Sacc. f. sp. *phaseoli* (Burkholder) W.C. Snyder & H.N. Hans., the cause of Fusarium root rot of bean (*Phaseolus vulgaris* L.). A year later, Nash and Snyder (66) introduced a semiselective medium particularly suitable for isolating the fungus from soil. They noted that: "The way is now open to investigate critically the influence on qualitative and quantitative changes in the soil population of *F. solani* f. *phaseoli*, directly in the field; of applications of organic and mineral amendments to soil; of application of fungicidal materials; cultural practices, such as crop rotations and cover cropping; and irrigation, cultivation, and all other practices and factors, such as soil type, which may influence biological changes in soils". Qualitative aspects of inoculum include its form, and the processes of germination, growth, reproduction, and infection; examples of quantitative features are the number of inoculum units, factors that affect inoculum density, and relation of inoculum density to disease. By reference to a limited selection of publications, I propose to examine progress made in determining the qualitative and quantitative aspects of inoculum dynamics of *F. solani* f. sp. *phaseoli* and the application of this information to the management of Fusarium root rot of bean.

IMPACT OF THE DISEASE

Fusarium solani f. sp. *phaseoli* causes root rot on plant species within several genera, including *Arachis*, *Dolichos*, *Onobrychis*, *Phaseolus*, *Pisum*, *Pueraria*, *Vicia*, and *Vigna* (1,18,29). It is economically important primarily as the cause of Fusarium root rot of *Phaseolus* bean (1,13,18).

Fusarium root rot of bean has attracted attention as a model system. It is easy to grow the host plant in controlled and field environments; to generate, recognize, and quantify inoculum of the fungus; and to produce symptoms of the disease within a few days. The pathosystem has therefore been used to explore basic issues such as nutritional regulation of disease, inoculum density-disease relationships, cultural management of disease, biological control, and the physiology of disease resistance. Some of these subjects are explored in this review.

The disease also has a significant impact on production of the bean crop. Kommedahl and Windels (41) classified Fusarium root rot of bean as a host-dominant disease caused by a cortex-specific macerative pathogen. They noted (p. 58) that such fungi are not "killers" and characterized the damage done by this group of diseases as "subtle and slow and the host generally retains control of the relationship by surviving but not with robust health." Nevertheless, *F. solani* f. sp. *phaseoli* can cause substantial losses in bean yield. Burkholder (18) considered that a 25% decrease in bean yield recorded over a 10-year period in New York State was largely attributable to the fungus. In Washington State, bean yields in soil not infested with *F. solani* f. sp. *phaseoli* were 1.7 to 3.6 times higher than yields in fields naturally infested with the pathogen at 260 to 285 colony-forming units (CFU) g^{-1} soil (15). More detailed documentation of the impact of root rot on bean yield was provided by studies in Colorado and Nebraska, in which *F. solani* f. sp. *phaseoli* was considered to be the main cause of the disease. Similar disease rating systems and methodology of estimating yield loss were used in both studies.

In Colorado, Keenan et al. (39) measured root rot severity and seed yield per plant in 25 fields in 1971 and in 31 fields in 1972. Yield per acre was calculated by assuming a stand of 16,000 plants $acre^{-1}$. Disease severity was rated on a scale of 1 (none or slight) to 5 (dead). About 60% of the fields were rated 2 or 3 (referred to as "heavy" root rot by the authors). Each unit increase in disease severity was associated with a 22% decrease in yield. Yield losses averaged 62% and 27% and ranged up to 89% and 66%, in 1971 and 1972 respectively.

Steadman et al. (78) examined 114 bean fields in Nebraska over 3 years. Root rot severity was rated on a scale of 0 (no disease) to 5 (dead plant). Seed yield per plant was correlated negatively with root rot severity in great northern bean but not in pinto bean. Coefficients of determination for multiple regression of seed yield against root rot rating, adventitious root rating, plant density, and year of survey were 0.43 and 0.44 for pinto and great northern bean, respectively. Significant partial regression coefficients were obtained for root rot rating, plant density, and year of survey (coefficients not given). Losses were estimated to be 9.0 bushels acre^{-1} per unit of root rot rating at an average yield of 48.2 bushels acre^{-1} for pinto bean, and 6.0 bushels acre^{-1} per unit of root rot for great northern bean at an average yield of 42.1 bushels acre^{-1}. If we assume that the average root rot severity was at the mid point of the observed range (0.8 to 4.0), i.e. 2.4, then yield at zero root rot in pinto bean is calculated to be $48.2 + (2.4 \times 9.0) = 69.8$ bushels acre^{-1}, and average yield loss would be $(69.8-48.2)/69.8 = 31\%$, i.e a 12.9% loss per unit of root rot. Similar calculations for great northern bean produce an estimated yield loss of 25.5%, i.e. 10.6% per unit of root rot. Maximum estimated losses would be 51.6% and 42.4% in pinto and great northern beans.

It is losses of this magnitude that justify and stimulate research aimed at controlling the disease.

CHLAMYDOSPORES: UNITS OF SURVIVAL AND INFECTION

Fusarium solani f. sp. *phaseoli* survives in soil as chlamydospores (65) produced within the cortex of infected stems and roots (18,23). The chlamydospores are released into the soil from decaying infected tissue and occur deeply embedded in organic matter and humus, singly, in pairs, or sometimes in chains. They tend to remain localized in the soil (11). For simplicity in this discussion, it is assumed that an inoculum unit consists of one chlamydospore.

Chlamydospores 1 mm or less from bean seeds, root tips, and injured mature roots germinate within 16 hours (72). In contrast, chlamydospores near the hypocotyl germinate slowly over a period of 48 hours (82). An extensive mycelium is produced over the hypocotyl surface and lesions appear on the hypocotyl within 48 hours (82). Sugars and amino acids are exuded from bean seeds and root tips, sugars and trace amounts of amino acids are released from hypocotyls, and no exudates are released from mature roots (72,73). It was thus concluded that germination of chlamydospores in soil requires exogenous sources of carbon and nitrogen. Limited exudation of nitrogen compounds from hypocotyls may explain the slow germination of chlamydospores near hypocotyls (82) and the increase in disease severity on hypocotyls in soil supplemented with inorganic nitrogen (54,84). Macroconidia may be produced on the stem at the soil surface (23). These spores also require exogenous sources of carbon and nitrogen for germination (22,24,31) and are quickly converted into chlamydospores in the soil (65).

Although the fungus requires both carbon and nitrogen for germination, Cook and Snyder (27) concluded that rapid germination of chlamydospores does not favor long-term survival of the fungus. Bean seeds stimulated 60% germination of chlamydospores in 16 hours but germ tubes lysed quickly. Hypocotyls of bean seedlings stimulated 20% germination of chlamydospores in 48 hours but germlings did not lyse and extensive mycelium formed on the surface of the hypocotyl, as shown by Toussoun and Snyder (82). Rapid germination and lysis were induced by addition of sugars and amino acids to soil. Slow germination and limited lysis were initiated by addition of sugars only. Cook and Snyder (27) thus suggested that carbon and nitrogen activated a microflora in the soil that led to lysis of *F. solani* f. sp. *phaseoli*. It was subsequently shown that soil bacteria can lyse the hyphae (34) and stimulate formation of chlamydospores (30). There is also evidence that formation of chlamydospores in soil is inhibited by high nutrient levels and stimulated by starvation and chemical factors (4).

Mycelium of the fungus enters roots directly or through natural or mechanical wounds, and especially through stomata in the hypocotyl (23). The mycelium colonizes the cortex but stops at the endodermis. The fungus does not spread from infected to uninfected plants over distances greater than 1.3 to 5 cm (9).

Chlamydospores also germinate in the soil adjacent to organic debris (71,81), and in the rhizosphere of nonhost plants, including barley, corn, cucumber, spinach, and tomato (71). The mycelium quickly produces new chlamydospores if no host is present (72). Nonhosts can therefore maintain inoculum of the fungus in soil.

NUMBERS OF CHLAMYDOSPORES

The population density of *F. solani* f. sp. *phaseoli* in the top 15 cm of naturally infested soils ranges from 0 to 5,500 CFU g^{-1}. In this section, I discuss the protocol and reliability of methods for estimating the population density of the fungus.

Isolation Of The Fungus

The fungus can be isolated and recognized on several media (69). Nash and Snyder (66) used a medium consisting of 15 g Difco peptone, 20 g agar, 1 g KH_2PO_4, 0.5 g $MgSO_4.7H_2O$, 300 mg streptomycin and 1 g pentachloronitrobenzene (PCNB) per liter of water. Papavizas (69) also recommended this medium if bacteria and spreading fungi are not a problem, or a modification in which 50 mg liter^{-1} chlortetracycline HCl and 0.5 g liter^{-1} oxgall are added, and streptomycin and PCNB are reduced to 100 mg liter^{-1} and 0.5 g liter^{-1}. The media of Nash and Snyder (66) and Papavizas (69) are useful for a broad range of species and formae speciales of *Fusarium*. Hall (32) rendered a variant of the Papavizas medium (300 mg liter^{-1} streptomycin, 100 mg liter^{-1} chlortetracycline, no oxgall) highly selective for *F. solani* f. sp. *phaseoli* by the addition of 5 mg liter^{-1} benomyl. Although *F. solani* f. sp. *phaseoli* is defined by its pathogenic specificity to *Phaseolus* (52,75), it can be easily recognized by its appearance in culture on all three media described above. For routine work on quantification of inoculum it is possible to identify the forma specialis directly on the isolation medium, without recourse to inoculation tests except as necessary to verify visual identification. The existence of highly selective media on which *F. solani* f. sp. *phaseoli* can be identified macroscopically has facilitated quantitative studies of the inoculum dynamics of the fungus.

The fungus is generally isolated by dilution plating techniques. Nash and Snyder (66) used 1:200 to 1:2,000 dilutions of soil in water and applied 1 ml of the suspension (0.5 to 5 mg soil) to the surface of the selective medium in a 9-cm-diameter petri plate. Lower dilutions tended to produce more colonies per plate and they chose the lowest dilution that provided stable readings. Hall and Phillips (33) suspended 2 g soil in 100 ml water (1:50 dilution), applied 0.1 ml (2 mg soil) per plate, and prepared 100 plates per suspension. By plating 200 mg soil they could detect ≥ 5 CFU g^{-1}. Similarly, Burke and Kraft (15) spread a 1:200 soil dilution on six to 18 replicate plates of Papavizas PCNB agar (69) and reported populations of ≥ 7 CFU g^{-1}.

Distribution In The Field

Nash and Snyder (66) took core samples from 37 approximately equally spaced stations in a field. They concluded that the distribution of *F. solani* f. sp. *phaseoli* was uniform throughout the plowed layer because there were no significant differences in counts obtained from single core samples (40 g), composites of core samples (200 g), or small samples from cores (43 mg). Average counts for the whole field did not vary seasonally; similar counts were obtained 16 September 1960 (after windrowing but before threshing), 18 January 1961 (during fallow), and 26 April 1961 (just after seeding a new crop). But the counts from separate stations differed widely, ranging from 391 to 1,975 CFU g^{-1}. The population density was correlated with the clay content of the soil (r $= 0.61, P = 0.01$).

Burke et al. (14) reported that *F. solani* f. sp. *phaseoli* was distributed throughout the plowed layer (0 to 30 cm) but the population density tended to decline progressively with distance from the surface. Average densities ranged from 368 to 1,546 CFU g^{-1} at 0 to 8 cm, from 20 to 587 CFU g^{-1} at 23 to 31 cm, and from 0 to 80 CFU g^{-1} below the plowed layer (33 to 41 cm). The fungus was not found by dilution plating at depths of 90 to 100 cm. The population density per location ranged from 100 to 5,500 CFU g^{-1}. Lateral distribution of the fungus in the field was not associated with other factors except that sometimes population density was higher in the row than between rows. There were some associations between vertical distribution of the fungus and other factors in the soil profile. It was noted, for example, that bean roots and *F. solani* f. sp. *phaseoli* occurred mostly within the plowed layer, and that population density of the fungus decreased as the bulk density of the soil increased with depth.

Dryden and Van Alfen (28) reported that 99% of the population of *F. solani* f. sp. *phaseoli* was distributed uniformly in the top 45 cm of soil (mean 528 to 780 CFU in 20 ml of soil), the zone where most roots occurred. The remaining 1% of the population was located between 45 and 105 cm. The population density in the top 45 cm differed among samples (397 to 1,112 CFU in 20 ml of soil) but did not change between July and September.

A brief report (55) on the effect of the sampling pattern and number of samples on population estimates is enlarged here. A 4.5-ha portion of a field that had been repeatedly cropped to bean was divided by a 6 × 12 grid into 72 squares each measuring 25 × 25 m (Figure 1). A 1-liter sample of soil was taken from the top 15 cm of the plowed layer from the centre of each square. A small portion (0.8 g) of soil from each sample was suspended in 100 ml of water and 50 ml of the suspension was distributed across 100 plates of selective medium (32). Colonies of *F. solani* f. sp. *phaseoli* were counted after incubation of the plates for 14 days in the laboratory. Population density per site averaged 253 and ranged from 30 to 635 CFU g^{-1}. In 20 subsamples (0.8 g) from one site

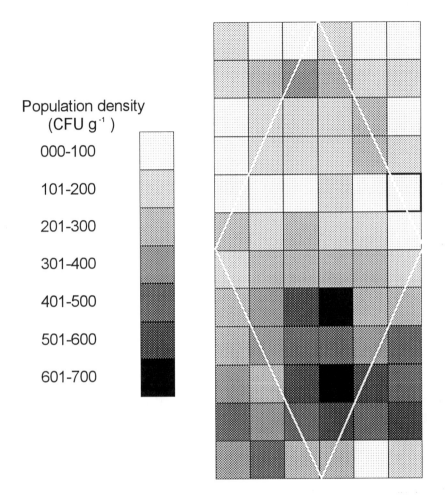

Figure 1. Distribution of *F. solani* f. sp. *phaseoli* in a 4.5-ha portion of a field divided into 72 sampling areas. The clear line represents the sampling path followed to determine the effect of number of samples on population estimates. Subsamples were taken from the plot outlined in bold.

(outlined in bold in Figure 1), counts averaged 200 and ranged from 85 to 1,050 CFU g^{-1}. The population appeared to be higher at one end of the field. From the total of 72 data points for the field, four to 24 were

Table 1. Effect of sampling pattern and number of samples on estimate of population density of *F. solani* f. sp. *phaseoli* in a bean field.

Source of samples	Sampling pattern	Samples	Population density (CFU g^{-1})	
			Mean	Confidence limits (95%)
Field	Total field	72	253	215-291
	Diamond	24	243	185-301
	Diamond	12	243	150-337
	Diamond	8	295	154-436
	Diamond	4	316	75-557
	Random	20	223	155-290
	Random	20	282	204-359
	Random	20	249	167-330
	Random	20	229	159-299
One site	Subsamples	20	200	102-299

selected either randomly or by sampling at regular intervals along a diamond-shaped sampling path (shown as a clear line in Figure 1). Average population densities estimated by all procedures were very similar but the confidence limits became wider as the number of samples decreased (Table 1). It was concluded that 10 to 20 samples collected either systematically or randomly would provide a reasonable estimate of the density of *F. solani* f. sp. *phaseoli* in the field.

INOCULUM DENSITY AND DISEASE

Low populations of *F. solani* f. sp. *phaseoli* can lead to high incidences of infected plants. Maloy and Burkholder (50), for example, reported from a greenhouse study that 5 conidia g^{-1} in sterilized soil led to infection of nearly 80% of plants and 0.1 CFU g^{-1} produced lesions on 20% of the plants. Low population densities have also been associated with moderately severe root rot. In a field study, Dryden and Van Alfen (28) reported that, on a scale of 1 to 5, the average hypocotyl disease severity was 1.9, equivalent to 20 to 40% necrosis of the cross-sectional area of the hypocotyl. The average population density in the top 45 cm of the soil was 680 CFU in 20 ml of soil, equivalent to 26 CFU g^{-1} of soil if the bulk density is assumed to be 1.3. In addition, roots in the top 45 cm of soil were uniformly colonized, yielding about one colony per 2 cm of root.

Short-term experiments in growth chambers have shown that the severity of Fusarium root rot of bean rises as the inoculum density of the pathogen is increased (5,6,74). For example, Baker (5) found that, at low carbon:nitrogen ratios, 4 conidia g^{-1} led to an average of less than one lesion per hypocotyl within 15 days whereas 4,096 conidia g^{-1} caused more than eight lesions per hypocotyl. Similarly, Sippell and Hall (74) showed that, within 10 days after inoculation, total root rot severity ratings (sum of disease severity on hypocotyl, tap root, and adventitious roots) on a cumulative scale of 0 to 12 were 1.5, 2.7, 3.9, and 5.3 at inoculum concentrations of 0, 10, 100, and 1,000 conidia g^{-1} pasteurized soil.

In tests more closely related to field conditions, Abawi and Cobb (2) tested *F. solani* f. sp. *phaseoli* at concentrations of 0, 200, 2,000, and 4,000 CFU g^{-1} in field microplots. Inoculum density was correlated with root weight (r = -0.54), and with disease severity on stems (r = 0.67) and on roots (r = 0.65), but not with yield.

There are also reports from field-scale experiments that higher population densities of *F. solani* f. sp. *phaseoli* are associated with more severe Fusarium root rot. Maloy and Burkholder (50) estimated populations of the fungus by applying the most probable number method (49) to colony counts on an agar medium containing rose bengal and streptomycin (51). Soil samples were taken from at least 10 randomly chosen sites per field. Root rot severity was rated on a scale of 0 to 100. They noted in one trial that root rot severity rose from 7 to a maximum near 60, 6 weeks after seeding. Among 39 fields, average root rot severities were <56.0 in 13, 56.1 to 65.0 in 15, and >65.1 in 11. Average populations in these three groups of fields were 19, 54, and 31 CFU g^{-1}. There was thus no direct relation between inoculum density and disease severity but the lowest populations were associated with the least severely affected fields.

Two later studies estimated population densities directly from the number of colonies produced by soil dilution techniques on versions of the medium of Nash and Snyder (66). In the first study, Burke and Kraft (15) reported that the severity of Fusarium root rot of bean and the population density of *F. solani* f. sp. *phaseoli* were higher in a field previously monocultured to bean for 15 years than in a field monocultured to pea for the previous 6 years.

In a more detailed analysis of the relation of population density to root rot severity, McFadden et al. (55) examined 48, 56, and 48 commercial fields of white bean in southern Ontario in 1984, 1985, and 1986. Seventeen samples of soil were collected along a W-shaped transect 144 m long and bulked to a total volume of 10 liter. Population density of *F. solani* f. sp. *phaseoli* was estimated by suspending 2 g or 0.8 g of soil in 100 ml water and distributing 50 ml of the suspension on 100 plates of selective medium (32). Severity of Fusarium root rot was measured on an index of 0 to 100 that represented visual symptoms ranging from none to plant dead or dying. Root rot indexes ranged from 1 to 68 and population densities ranged from 0 to 1,420 CFU g^{-1}. Correlation coefficients relating severity of root rot to population density of *F. solani* f. sp. *phaseoli* ranged from 0.52 to 0.66, depending on the year and the scale used, and were all significant at $P = 0.01$. The intercepts of regression lines on the disease severity axis ranged from 13 to 24. The positive intercepts on the disease severity axis and the low coefficients of determination of the regression lines (0.27 to 0.43) indicate that factors other than population density of the pathogen were contributing to root rot severity. Despite these unknown interfering influences, the significant correlation coefficients provide evidence that an increase in the population density of *F. solani* f. sp. *phaseoli* tends to increase the severity of root rot under field conditions. The correlation coefficients relating field root rot index to seed yield were not significant, possibly because field root rot indexes were generally less than 50, the value that Kobriger and Hagedorn (40) associated with appreciable yield losses.

INOCULUM TYPE AND DISEASE

Fusarium solani f. sp. *phaseoli* occurs as conidial and mycelial types (21). The conidial type produces abundant macroconidia and few microconidia, and lacks aerial mycelium; it is the type commonly isolated from diseased bean. The mycelial type produces aerial mycelium, few macroconidia and abundant microconidia. The mycelial type is less pathogenic than the conidial type. Maloy (48) examined the possibility that loss of virulence explains the reduction in bean root rot severity in certain rotations. Soil sterilized by steaming was infested with a mixture of the two types at a rate of 1,000 spores g^{-1} soil. In unamended soil, the mycelial type constituted 54% of the population at the beginning of the experiment, fell to 16% after 1 week, then rose to 70% after 20 weeks. In soil amended with 1% soybean meal, the mycelial type made up 54% of the population at the beginning of the experiment and 38% 20 weeks later. The total number of propagules was 1,000 g^{-1} at the beginning of the experiment and 11,520 and 13,520 after 20 weeks in unamended and amended soil respectively. It was suggested that the predominance of the mycelial type in unamended soil would help explain the reduction in the severity of bean root rot after a period of absence of bean in the rotation. However, no data were provided on population changes in unsteamed soil or on root rot severity in soil infested with the two types. Moreover, Nash et al. (65) reported that none of 3,700 isolates from field soil or of 350 isolates from lesions from plants infected in the field yielded the mycelial type. They concluded that it was unlikely that the mycelial type was a significant component of the population of *F. solani* f. sp. *phaseoli* in soil.

CROP HISTORY AND INOCULUM DENSITY

Nash and Snyder (68) reported that the population of *F. solani* f. sp. *phaseoli* varied by a factor of less than three within a field, and remained constant even when the host plant was not grown. Their opinion (68, p.944) was that "this fungus survives a long time in a resting state, seldom germinating, and only slowly drying out". The fungus was considered to be an example of fusaria that germinate readily, use a variety of food sources, and maintain their population by repeated production of propagules. This section examines the effects of crop history on inoculum density of the fungus.

Population densities of *F. solani* f. sp. *phaseoli* are zero or low in noncultivated soils and in soils not previously cropped to bean. The fungus could not be found in noncultivated soils or in fields where only cereals had been grown in California (68); nor was it found in sagebrush land or in fields cropped to bean for the first time in Washington State (14). However, the fungus is likely to occur in all or most soils within a geographic area where bean has been grown, even in fields that have not been sown to the crop, because the fungus can be introduced in the dust on seeds (67) and can be dispersed in airborne or waterborne soil or infested debris. I have isolated the fungus from soil in southern Ontario that had been under sod for 30 years but was located near research plots and commercial fields in which bean had been grown (R. Hall, unpublished). Species of *Phaseolus* have been grown for at least 600 years in the Great Lakes region of North America (38), including portions of southern Ontario (56). The fungus was therefore probably introduced into the region hundreds of years ago and is now undoubtedly widespread. The origin of bean in Central and South America; the cultivation of the crop for hundreds or thousands of years in Central, South, and North America (38); and the consequent widespread distribution of *F. solani* f. sp. *phaseoli*, would explain why Menzies observed that Fusarium root rot behaved in the Columbia Basin "as though the pathogen were native in the area" (57, p. 44).

The population density of *F. solani* f. sp. *phaseoli* can increase quickly in the presence of bean. Hall and Phillips (33) noted that the population density rose from < 1.7 CFU g^{-1} to about 660 CFU g^{-1} after seven successive bean crops. Burke and Kraft (15) reported that the fungus was not detected in soil in which bean had not been grown, or in bean rhizosphere soil in a field that had not previously produced the crop, and was barely detectable (7 to 13 CFU g^{-1}) in bean rhizosphere soil in the second crop of bean in a field previously cropped to pea for 6 years. However, the population density of the fungus was 285 CFU g^{-1} in a field cropped to bean for the previous 15 years. In the 17th year of bean monoculture, the population density of *F. solani* f. sp. *phaseoli* was 260 CFU g^{-1} in the field before seeding and 560 and 276 CFU g^{-1} in bean rhizosphere soil 2 and 6 weeks after seeding.

Crop sequence affects population density of the fungus. Nash and Snyder (66) reported that in a bean-barley-bean rotation the mean population density was $> 3,000$ CFU g^{-1} after barley whereas two adjoining fields that had experienced a bean-sugar beet-bean rotation had counts of 578 and 1,220 CFU g^{-1}. There was thus some evidence that barley increases the population of the fungus, even though the crop reduces losses from Fusarium root rot, possibly because barley straw deprives *F. solani* f. sp. *phaseoli* of nitrogen needed for infection (79). However, the effects of crop sequence are variable. Huber and Andersen (34) reported that the incidence of bean root rot was low when the crop was preceded by corn and high when the crop was preceded by barley. Converse results have been obtained for barley (66) and corn, as described below (R. Hall, unpublished).

Schroth and Hendrix (71) showed that the population of the fungus could increase in the rhizosphere of tomato, lettuce, and corn, and in the presence of certain chopped materials. This suggested that some rotations could be ineffective in reducing the population of the fungus in soil. Similarly, Maier (47) observed that many rotations increased the populations of the fungus. However, populations were reduced in some rotations, especially following a fallow period, suggesting that the fungus cannot maintain its population for prolonged periods without some parasitic or saprophytic activity.

Several reports document a reduction in the population of *F. solani* f. sp. *phaseoli* in the absence of the bean crop. For example, Nash and Snyder (66) noted a decline from 1,497 CFU g^{-1} in January 1959 after a bean crop to 373 CFU g^{-1} in October 1959 after a tomato crop. Burke and Kraft (15) reported that in the second year of pea after 15 years of bean, counts of the fungus in the rhizosphere of a pea crop were 233 and 0 CFU g^{-1} 2 and 6 weeks after seeding compared to 260 in soil before seeding. In a bean-soybean-corn rotation trial conducted between 1984 and 1994, Hall and Phillips (33) showed that populations of *F. solani* f. sp. *phaseoli* in the fall ranged from 443 to 723 CFU g^{-1} after a bean crop and from 163 to 393 CFU g^{-1} after production of soybean or corn. However, population density of the pathogen also fluctuated when bean was grown each year, ranging from 310 to 1,165 CFU g^{-1}. Expressed as a percentage of populations in the repeated bean treatment, population density of the fungus in the rotation treatment ranged from 73% to 86% in years in which bean was grown, and declined to levels ranging from 30% to 68% in years when soybean or corn was grown (Figure 2). Thus, populations in the rotation treatment, both in absolute numbers and relative to the repeated bean treatment, consistently declined in the absence of the bean crop and increased in its presence.

Similar data were obtained with other crop sequences grown in microplots (1 m × 1 m) in the field (R. Hall, unpublished). A high population (about 1,100 CFU g^{-1}) of *F. solani* f. sp. *phaseoli* was established by infestation of the plots with macroconidia and repeated cultivation of bean for several years. Beginning in 1992, crop sequences tested were repeated bean, corn-soybean-corn, wheat-soybean-wheat, repeated red clover, and fallow. By the fall of the first year without bean, population densities had declined to 60 to 70% of those in repeated bean plots (Figure 3). After 3 years without bean, population density in the most effective treatment (red clover) had declined to 50% of that in the repeated bean check. Populations that were low (< 100 CFU g^{-1}) at the beginning of the trial remained constant and low under fallow throughout the 3 years of the trial. The results indicate that populations of the fungus in soil can be lowered over 3 years, but not eliminated, by omitting bean from the rotation.

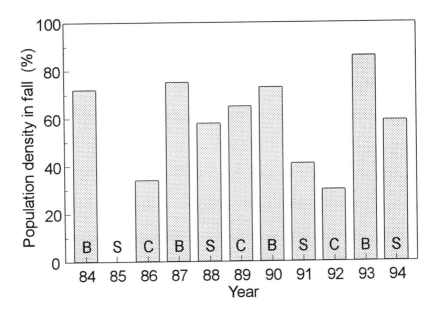

Figure 2. Population density of *F. solani* f. sp. *phaseoli* in soil in the fall in a rotation of bean (B), soybean (S), and corn (C) as a percentage of population density in a repeated bean treatment during the years 1984 to 1994.

Another way of showing how the intensity of bean production affects populations of *F. solani* f. sp. *phaseoli* is to use historical data on bean production in a geographical area (R. Hall, unpublished). The area cropped to bean in each county in Ontario has been recorded since 1882. For each county we can sum the annual areas in bean to produce a measure of the intensity of bean production for the county over a number of years, with units of hectare-years. Soil samples were obtained from 310 cultivated fields in seven counties in southern Ontario in 1978. Bean was being grown in 170 of these fields at the time of the survey; most other fields sampled contained corn, soybean, or small-grain cereals. One field was sampled for every 3,600 ha of land considered to be available for bean production (i.e. the area in hay and field crops). A small quantity (100 g) of soil was obtained from the upper 8 cm of the plowed layer at each of 20 sites located along a 150-m transect through the field.

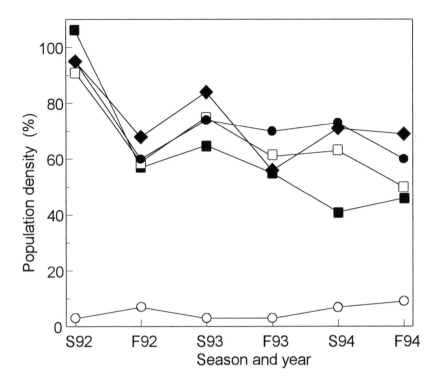

Figure 3. Effect of absence of bean cultivation on population density of *F. solani* f. sp. *phaseoli* in soil in field microplots in spring (S) and fall (F) as a percentage of population density in a repeated bean treatment. The trial started in spring 1992 (S92) and ended in fall 1994 (F94). The crop histories during the trial were corn-soybean-corn (●), wheat-soybean-wheat (♦), repeated red clover (■), high inoculum fallow (□), and low inoculum fallow (○). Initial population densities were about 1,100 CFU g^{-1} in all treatments except low inoculum fallow (<100 CFU g^{-1}).

The samples were mixed to form one bulk sample per field. To determine the population density of the pathogen, the equivalent of 0.8 g of air dry soil per bulk sample was suspended in 100 ml or 200 ml of water, and 0.25-ml aliquots were spread over a selective medium (32) in petri plates. Colonies were identified after incubation in the laboratory for 2 to 4 weeks. The procedure could detect ≥ 5 CFU g^{-1}. The fungus was detected in 42 (13.5%) of the 310 fields at population densities ranging from 10 to 600 CFU g^{-1}. Average population densities per county ranged from 0 to 56 CFU g^{-1} and the percentage of infested fields per county ranged from 0 to 28%. The average population density of *F. solani* f. sp. *phaseoli* per county (Figure 4) was positively correlated with the intensity of bean production over the previous 97 years.

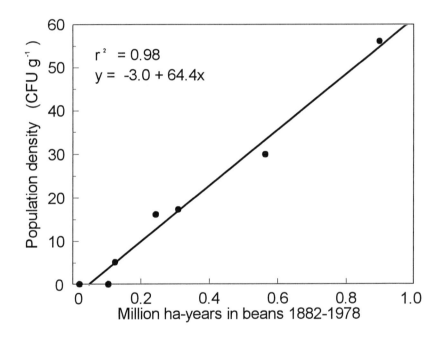

Figure 4. Relation between intensity of bean cultivation (million ha-years) in seven counties in Ontario in the period 1882 to 1978 and average population density of *F. solani* f. sp. *phaseoli* (CFU g^{-1}) per county in 1978.

CROP HISTORY, INOCULUM DENSITY, AND DISEASE

Fusarium root rot and *F. solani* f. sp. *phaseoli* occur where bean has been grown repeatedly (66). For example, Nash and Snyder (68) observed the disease in cultivated soils of the Salinas Valley in California whenever the crop was grown. The population density of the fungus in cultivated soils ranged from 100 to 3,000 CFU g^{-1}. *Fusarium solani* f. sp. *phaseoli* was numerous in all fields where bean was part of the crop sequence, often composing over half of the population of *F. solani*. Numerous reports show that the disease tends to be more severe where bean has been grown repeatedly (16,50,57).

Results from a trial established in 1976 at the Centralia Research Station of the Ontario Ministry of Agriculture, Food and Rural Affairs illustrate relations between crop history, inoculum density, and disease. The trial examined the effects of 12 crop sequences on yield and root rot severity in white bean and population density of *F. solani* f. sp. *phaseoli*. The data on yield were collected by J. O'Toole and published in annual reports of the research station. The data on root rot and population density were obtained by the author (R. Hall, unpublished). The first crop was sown into land that had been maintained under grass for the previous 30 years. The crop sequences followed a 4-year cycle, and bean yields were measured in 1977, 1981, and 1985. In 1977, bean yields were similar in all rotations but were lowest in the repeated bean treatment, which was in its second year of bean (Figure 5). In 1981, bean yields were about 50% of those recorded in 1977 and were again lowest in the repeated bean treatment. By 1985, bean yields were one-third of those observed in 1977, except in repeated bean, where they had fallen to one-fifth.

At the beginning of the trial, *F. solani* f. sp. *phaseoli* was not detected in the soil. In 1986, soil was removed from the plots and assayed for the fungus and for root rot potential using methods described by McFadden et al. (55). Again, *F. solani* f. sp. *phaseoli* was not detected in uncultivated soil adjacent to the cultivated area. Low population densities (50 CFU g^{-1}) were detected where bean had been grown once every 4 years. The highest populations (600 CFU g^{-1}) occurred where bean had been grown every year, or twice every 4 years in rotation with corn. Intermediate population densities occurred where bean had been grown twice in 4 years in rotations with barley or wheat. The root rot potential of the soil was lowest (35) in untilled soil, intermediate (60) in soil where bean had been grown once every 4 years, and high (80 to 90) in all other treatments.

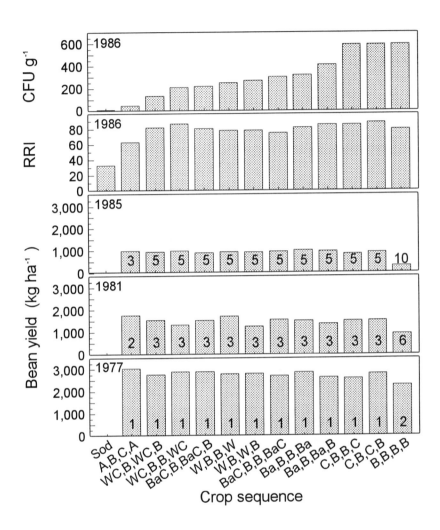

Figure 5. Effect of crop sequence on bean seed yield (kg ha^{-1}) in 1977, 1981, and 1985, and on root rot index (RRI) and population density (CFU g^{-1}) of *F. solani* f. sp. *phaseoli* in soil collected in spring 1986. The trial was started in 1976. Crops grown were alfalfa (A), bean (B), corn (C), wheat (W) alone or seeded down to red clover (WC), and barley (Ba) alone or seeded down to red clover (BaC). Population density and root rot index were also determined in soil from a nearby area maintained under sod for 30 years. Numbers associated with bars indicate the number of bean crops grown.

The experiment showed that bean yields declined with repeated cultivation of crops, regardless of crop sequence; that *F. solani* f. sp. *phaseoli* could not be detected in uncultivated soil but was detected after three bean crops (grown once every 4 years); that population density of the fungus tended to increase as the frequency of bean production in the rotation increased; that soils with low root rot potentials had low population densities of *F. solani* f. sp. *phaseoli* and soils with high root rot potential had intermediate to high populations of the fungus; that the highest populations of the pathogen were associated with the lowest yields; but that across the twelve rotations there was no correlation between bean yield and population density of the fungus. *Fusarium solani* f. sp. *phaseoli* has the capacity to reduce bean yield but may not be the only, or even major, factor contributing to yield reductions following repeated cultivation of the crop, even at relatively high population densities.

SUPPRESSIVE SOILS, INOCULUM, AND DISEASE

There is evidence that some soils are highly suppressive to Fusarium root rot of bean. In Washington State, the disease spread rapidly in most new soils brought under irrigation in the Columbia Basin but was not a problem in some longer-irrigated fields of similar soil types in Yakima Valley (10,57). Burke (8,10) concluded that the difference in the soils could be attributed to biological activity of soil but he could not identify the organisms responsible for the disease-suppressive effect. In soils favorable to persistence of the pathogen the conidia are converted rapidly to chlamydospores. In unfavorable soils, conidia germinate but most conidia and germ tubes lyse and few persistent chlamydospores form (8,10). A degree of biological suppression may occur in many soils not highly suppressive to the disease. Nash and Snyder (66) reported that estimates of soil populations were lower when soil dilutions were placed on stems than when placed on selective medium. Some of the competing organisms appear to be bacteria, since Cook and Schroth (26) increased the germination of chlamydospores by adding antibiotics to soil. Other competitors may be fungi; Lindsay (42) observed competition amongst species of *Fusarium*, including *F. solani* f. sp. *phaseoli*. Huber and Watson (35) suggested that fumigation and irrigation with water containing crop residue increased bean root rot in Idaho by reducing biological buffering of soil and by recontaminating soil with the pathogen. Adams et al. (3) noted that adding spent coffee grounds temporarily removed the fungistatic activity of the soil towards *F. solani* f. sp. *phaseoli* then increased it above levels in the untreated check. Using microbial activity to eliminate the pathogen may be difficult. However, Watson (83) reported that decomposition of crop residues in flooded soil under anaerobic conditions eliminated all *Fusarium* species.

ENVIRONMENT, INOCULUM, AND DISEASE

Temperature

Fusarium root rot, and its impact on plant productivity, are generally reported to be most severe at temperatures of 16 to 24°C (19,45,70). Baker (5) found that disease severity was similar at 18°C and 23°C and lower at 28°C. Sippell and Hall (74) observed that root rot ratings were greatest at 21°C and less at 14°C and 28°C.

Moisture

Fusarium root rot tends to be more severe in moist (-5 kPa) soil than in drier soils (-17 to -80 kPa) (59,74), but may reduce yields more in a drier year (18) or drier soil (20).

In the repeated-bean treatment of the rotation trial started in 1978 at the Arkell Research Station (33), described earlier in this chapter, it was noted that, even though bean was grown every year, annual population densities of *F. solani* f. sp. *phaseoli* in the period 1984 to 1991 fluctuated widely around a mean of about 600 CFU g^{-1}, ranging from 442 to 1,013 CFU g^{-1} in June before seeding and from 310 to 718 CFU g^{-1} in September after harvest. This suggested that something besides production of bean was influencing population density of the fungus. Correlations were determined between population density of *F. solani* f. sp. *phaseoli* and temperature and rainfall data for different periods of the year. The only significant correlation obtained was between population density in June and total rainfall during the preceding growing season (June to August). When effects of temperature were excluded, the partial correlation coefficient was 0.84, significant at $P = 0.04$. The coefficient of determination was 0.59 (Figure 6). The mechanism by which rainfall affects the population density of the fungus is not known but could involve effects on the size or susceptibility of the root system, the amount or nature of root exudates, and the activity of *F. solani* f. sp. *phaseoli* or other microorganisms in the soil.

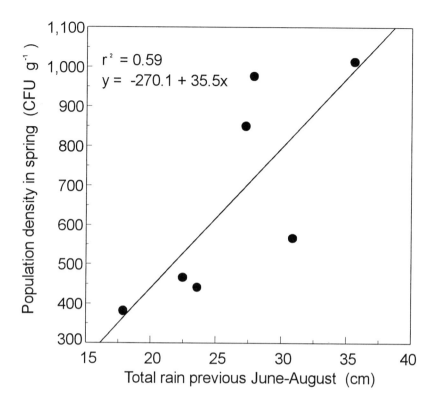

Figure 6. Relationship between population density of *F. solani* f. sp. *phaseoli* (CFU g^{-1}) in spring in a repeated bean plot and total rainfall (cm) during the preceding summer (June to August).

Organic Amendments To Soil

Addition of organic material to soil may increase or decrease bean root rot (36). Amendments that have been tested include alfalfa, alfalfa hay, barley straw, green barley hay, bean seed meal, bean straw, cellulose, chitin, corn shucks, corn stover, glucose, lettuce, lignin, oat straw, pine shavings, sawdust, sorghum, soybean hay, soybean straw, tomato, wheat bran, and wheat straw. Certain amendments, such as barley straw, have been reported to both reduce and increase the disease. Fusarium root rot, for example, has been reduced by the addition of clean barley straw (77), lignin (53), chitin (53), and cellulose (31,54). Other organic materials, such as soybean and alfalfa residues (77), may increase the disease. It has been suggested that organic materials affect bean root rot by altering the carbon:nitrogen ratio in the soil, by regulating microbial activity in the soil, or by affecting the form of nitrogen in the soil.

The effect of the carbon:nitrogen ratio on Fusarium root rot was shown in experiments in controlled environments by Baker (5); disease severity declined as the carbon:nitrogen ratio of the soil was increased from 10 to greater than 45. This confirmed previous reports that the disease may not be suppressed if a nitrogen source, such as ammonium nitrate, is added with the carbon source (77), or if the amendments, such as alfalfa and soybean residues, have a low carbon:nitrogen ratio (77). Similarly, Maier (43,44,45,46) reported that the severity of a root rot complex in pinto bean was reduced by the addition to soil of organic matter with a high carbon:nitrogen ratio, but the materials were ineffective if supplemented with ammonium nitrate. It has been suggested that high carbon:nitrogen ratios in field soil reduce disease because the nitrogen is rapidly immobilized by soil microorganisms (77), thereby depriving *F. solani* f. sp. *phaseoli* of this essential nutrient. Direct observation supported the nitrogen starvation hypothesis. The infection thalli of the fungus were smaller and produced fewer lesions on bean hypocotyls in soil amended with barley straw than in unamended, nitrogen-rich soil (25,79). Although sugars stimulated chlamydospore germination in soil, nitrogen was needed for maximum infection (79).

Huber et al. (36) suggested that amendments influenced root rot through their effects on the form of nitrogen in the soil. They found that organic materials that had been reported to reduce root rot, such as alfalfa and corn, stimulated biological oxidation of ammonium to nitrate. Materials that had been reported to increase root rot, such as barley straw and glucose, suppressed nitrification. They therefore suggested that the ammonium form of nitrogen favored root rot. However, they relied on literature reports for assessing the effects of organic amendments on disease. These effects, unfortunately, are variable; for example, root rot may increase (53) or decrease (54) after addition of glucose.

The difficulty of interpreting the effects of organic amendments on bean root rot is further shown by the report of Burke et al. (14) that soil containing undecomposed barley residue had a lower bulk density than soil with bean residue. Thus, barley residues may reduce root rot by relieving restrictions on root growth (16).

Disease suppression by organic amendments may be related more to qualitative than quantitative aspects of inoculum, at least in the short term. Crop residues that suppress root rot may not have immediate effects on the population density of *F. solani* f. sp. *phaseoli*. For example, cellulose reduced root rot (54) but not the inoculum density of the fungus (7). In addition, the disease suppressiveness of residues may be effective against unusually high populations of the fungus. Maier (46) reported that inoculum densities four to five times the levels of natural infestation were needed to overcome the suppression of symptoms by crop residues.

Stress

Fusarium solani f. sp. *phaseoli* has little effect on yield unless the plant is under stress (16). Populations as high as 4,000 propagules g^{-1} did not reduce yield of nonstressed plants in field microplots (2). Stress factors that increase the damage done to bean by the fungus include soil compaction (12,14,16,17,62), poor aeration (59,61), excess soil moisture (60), drought (18), herbicide toxicity (85), high plant density (9), and toxic products from decomposing crop residues (80). Plant-parasitic nematodes increased the incidence but not the severity of Fusarium root rot (37).

A few reports on stress document the quantitative aspects of inoculum. In controlled environments, *F. solani* f. sp. *phaseoli* added to fumigated soil (inoculum density not reported) reduced fresh shoot weight of three 4-week-old plants from 33.9 g to 19.1 g (44%) and total fresh root weight from 25.8 g to 14.9 g (42%). When an oxygen deficit was applied to the soil 5 to 8 days after seeding, *F. solani* f. sp. *phaseoli* reduced fresh shoot weight from 27.3 g to 4.3 g (84%) and fresh root weight from 23.8 g to 4.1 g (83%). Twenty-five days after seeding, the fungus had reduced water use rates about 50% in normally aerated plants and by about 90% in plants subjected to low oxygen stress for 3 days (64). Similar results were obtained by comparing fumigated soil with soil naturally infested with *F. solani* f. sp. *phaseoli*, and reported previously (15) to contain 260 to 285 CFU g^{-1}.

In examining the effects of drought, soil compaction, and the fungus on bean growth, Miller and Burke (58) used surface soil (0 to 15 cm) and subsoil (60 to 75 cm) from a field cropped to bean for about 15 years. *Fusarium solani* f. sp. *phaseoli* was present in the surface soil at population densities of 200 to 500 CFU g^{-1} but was essentially absent from the subsoil. Bean plants were grown in a slab of soil consisting of a bottom layer of subsoil 14 cm deep, a central layer 4 cm deep of top soil at bulk densities of 1.2, 1.4 or 1.55 g cm^{-3}, and an upper layer of topsoil.   Three bean seedlings were planted in each container. Measurements were taken 4 weeks after planting. Root weights in the middle layer were lower in drier soil, in more compact soil, and in infested soil. Shoot weight was affected by water potential and by the pathogen but not by bulk density. In this experiment, there were no interactions among water potential, the fungus, and bulk density in effects on root or shoot weight. However, the effects of the treatments were additive; the greatest reduction in plant productivity occurred in plants exposed to the fungus and to the stresses of drought and soil compaction. A similar combination of effects of stress and the pathogen was reported by Wyse et al. (85). They noted that the herbicide EPTC increased the severity of root rot and reduced yield, and that plant growth was reduced as inoculum density was increased in the presence of EPTC.

IMPLICATIONS OF INOCULUM DYNAMICS FOR DISEASE MANAGEMENT

Ever since Fusarium root rot of bean was described (18), an integrated approach has been taken to managing the disease. Nevertheless, at different times, different themes have received special attention. Through the 1950s and 1960s, some emphasis was placed on managing the disease by restricting the activity of the fungus in soil, principally through organic amendments, with the hope that this would lead to manageable biological control.   A later theme was that of controlling the disease by reducing abiotic stress on the plant. In this chapter, it is suggested that lowering the density of the inoculum of *F. solani* f. sp. *phaseoli* should also be a deliberate component of managing Fusarium root rot of bean. This view is based on the following argument. The fungus is essential to the development of the disease. An increase in inoculum density has been associated with a rise in disease severity. Higher disease severity has been associated with lower yields.   The severity of disease at a given population density depends on environmental factors such as carbon:nitrogen ratio, moisture, temperature, drought, soil compaction, and crop residue. Stress factors increase disease severity and cannot always be avoided. Populations of the fungus can be reduced by crop rotation and by addition of organic amendments to soil.   It is therefore desirable, feasible, and practical to incorporate inoculum management into integrated control of the disease.

Monitoring the density and activity of inoculum of *F. solani* f. sp. *phaseoli* is also important to the control of Fusarium root rot of bean. Monitoring inoculum can assist in identifying fields with elevated risk of the disease, in recognizing practices that increase or decrease the risk or impact of the disease, in verifying inoculum reduction in soil, and in elucidating mechanisms of disease increase or reduction. Suppressive soils might be identified rapidly by observing the behavior of chlamydospores in soil samples. In addition, the fungus is genetically variable, existing as multiple clonal types in soil and in plants. For example, Snyder et al. (76) observed 21 distinct wild-type clones among 3,700 isolates from one field. Monitoring inoculum could therefore be useful in determining how the fungus varies in characteristics such as survival, infectivity, susceptibility to microbial suppression, and sensitivity to control measures.

Knowing that the fungus is present is important in developing recommendations for integrated management of bean diseases. Burke and Miller (16) proposed that management practices for controlling Fusarium root rot should be directed at relieving or avoiding plant stress. They emphasized practices that provide an environment favorable to root growth, such as crop rotation, subsoil chisel plowing, sowing in warm soil, addition of organic matter to soil, and avoiding excessive or inadequate soil moisture. Nevertheless, because the severity of root rot depends on the effects of *F. solani* f. sp. *phaseoli* and abiotic features of the environment, they noted that the presence of the fungus is an important criterion for developing recommendations on the frequency of irrigation and the use of deep tillage for integrated management of root rot and Sclerotinia wilt (caused by *Sclerotinia sclerotiorum*) in the bean crop (63).

As originally suggested, the recognition of chlamydospores as survival and infectious units of *F. solani* f. sp. *phaseoli* (65), and the introduction of a selective medium (66), have been vital to understanding factors that affect qualitative and quantitative changes in the soil population of the fungus, and to developing methods of managing Fusarium root rot of bean. Managing both the activity and the number of propagules of *F. solani* f. sp. *phaseoli* in soil can contribute to sustainable production of the bean crop and to maintenance of the productivity of the soil. The latter is especially important where the area of land available for bean production is limited.

LITERATURE CITED

1.   Abawi, G.S. 1989. Root rots. Pages 105–157 in: Bean Production in the Tropics. H.F. Schwartz and M.A. Pastor-Corrales, eds. Centro Internacional de Agricultura Tropical, Cali, Colombia.

2.   Abawi, G.S., and Cobb, A.C. 1984. Relating soil densities of *Fusarium*, *Pythium*, *Rhizoctonia*, and *Thielaviopsis* to disease severity and yield of snap beans in field microplots. Phytopathology 74:813. (Abstr.).

3.   Adams, P.B., Lewis, J.A., and Papavizas, G.C. 1968. Survival of root-infecting fungi in soil. IX. Mechanism of control of Fusarium root rot of bean with spent coffee grounds. Phytopathology 58:1603–1608.

4.   Alexander, J.V., Bourret, J.A., Gold, A.H., and Snyder, W.C. 1966. Induction of chlamydospore formation by *Fusarium solani* in sterile soil extracts. Phytopathology 56:353–354.

5.   Baker, R. 1970. Use of population studies in research on plant pathogens in soil. Pages 11–15 in: Root Diseases and Soil-Borne Pathogens. T.A. Toussoun, R.V. Bega, and P.E. Nelson, eds. University of California Press, Berkeley, CA.

6.   Baker, R., and Maurer, C.L. 1967. Interaction of major factors influencing severity of bean root rot. Phytopathology 57:802. (Abstr.).

7.   Baker, R., and Nash, S.M. 1965. Ecology of plant pathogens in soil. VI. Inoculum density of *Fusarium solani* f. sp. *phaseoli* in bean rhizosphere as affected by cellulose and supplemental nitrogen. Phytopathology 55:1381–1382.

8.   Burke, D.W. 1954. Pathogenicity of *Fusarium solani* f. *phaseoli* in different soils. Phytopathology 44:483. (Abstr.).

9.   Burke, D.W. 1965. Plant spacing and Fusarium root rot of beans. Phytopathology 55:757–759.

10.   Burke, D.W. 1965. Fusarium root rot of beans and behavior of the pathogen in different soils. Phytopathology 55:1122–1126.

11.   Burke, D.W. 1965. The near immobility of *Fusarium solani* f. *phaseoli* in natural soils. Phytopathology 55:1188–1190.

12.   Burke, D.W. 1968. Root growth obstructions and Fusarium root rot of beans. Phytopathology 58:1575–1576.

13.   Burke, D.W., and Hall, R. 1991. Fusarium root rot. Pages 9–10 in: Compendium of Bean Diseases. R. Hall, ed. APS Press, St. Paul, MN.

14. Burke, D.W., Holmes, L.D., and Barker, A.W. 1972. Distribution of *Fusarium solani* f. sp. *phaseoli* and bean roots in relation to tillage and soil compaction. Phytopathology 62:550–554.

15. Burke, D.W., and Kraft, J.M. 1974. Responses of beans and peas to root pathogens accumulated during monoculture of each crop species. Phytopathology 64:546–549.

16. Burke, D.W., and Miller, D.E. 1983. Control of Fusarium root rot with resistant beans and cultural management. Plant Dis. 67:1312–1317.

17. Burke, D.W., Miller, D.E., Holmes, L.D., and Barker, A.W. 1972. Counteracting bean root rot by loosening the soil. Phytopathology 62:306–309.

18. Burkholder, W.H. 1919. The dry root-rot of the bean. Cornell Univ. Agric. Exp. Sta. Mem. 26:999–1033.

19. Burkholder, W.H. 1920. The effect of two soil temperatures on the yield and water relations of healthy and diseased bean plants. Ecology 1:113–123.

20. Burkholder, W.H. 1924. The effect of varying soil moistures on healthy bean plants and on those infected by a root parasite. Ecology 5:179–187.

21. Burkholder, W.H. 1925. Variation in a member of the genus *Fusarium* grown in culture for a period of five years. Am. J. Bot. 12:245–253.

22. Byther, R. 1965. Ecology of plant pathogens in soil. V. Inorganic nitrogen utilization as a factor of competitive saprophytic ability of *Fusarium roseum* and *F. solani*. Phytopathology 55:853–858.

23. Christou, T., and Snyder, W.C. 1962. Penetration and host-parasite relationships of *Fusarium solani* f. *phaseoli* in the bean plant. Phytopathology 52:219–226.

24. Cochrane, J.C., Cochrane, V.W., Simon, F.G., and Spaeth, J. 1963. Spore germination and carbon metabolism in *Fusarium solani*. I. Requirements for spore germination. Phytopathology 53:1155–1160.

25. Cook, R.J. 1962. Influence of barley straw on the early stages of pathogenesis in Fusarium root rot of bean. Phytopathology 52:728. (Abstr.).

26. Cook, R.J., and Schroth, M.N. 1965. Carbon and nitrogen compounds and germination of chlamydospores of *Fusarium solani* f. *phaseoli*. Phytopathology 55:254–256.

27.   Cook, R.J., and Snyder, W.C. 1965. Influence of host exudates on growth and survival of germlings of *Fusarium solani* f. *phaseoli* in soil. Phytopathology 55:1021–1025.

28.   Dryden, P., and Van Alfen, N.K. 1984. Soil moisture, root system density, and infection of pinto beans by *F. solani* f. sp. *phaseoli* under dryland conditions. Phytopathology 74:132–135.

29.   Farr, D.F., Bills, G.F., Chamuris, G.P., and Rossman, A.Y. 1989. Fungi on Plants and Plant Products in the United States. APS Press, St. Paul, MN. 1252 pp.

30.   Ford,, E.J., and Trujillo, E.E. 1967. Bacterial stimulation of chlamydospore production in *Fusarium solani* f. sp. *phaseoli*. Phytopathology 57:811. (Abstr.).

31.   Griffin, G.J. 1964. Long-term influence of soil amendments on germination of conidia. Can. J. Microbiol. 10:605–612.

32.   Hall, R. 1981. Benomyl increases the selectivity of the Nash-Snyder medium for *Fusarium solani* f. sp. *phaseoli*. Can. J. Plant Pathol. 3:97–102.

33.   Hall, R., and Phillips, L.G. 1992. Effects of crop sequence and rainfall on population dynamics of *Fusarium solani* f. sp. *phaseoli* in soil. Can. J. Bot. 70:2005–2008.

34.   Huber, D.M., and Andersen, A.L. 1966. Necrosis of hyphae of *F. solani* f. *phaseoli* and *Rhizoctonia solani* induced by a soil-borne bacterium. Phytopathology 56:1416–1417.

35.   Huber, D.M., and Watson, R.D. 1970. Effect of organic amendment on soil-borne plant pathogens. Phytopathology 60:22–26.

36.   Huber, D.M., Watson, R.D., and Steiner, G.W. 1965. Crop residues, nitrogen, and plant disease. Soil Sci. 100:302–308.

37.   Hutton, D.G., Wilkinson, R.E., and Mai, W.F. 1973. Effect of two plant-parasitic nematodes on Fusarium dry root rot of beans. Phytopathology 63:749–751.

38.   Kaplan, L. 1965. Archeology and domestication in American *Phaseolus* (beans). Econ. Bot. 19:358–368.

39.   Keenan, J.G., Moore, H.D., Oshima, N., and Jenkins, L.E. 1974. Effect of bean root rot on dryland pinto bean production in southwestern Colorado. Plant Dis. Rep. 58:890–892.

40.   Kobriger, K.M., and Hagedorn, D.J. 1983. Determination of bean root rot potential in vegetable production fields of Wisconsin's Central Sands. Plant Dis. 67:177–178.

41. Kommedahl, T., and Windels, C.E. 1979. Fungi: pathogen or host dominance in disease. Pages 1–103 in: Ecology of Root Pathogens. S.V. Krupa and Y.R. Dommergues, eds. Elsevier, Amsterdam, The Netherlands.

42. Lindsay, D.L. 1965. Ecology of plant pathogens in soil. III. Competition between soil fungi. Phytopathology 55:104–110.

43. Maier, C.R. 1959. Effect of certain crop residues on bean root-rot pathogens. Plant Dis. Rep. 43:1027–1030.

44. Maier, C.R. 1961. Selective effects of barley residue on fungi of the pinto bean root-rot complex. Plant Dis. Rep. 45:808–811.

45. Maier, C.R. 1961. Effects of soil temperature and selected crop residues on the development and severity of Fusarium root-rot of bean. Plant Dis. Rep. 45:960–964.

46. Maier, C.R. 1965. Effects of Crop Residues on the Pinto Bean Root Rot Complex. New Mexico Agric. Exp. Sta. Bull. 491. 29 pp.

47. Maier, C.R. 1965. Effects of Crop Sequences on Cotton Seedling Diseases and Pinto Bean Root Rot. New Mexico Agric. Exp. Sta. Bull. 492. 14 pp.

48. Maloy, O.C. 1960. Physiology of *Fusarium solani* f. *phaseoli* in relation to saprophytic survival in soil. Phytopathology 50:56–61.

49. Maloy, O.C., and Alexander, M. 1958. The "most probable number" method for estimating populations of plant pathogenic organisms in the soil. Phytopathology 48:126–128.

50. Maloy, O.C., and Burkholder, W.H. 1959. Some effects of crop rotation on the Fusarium root rot of bean. Phytopathology 49:583–587.

51. Martin, J.P. 1950. Use of acid, rose bengal, and streptomycin in the plate method for estimating soil fungi. Soil Sci. 69:215–232.

52. Matuo, T., and Snyder, W.C. 1973. Use of morphology and mating populations in the identification of formae speciales in *Fusarium solani*. Phytopathology 63:562–565.

53. Maurer, C.L., and Baker, R. 1964. Ecology of plant pathogens in soil. I. Influence of chitin and lignin amendments on development of bean root rot. Phytopathology 54:1425–1426.

54. Maurer, C.L., and Baker, R. 1965. Ecology of plant pathogens in soil. II. Influence of glucose, cellulose, and inorganic nitrogen amendments on development of bean root rot. Phytopathology 55:69–72.

55.   McFadden, W., Hall, R., and Phillips, L.G. 1989. Relation of initial inoculum density to severity of fusarium root rot of white bean in commercial fields. Can. J. Plant Pathol. 11:122–126.

56.   McGill, J. 1979. Layered lakes reveal much about local history. Canadian Geographic 99:56–59.

57.   Menzies, J.D. 1952. Observations on the introduction and spread of bean diseases into newly irrigated areas of the Columbia Basin. Plant Dis. Rep. 36:44–47.

58.   Miller, D.E., and Burke, D.W. 1974. Influence of soil bulk density and water potential on Fusarium root rot of beans. Phytopathology 64:526–529.

59.   Miller, D.E., and Burke, D.W. 1975. Effect of soil aeration on Fusarium root rot of beans. Phytopathology 65:519–523.

60.   Miller, D.E., and Burke, D.W. 1977. Effect of temporary excessive wetting on aeration and Fusarium root rot of beans. Plant Dis. Rep. 61:175–179.

61.   Miller, D.E., and Burke, D.W. 1985. Effects of low soil oxygen on Fusarium root rot of beans with respect to seedling age and soil temperature. Plant Dis. 69:328–330.

62.   Miller, D.E., and Burke, D.W. 1985. Effects of soil physical factors on resistance in beans to Fusarium root rot. Plant Dis. 69:324–327.

63.   Miller, D.E., and Burke, D.W. 1986. Reduction of Fusarium root rot and Sclerotinia wilt in beans with irrigation, tillage, and bean genotype. Plant Dis. 70:163–166.

64.   Miller, D.E., Burke, D.W., and Kraft, J.M. 1980. Predisposition of bean roots to attack by the pea pathogen, *Fusarium solani* f. sp. *pisi*, due to temporary oxygen stress. Phytopathology 70:1221–1224.

65.   Nash, S.M., Christou, T., and Snyder, W.C. 1961. Existence of *Fusarium solani* f. *phaseoli* as chlamydospores in soil. Phytopathology 51:308–312.

66.   Nash, S.M., and Snyder, W.C. 1962. Quantitative estimations by plate counts of propagules of the bean root rot *Fusarium* in field soils. Phytopathology 52:567–572.

67.   Nash, S.M., and Snyder, W.C. 1964. Dissemination of the root rot *Fusarium* with bean seed. Phytopathology 54:880.

68.   Nash, S.M., and Snyder, W.C. 1965. Quantitative and qualitative comparisons of *Fusarium* populations in cultivated fields and noncultivated parent soils. Can. J. Bot. 43:939–945.

69. Papavizas, G.C. 1967. Evaluation of various media and antimicrobial agents for isolation of *Fusarium* from soil. Phytopathology 57:848–852.

70. Reddick, D. 1917. Effect of soil temperature on the growth of bean plants and on their susceptibility to a root parasite. Am. J. Bot. 4:513–519.

71. Schroth, M.N., and Hendrix Jr., F.F. 1962. Influence of nonsusceptible plants on the survival of *Fusarium solani* f. *phaseoli* in soil. Phytopathology 52:906–909.

72. Schroth, M.N., and Snyder, W.C. 1961. Effect of host exudates on chlamydospore germination of the bean root rot fungus, *Fusarium solani* f. *phaseoli*. Phytopathology 51:389–393.

73. Schroth, M.N., Toussoun, T.A., and Snyder, W.C. 1963. Effect of certain constituents of bean exudate on germination of chlamydospores of *Fusarium solani* f. *phaseoli* in soil. Phytopathology 53:809–812.

74. Sippell, D.W., and Hall, R. 1982. Effects of pathogen species, inoculum concentration, temperature, and soil moisture on bean root rot and plant growth. Can. J. Plant Pathol. 4:1–7.

75. Snyder, W.C., and Hansen, H.N. 1941. The species concept in *Fusarium* with reference to section Martiella. Am. J. Bot. 28:738–742.

76. Snyder, W.C., Nash, S.M., and Trujillo, E.E. 1959. Multiple clonal types of *Fusarium solani phaseoli* in field soil. Phytopathology 49:310–312.

77. Snyder, W.C., Schroth, M.N., and Christou, T. 1959. Effect of plant residues on root rot of bean. Phytopathology 49:755–756.

78. Steadman, J.R., Kerr, E.D., and Mumm, R.F. 1975. Root rot of bean in Nebraska: primary pathogen and yield loss appraisal. Plant Dis. Rep. 59:305–308.

79. Toussoun, T.A., Nash, S.M., and Snyder, W.C. 1960. The effect of nitrogen sources and glucose on the pathogenesis of *Fusarium solani* f. *phaseoli*. Phytopathology 50:137–140.

80. Toussoun, T.A., and Patrick, Z.A. 1963. Effect of phytotoxic substances from decomposing plant residues on root rot of bean. Phytopathology 53:265–270.

81. Toussoun, T.A., Patrick, Z.A., and Snyder, W.C. 1963. Influence of crop residue decomposition products on the germination of *Fusarium solani* f. *phaseoli* chlamydospores in soil. Nature (Lond.) 197:1314–1316.

82.   Toussoun, T.A., and Snyder, W.C. 1961. Germination of chlamydospores of *Fusarium solani* f. *phaseoli* in unsterilized soils. Phytopathology 51:620–623.

83.   Watson, R.D. 1964. Eradication of soil fungi by a combination of crop residues, flooding, and anaerobic fermentation. Phytopathology 54:1437. (Abstr.).

84.   Weinke, K.E. 1962. The influence of nitrogen on the root disease of bean caused by *Fusarium solani* f. *phaseoli*. Phytopathology 52:757. (Abstr.).

85.   Wyse, D.L., Meggitt, W.F., and Penner, D. 1976. Effect of herbicides on the development of root rot on navy bean. Weed Sci. 24:11–15.

Chapter 14

THEORY AND PRACTICE OF INNOVATION IN MANAGING SOILBORNE PLANT PATHOGENS

R. Hall

Progressive improvement in management of soilborne plant pathogens will require continued innovation. How can innovation be facilitated? Where do new ideas come from? A recurring theme in this book is that an important source of innovation is the process of abstraction, i.e. the formation of general statements, labeled by terms such as principles, concepts, theories, precepts, paradigms, beliefs, models, hypotheses, or strategies. Abstractions then become the motor for innovation in knowledge, understanding, and management. Other sources of innovation considered are new methods, new uses of methods, and comparisons between and within levels of abstraction. In this final chapter, I present an innovation model and consider its application to managing soilborne plant pathogens.

SOURCES OF INNOVATION

Innovation Model

Abstractions and facts form an integrated, interacting system in which each informs the other. Each is the catalyst for innovation in the other. Abstractions derive from the facts, and in turn explain the facts. Abstractions lead to new facts. New facts enlarge and improve abstractions. Practices in research, problem solving, and management can be developed from the abstractions and the facts. Improved management systems emerge from the formulation and improvement of abstractions. In short, the system of abstractions and facts is a source of the innovation that is required for progress in understanding and managing soilborne plant pathogens.

Various terms are used throughout this book as labels for generalized statements. For the sake of consistency in the following discussion, I will use the term "principle" to refer to any generalized statement. Principles at high, intermediate, and low levels of abstraction are referred to as theories, strategies, and tactics, respectively.

Concepts of how principles relate to practice in scientific method, problem solving, and management (1,2,5) are brought together in the innovation model proposed here (Table 1). Key elements of problem solving models are definition of the problem, development of a plan of action, implementation of the plan, observation of the results, and evaluation of the results. Scientific method may involve induction from observation to theory, deduction of a hypothesis from theory, planning and conducting experiments, collecting and analyzing results, and drawing conclusions from the results. Management practices recognize the value of thinking and acting at general (strategic) and specific (tactical) levels. The innovation model (Table 1) thus recognizes four kinds of activity: planning, implementing, observing, and evaluating. It then postulates that each of these activities can be conducted at high (theoretical), intermediate (strategic), and low (tactical) levels of abstractness. I thus recognize three modes of activity: theoretical, strategic, and tactical. The innovation model facilitates and organizes innovative solutions to problems by inviting us to think creatively in each cell of the table.

The innovation model presented can be applied to research, to problem solving, and to management. Indeed, it emphasizes the common features of these activities. For example, the experimental approach characterized by a cycle of steps from observation, to induction of theory, to deduction of hypothesis, to experimentation, has its counterpart in the innovation model that allows for progression from observation, through evaluation, to theoretical, strategic, and tactical planning, and finally to implementation. The problem solving steps of planning, implementation, observation, and evaluation, and the management levels of strategies and tactics, are also explicitly recognized.

An example (Table 2) may help to explain one use for the model. Having defined the problem, say Verticillium wilt of potato, the first step (Table 1) is to plan the action. This takes place in three modes: theoretical planning, strategic planning, and tactical planning. Planning can be rationalized by consideration of relevant generalized statements. Each mode of planning has a level of abstractness, and therefore can be associated with principles. There is a hierarchy of principles ranging from high to low degrees of abstractness. In the hierarchy chosen in the

Table 1. Innovation model.

Mode	Activity			
	Planning	Implementing	Observing	Evaluating
Theoretical				
Strategic				
Tactical				

example (Table 2), we start at high levels of abstraction (theories) that relate inoculum density to disease. At intermediate levels of abstraction (strategies), we relate solarization to inoculum and disease. At low levels of abstraction (tactics), we consider how to conduct solarization. The planning process leads from theories to the identification of tactics that can be used.

The next step in the model (Table 1) is implementation. Implementation of the tactics implies that the strategies and theories are also being implemented and that the effects of the tactics, strategies, and theories can be observed. In the example, implementation of specific tactics of solarization implies implementation of strategies of solarization and theories of inoculum and disease. The third step in the model is to collect the data. These also relate to each level of abstractness. At the tactical level, the data relate to particular plastic sheeting under particular conditions. At the strategic level, they connect with the ability of solarization to reduce inoculum density or disease severity. At the theoretical level, they pertain to relations between inoculum density and disease. The fourth step in the model, evaluation of the data, also occurs at all levels of abstraction and leads to support for, or modification of, the original theories, strategies, and tactics. Having run the experiment, or employed the tactics as a management practice, we are ready to begin the process again.

Ideas are central to research, problem solving, and management, and permeate the entire innovation model. Planning and evaluation are clearly thinking activities. So are implementation and observation, because they are mental constructs of the observer, and are conditioned by the methods of observation and the framework of principles. In addition, we have to keep our wits about us to conduct the research or apply the plan.

Table 2. Example of relation of modes of planning to principles.

Mode of planning	Principles
Theoretical	Inoculum density is related to disease severity
	Reducing inoculum density reduces disease severity
Strategic	Soil solarization reduces inoculum density
	Soil solarization reduces disease severity
Tactical	Soil can be solarized using plastic sheets
	Characteristics of sheets can be specified

Ideas come in a multitude of ways, many of which we do not understand. This discussion has centred on the generation of ideas from an organized structure. A classic example of innovation through organization is the use of the periodic table of elements to predict the existence and properties of undiscovered elements. So how might the innovation model be used to generate new ideas from the material presented by the authors?

Application Of The Innovation Model

Examples of management principles, themes, and methods that the authors considered to be important to the understanding and control of soilborne plant pathogens are listed below. The principles are classified as management theories or management strategies according to my assessment of their level of abstractness.

Management theories

1.   Reduce rate of primary or secondary infection.
2.   Reduce production of inoculum.
3.   Reduce amount or density of inoculum.
4.   Reduce survivability of inoculum.
5.   Reduce pathogenicity of inoculum units.

6. Disinfest soil to improve plant health.
7. Provide a physical, chemical, or biological environment unfavorable to one or more stages of the life cycle of the pathogen.
8. Exclude inoculum.
9. Set decompostion level for maximum suppressiveness of organic amendments to soil.
10. Increase resistance of the plant to the pathogen.
11. Adjust host density.
12. Reduce stress on plant.
13. Predict pathogen behavior from environmental parameters such as temperature.

Management strategies

1. Use natural processes to regulate inoculum.
2. Starve inoculum, e.g. nitrogen starvation through high carbon:nitrogen ratio, iron starvation through action of siderophores.
3. Protect infection courts with biological control agents possessing high rhizosphere competence.
4. Promote biological antagonism of inoculum through antibiosis, parasitism, predation, or competition for space or food.
5. Protect infection sites on roots.
6. Breed plants with improved resistance to the pathogen.
7. Induce resistance in the plant.
8. Avoid infection of the plant.
9. Develop transgenic plants.
10. Use cultural practices.
11. Identify the pathogen correctly.
12. Link control practices to new production techniques such as precision farming.
13. Develop, register, and use biological control agents.
14. Use pathogen-free plants.
15. Heat plant to kill pathogens.
16. Heat soil to kill pathogens.
17. Fumigate soil to kill pathogens.
18. Use added antagonists where amount of soil is limited.
19. Use organic amendments with an intermediate decomposition level.
20. Monitor the environment.

Themes and methods

1.   Identification, detection, monitoring of pathogens; cultural, serological, molecular techniques; need for qualified personnel.
2.   Quantification of inoculum.
3.   Assessment of disease.
4.   Design of experiments.
5.   Modeling of epidemics.
6.   Comparison and classification of epidemics.
7.   Sampling soil and roots.
8.   Inoculum density-disease-yield relations.
9.   Effects of organic amendments on microbial activity and disease.
10.   Fumigant use.
11.   Heat, including solarization.
12.   Induced resistance.
13.   Rotation.
14.   Biological control.
15.   Selective media.
16.   Suppressive soil.
17.   Stress reduction.
18.   Commercial biological control products.
19.   Integrated pest management.
20.   Seed treatments.
21.   Irrigation.
22.   Use of molecular biology to determine role of antibiosis in biological control and to improve biological control agents.

This list and other comments from the authors illustrate several characteristics of the innovation model. For example, whether authors considered a statement to be a principle, methodology, strategy, or tactic depended on the context and the level of abstraction. Thus, Van Vuurde considered general methods for detecting bacteria and fungi (plating, serology, molecular biology) as principles of detection. Similarly, Katan proposed soil solarization as a management principle, although it could also be considered a method. The innovation model recognizes that hierarchies of abstractions (principles) can be erected in each area of activity, whether it be planning, implementation, observation, or evaluation.

The hierarchical level and wording of principles affect their utility and breadth of application. A principle at a low level of abstraction might be that incorporation of barley straw into soil reduces Fusarium root rot of bean. A more general principle that encompasses the former but derives from a greater array of data might be that incorporation into soil of organic matter with a high carbon:nitrogen ratio reduces Fusarium root rot of bean. The second principle is more useful than the first because it indicates that effective organic amendments are not limited to barley straw, it gives guidance on how to select organic amendments for disease control, and it provides more insight into mechanisms of suppression of this disease. On the other hand, another highly abstract principle, that temperature influences pathogen activity, is less immediately useful than related lower order, more specific, principles such as: (i) high temperatures kill pathogens, (ii) high temperatures in soil can be generated by the process of solarization, (iii) solarization can become more effective in ways described by Katan (Chapter 12).

Table 3 shows my interpretation of some relationships between the management theories and strategies listed above. A darkened cell means that I perceive a relationship between the respective theory and strategy. The analysis is not exhaustive but serves to demonstrate several reasons why the process of abstraction is useful to disease management. A given strategy can be derived from one or more theories, and one theory can give rise to one or more strategies. Theories associated with many strategies indicate well studied areas. Theories associated with few strategies represent opportunities for further development of management options. Relationships not identified may suggest new possibilities for research or management. Thus the interaction of the known with the unknown becomes a source of innovation. A similar analysis can be used to link strategies to tactics. Our objective is to fill in the blanks at the levels of theories, strategies, and tactics. We do this by innovative reflection at all levels of abstraction.

Elaboration of principles helps us to focus on the common features of disease and to identify possible new management practices. Theories, strategies, and tactics form an interacting system with vertical and horizontal connections. Strategies can flow from theories, and tactics from strategies. One principle of managing plant diseases caused by soilborne pathogens might be to reduce inoculum density. A strategy based on this principle might therefore be to fumigate soil. The tactics employed would be the specific operational details such as choice of fumigant, and time, rate, and method of application. By observing the effects of the tactics employed, we might develop another principle; that disease can be managed by reducing pathogenicity per unit of inoculum. The flow of innovation is now in the opposite direction, with new observations leading to new principles. In addition, consideration of other

Table 3. Relation of theories to strategies of managing soilborne plant pathogens.

Strategy	Theory												
	1	2	3	4	5	6	7	8	9	10	11	12	13
1		■	■	■	■	■			■				
2			■	■	■	■			■				
3	■			■									
4	■	■	■	■	■	■			■				
5	■			■									
6	■	■	■										
7	■	■	■										
8	■	■		■		■		■	■				
9	■	■								■			
10	■	■	■	■	■	■	■	■		■	■	■	
11	■	■	■	■	■	■	■	■	■	■			■
12	■		■		■	■		■			■	■	
13	■	■	■	■	■	■				■			
14	■	■	■					■					
15	■	■	■				■						
16	■	■	■		■	■		■					
17	■	■	■	■	■	■		■					
18	■	■	■	■	■	■	■	■	■				
19	■	■	■	■	■	■	■	■	■				
20													■

Theories 1 to 13 and strategies 1 to 20 refer to the list presented earlier.

principles of disease management, such as reducing the production, survivability, or infectivity of inoculum, may contribute to new tactics for using fumigation as a management tool, or for integrating fumigation with other approaches to managing inoculum. It is this network of interrelationships between ideas and data, and the importance of extracting principles from observations, that the authors of the present volume have striven to emphasize.

CONCLUSIONS

In the rush to solve problems there may be a tendency to clutch at the closest and most familiar ideas and techniques, and to stay at the tactical level of abstraction. We may indeed prefer to avoid biological generalizations; they can be limited and unreliable in their application. On the other hand, plant pathologists can contribute substantially to discussions on the philosophy of biology (3,4), particularly on the theme of intimate symbiotic relationships among organisms. Moreover, with continued refinement, principles of disease management can have substantial practical benefits. Because they are abstractions of relationships among observations, principles of disease provide: (i) a summary of observations, (ii) an explanation of disease, (iii) a rational basis for developing management strategies and tactics, (iv) a guide to managing diseases where knowledge is limited, (v) a method of predicting disease and disease control, and (vi) a source of innovation in disease management. Progressive innovations in managing soilborne plant pathogens will depend on the continuing and concomitant development of principles and practice.

LITERATURE CITED

1. Albrecht. K. 1992. Brain Power. Simon & Schuster Inc., New York, NY. 312 pp.
2. Basadur, M. 1995. The Power of Innovation. Pitman, London, UK. 330 pp.
3. Mayr, E. 1988. Towards a New Philosophy of Biology: Observations of an Evolutionist. The Belknap Press of Harvard University Press, Cambridge, MA. 564 pp.
4. Moore, J.A. 1993. Science as a Way of Knowing: The Foundations of Modern Biology. Harvard University Press, Cambridge, MA. 530 pp.
5. Quinlivan-Hall, D., and Renner, P. 1994. In Search of Solutions: 60 Ways to Guide Your Problem-Solving Group. Pfeiffer, Toronto, Canada. 177 pp.

INDEX